European Consortium for
Mathematics in Industry 8

Frank Hodnett (Ed.)

Proceedings of the Sixth European
Conference on Mathematics in Industry

European Consortium for Mathematics in Industry

Edited by

Michiel Hazewinkel, Amsterdam
Helmut Neunzert, Kaiserlautern
Alan Tayler, Oxford

ECMI Vol. 8

Within Europe a number of academic groups have accepted their responsibility towards European industry and have proposed to found a European Consortium for Mathematics in Industry (ECMI) as an expression of this responsibility.

One of the activities of ECMI is the publication of books, which reflect its general philosophy; the texts of the series will help in promoting the use of mathematics in industry and in educating mathematicians for industry. They will consider different fields of applications, present casestudies, introduce new mathematical concepts in their relation to practical applications. They shall also represent the variety of the European mathematical traditions, for example practical asymptotics and differential equations in Britian, sophisticated numerical analysis from France, powerful computation in Germany, novel discrete mathematics in Holland, elegant real analysis from Italy. They will demonstrate that all these branches of mathematics are applicable to real problems, and industry and universities in any country can clearly benefit from the skills of the complete range of European applied mathematics.

Proceedings of the

Sixth European Conference on Mathematics in Industry

August 27–31, 1991 Limerick

Edited by

Frank Hodnett
Department of Mathematics and Statistics
University of Limerick
Limerick, Republic of Ireland

Springer Fachmedien Wiesbaden GmbH 1992

Die Deutsche Bibliothek – CIP-Einheitsaufnahme

European Conference on Mathematics in Industry <06, 1991, Limerick>:
Proceedings of the Sixth European Conference on Mathematics in
Industry: August 27–31, 1991, Limerick / ed. by Frank Hodnett.
Department of Mathematics and Statistics, University of Limerick,
Limerick, Republic of Ireland.
  (European Consortium for Mathematics in Industry ; Vol. 8)
  ISBN 978-3-663-09835-5   ISBN 978-3-663-09834-8 (eBook)
  DOI 10.1007/978-3-663-09834-8
NE: Hodnett, Frank [Hrsg.] ; University <Limerick> / Department of
  Mathematics and Statistics ; European Consortium for Mathematics
  in Industry: European Consortium for...

© Copyright 1992 by Springer Fachmedien Wiesbaden
Originally published by B. G. Teubner Stuttgart in 1992

Preface

The sixth European Conference on Mathematics in Industry (ECMI 91) took place at Limerick, Ireland during August 27-31, 1991. This series of ECMI Conferences is clearly satisfying the need to provide a forum, in a European wide context, for the presentation of material related to the application of mathematics to industrial problems. The result, at Limerick, was the presence of approximately 170 delegates representing most countries of Western and Eastern Europe (with a small number from the U. S. and Canada) participating in this information exchange on a wide range of types of problems. The only identifiable common characteristic of the presentations was the element of application of mathematics to an industrially related problem.

The conference programme consisted of 8 invited lectures with approximately 80 contributed papers. These proceedings consist of 7 invited lectures and 63 contributed papers. The range of topics is very wide including areas such as optical fibres, biomechanics, electric power optimization, heat conduction, cryptography, catalytic reactors, dynamics of non-Newtonian fluids, fluid dynamics in a variety of applications, continuing education, epidemiology, numerical methods and software, optimal control, robotics etc. It is not possible to group the presentations under identifiable topic areas and the contents have therefore been organized in sections devoted to the invited lectures and the contributed papers. Within the contributed papers there are two small subsections devoted to (a) industrial design and (b) mathematical education for industry. In each section papers are listed in alphabetical order by name of first author.

As editor I wish to thank the following ECMI members V. Capasso, H. Engl, A. Fasano, S. McKee, R. Mattheij, H. Martens, H. Neunzert, A. Tayler for their generous help with the editorial process; the many referees whose work has helped to improve the quality of the material in these proceedings; and the authors for their cooperation and patience during the editing stage.

Limerick, July 1992   Frank Hodnett

Table of Contents

VIII

Industrial Design Papers

Mathematical Education and Industry Papers

Invited Lectures

Diffusion approximation on neural networks and applications for image processing

G.-H. Cottet*
LMC-IMAG, Université Joseph Fourier
BP 53x, 38041 Grenoble Cédex
France

February 16, 1992

Abstract

We derive reaction-diffusion approximations for simple models of neural networks. On the basis of this formal link we construct a non linear diffusion operator which combined with reaction allows to obtain non trivial asymptotic states. The efficiency of this model in terms of image processing is discussed on test images as well as medical images.

1 Neural networks in biomodelling

Briefly speaking, neural networks consist in discrete sets of points distributed on a lattice which interact between each other in the following fashion: at each point (or neuron) the output is computed iteratively by integrating the output of the neighbouring points. A new output is then determined in either a deterministic or stochastic way, depending on the relative value of the result and a given threshold. Let V_i be the output of the neuron i ($i \in Z^2$) and T_{ij} a matrix measuring the interactions of the neurons (called the synaptic matrix), then the simplest dynamics is given by the following iteration:

$$V_i \rightarrow \left\{ \begin{array}{ll} 1 & \text{if } \sum_j T_{ij} V_j \geq \bar{U} \\ 0 & \text{if not} \end{array} \right. \tag{1}$$

Hopfield ([6]) has proposed a continuous alternative to this boolean model by introducing, besides the outputs V_i, state variables U_i which are linked to the former by a sigmoid law.

$$V = g(\lambda U) \tag{2}$$

where λ is a gain parameter. The state of each neuron evolves according to an ODE of the following type

$$\frac{dU_i}{dt} = \sum_j T_{ij} V_j - \alpha_i U_i. \tag{3}$$

---

*Research partially supported by a Grant from Digital Equipment Corporation

Another natural model is to replace the $(0,1)$ law in (1) by a stochastic iteration. If $\tilde{H}$ denotes a regularized Heaviside function and if η is a small parameter, the iteration is then defined by

$$V_i \rightarrow \begin{cases} 1 & \text{with probability } \tilde{H}\left(\frac{1}{\eta}(\sum_j T_{ij} V_j - N\bar{U})\right) \\ 0 & \text{with probability } 1 - \tilde{H}\left(\frac{1}{\eta}(\sum_j T_{ij} V_j - N\bar{U})\right) \end{cases} \tag{4}$$

Let us immediately observe that (4) can be viewed, after renormalizations, as a time discretization of (2)-(3). The interesting features of these models are their flexibility in modelling a large range of neural activities and their ability of being parallely implemented. Let us comment a little bit on the first point. We have considered above isolated neural networks. In realistic models it is important to feed the network with external inputs representing the discharges of afferent fibers. Although it is difficult to figure out in which way the incoming information is actually coded, this can be rather simply modeled by choosing inputs obeying simple stochastic laws. For instance in the simulation of an intermediate neural network involved in the auditory system, choosing translated geometric laws governed by only 2 parameters has proved to mimic with a good agreement very specific responses corresponding to different cell populations ([1]). The role of the neural network is then to filter the information and enhance contrasts. Another crucial factor, which so far did not appear in our models, is to allow the synaptic weights T_{ij} to obey plasticity laws (the so called Hebbian rules), that is to learn during a transient phase. For details on how to choose theses rules and how they lead to a reinforcement of correlations between neurons showing coherence in their activity, we refer to [1] and the references therein. In this reference it is also shown that such models can provide effective tools in image processing.

Our goal here is to establish a link between neural networks in their simplest forms (that is without inputs and hebbian rules) and other popular models in biology, namely reaction-diffusion models, and then to derive new reaction-diffusion models involving non-linear diffusion for image processing.

2 Diffusion approximation on neural networks and related PDE models

In the field of neural networks, a common problem is to study the behavior of the system when the size of the network grows to infinity. In particular, phase transition is a term to denote the possibility that the limit system may not have the same asymptotic behavior as the finite size models. We want to address this problem by deriving a partial differential equation the solution of which is an approximation of the outputs produced by the finite size networks. Related works have already been done in the context of statistichal physics of particle systems (see [4]), but our view point here is somewhat different and yields precise estimates. Let us first observe that, by properly rescaling the network, it is equivalent to consider a growing domain or a neural network occupying a fixed box with a growing number of points. We will use this approach and assume for simplicity periodic boundary conditions. Let $x_i = ih, i \in Z^2$, the neurons. For simplicity let us assume that the synaptic matrix (T_{ij}) is symmetric and translation invariant (which is actually a very crude assumption, regarding the possibility of plasticity rules for these coefficients) and that correlations are limited to a neighboorhood of size ε, so that we can write

$$T_{ij} = T(\frac{x_i - x_j}{\varepsilon})$$

where T is a symmetric function with support in the unit ball. Observe that in the limiting process we have $\varepsilon, h \to 0$. Let

$$\tau_0 = \int T(y)dy$$

and assume $\tau_0 > 0$ (that is the synapses are globally excitatory). Also assume that

$$\int y_1^2 T(y)dy = \int y_2^2 T(y)dy = 2\tau_2 > 0.$$

Since g is strictly increasing we can set $G = g^{-1}$ and introduce the additional notations:

$$\kappa(x) = \frac{\alpha(x)}{\mu}, \quad a(V) = \frac{\tau_2}{G'(V)}, \quad b(V, x) = \frac{1}{G'(V)}(-\kappa(x)G(V) + \tau_0 V).$$

Another important paramer is the number of neighbours $N = \varepsilon^2 h^{-2}$. We then consider the reaction diffusion equation

$$\begin{cases} \frac{\partial V}{\partial t} - \epsilon^2 \mu a(V)\Delta V &= \mu b(V) \\ V(0) &= V_0. \end{cases} \tag{5}$$

We can prove

Theorem 1. *Assume the solution of (5) smooth enough on the time interval $[0, T]$; then, if V_i denote the solution of (2)-(3), the following estimate holds for all $t \in [0, T]$:*

$$\max_i |V_i(t) - V(x_i, t)| \leq C\epsilon^2 \mu(\epsilon^2 + N^{-m}).$$

The proof (see [2]) consists in observing that the discrete convolution in the right hand side of (3) is a quadrature for a continous convolution which can be rewritten as follows:

$$h^{-2} \int T(\frac{x - y}{\varepsilon})V(y)dy = h^{-2} \int T(\frac{x - y}{\varepsilon})(V(y) - V(x))dy + h^{-2}\varepsilon^2 \tau_0 V(x).$$

By a Taylor expansion the integral in the right hand side above is an integral approximation of the diffusion with coefficient $h^{-2}\varepsilon^4 \tau_2$, while the second term contributes to the reaction function.

If we consider the large time behavior of the neural networks or reaction-diffusion models, we note that, for positive values of ε, the impossibilty to reach a non-trivial steady state solution for (5) reflects the uniqueness of an invariant measure established in a more general framework for stochastic networks of the type (4) (see [5]). On the other hand for $\varepsilon = 0$ if $\tau_0 > \kappa G'(0)$ (that is if the mean value of the correlations is large enough) the reaction function has a positive slope at the origin and it will have 2 symmetric zeros which will attract respectively positive and negative parts of the initial condition.

We now turn to the second part of this discussion. On the basis of the above analysis and the ability of neural networks (of a more complex type than the one considered above) to locally adapt thei synaptic weigths in order to improve the way information is processed, it is natural to try to derive reaction-diffusion models that could at the same time eliminate the noisy part of the signal and produce non-trivial asymptotic states. It is clear that the shape of the reaction function does not play an essential role in this context. Therefore one must seek non-linear diffusion operators involving a scale parameter which would determine the minimum scale not affected by the diffusion. Diffusion operators of this type have been derived in [7] but it does not seem that, even when associated to reaction, they could lead to non trivial asymptotic states. The diffusion operators we propose have the form

$$\Delta_\varepsilon u = \text{div}\left([A_\varepsilon(u)][\nabla u]\right)$$

where $A_\varepsilon(u)$ is a 2×2 matrix representing the orthogonal projection onto the direction orthogonal to the gradient of a regularization u_ε of u:

$$A_\varepsilon(u)_{i,j} = \frac{\tilde{\partial}_i u_\varepsilon \tilde{\partial}_j u_\varepsilon}{|\nabla u_\varepsilon|^2 + \varepsilon^2},$$

where $\tilde{\partial}_1 = \frac{\partial}{\partial x_2}$ and $\tilde{\partial}_2 = -\frac{\partial}{\partial x_1}$. One motivation for considering this kind of diffusion operator is that we wish to be able to detect one dimensional objects, so that it is natural to try to penalize the diffusion in the direction orthogonal to the local value of the gradient. We then consider the following reaction-diffusion model:

$$\frac{\partial u}{\partial t} - \sigma \varepsilon^2 \text{div}([A_\varepsilon(u)][\nabla u]) = \quad f(u) \quad \text{in } \Omega \tag{6}$$

$$u(\cdot, 0) = \quad u_0 \quad \text{in } \Omega \tag{7}$$

$$u = \quad 0 \quad \text{on } \Gamma. \tag{8}$$

where Ω is the unit square, Γ its boundary and f is a C^1 reaction function satisfying $f(u)u > 0$ if $u \neq 0$ and $f(\pm 1) = 0$. We can prove

Theorem 2.

i) *Let $u_0 \in H_0^1(\Omega)$ and $\varepsilon > 0$. The system (6)-(8) has at least one solution in $H_0^1(\Omega)$. If $-1 \leq u_0 \leq 1$ then $-1 \leq u(\cdot, t) \leq 1$ for all time $t > 0$.*

ii) *Let D a domain which is locally on one side of its boundary ∂D supposed to be of class C^3. Let $\bar{u}$ the characteristic function of D. Then we have for ε small enough*

$$[A_\varepsilon(\bar{u})][\nabla \bar{u}] = \alpha \delta_{\partial D}$$

where α is a bounded vector function defined on ∂D and satisfying

$$\|\alpha\|_{L^\infty} \leq C\varepsilon^2$$

We refer to [3] for a proof. Let us only mention that the key ingredient in the proof of the existence of a solution is the following positivity property of Δ_ε

$$\int_\Omega <\Delta_\varepsilon u, u> dx = \int_\Omega \frac{|<(\nabla u_\varepsilon)^\perp, \nabla u>|^2}{\varepsilon^2 + |\nabla u|^2} dx > 0$$

Therefore the maximum principle holds for our problem. As for the second assertion its meaning is that the diffusion will not affect details on a scale larger than ε. [3] gives a discussion of the consequence of this fact on the attractivity of piecewise constant states for a finite element discretizaton of (6)-(8).

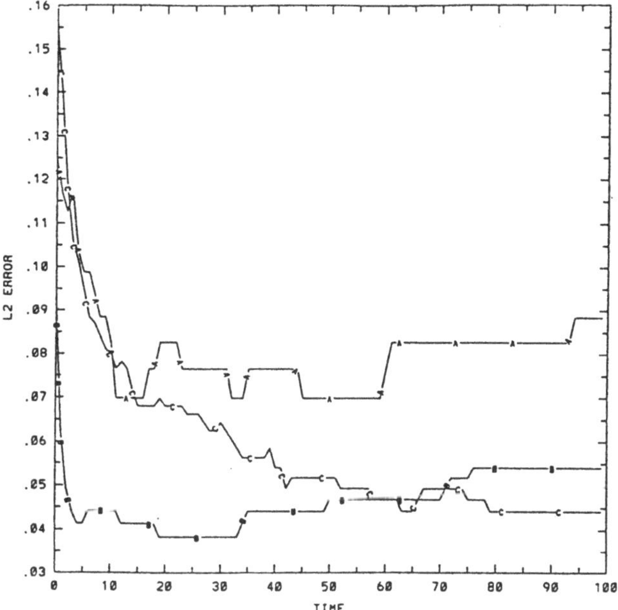

Figure 1: Convergence to the exact image as a function of the refinement (Curve A: 32 × 32, curves B and C:64 × 64; $\varepsilon = 2$ pixels for A and C, 4 pixels for B)

3 Numerical experiments

In the first experiment we wish to illustrate the attractivity of piecewise constant states. The image we want to recover consists in a black triangle above a black rectangle on a white background. It has been slightly perturbed by assigning to 10% of the pixels a random value. The curves A and B in Figure 1 correspond respectively to a 32 × 32 and 64 × 64 image. The value of ε has been kept constant ($\varepsilon = 2h$ for A, $\varepsilon = 4h$ for B, h being the distance between nearest pixels). The curve C corresponds to a 64 × 64 image where in addition the value of ε has been decreased to $2h$. All curves give the L^2 norm of the difference between the computed and the exact images in function of time. The successive improvements in the convergence illustrate the fact that the computed images fall for time large enough into a neighboorhood of the exact image with size proportional to εh. Figure 2 shows the effect of our image processing in the more sever case where 70% of the pixels have been destroyed (left picture). On the right picture the processed image after 150 iterations has undergone a thresholding splitting it between black and white. In this experiment the regualrization has been done on 14 pixels in each direction, while the rectangel is only 7 pixels wide. The alterations of the corners reflect the large values of the second and third order derivatives of the boundary at these points. We now come to a real image which we believe illustrate the possibilites of our model. It is a 512 × 512 medical image obtained through Magnetic Resonance Imaging. This technique allows to obtain images where homogoneous tissues give coherent signals. Therefore the goal of image processing is essentially to average the other parts of the signal without affecting its significant details, in such a way that the image will be more adapted to a post processing such as a segmentation. In our example the significant structures are monodimensional which obviously makes it difficult to deal with.

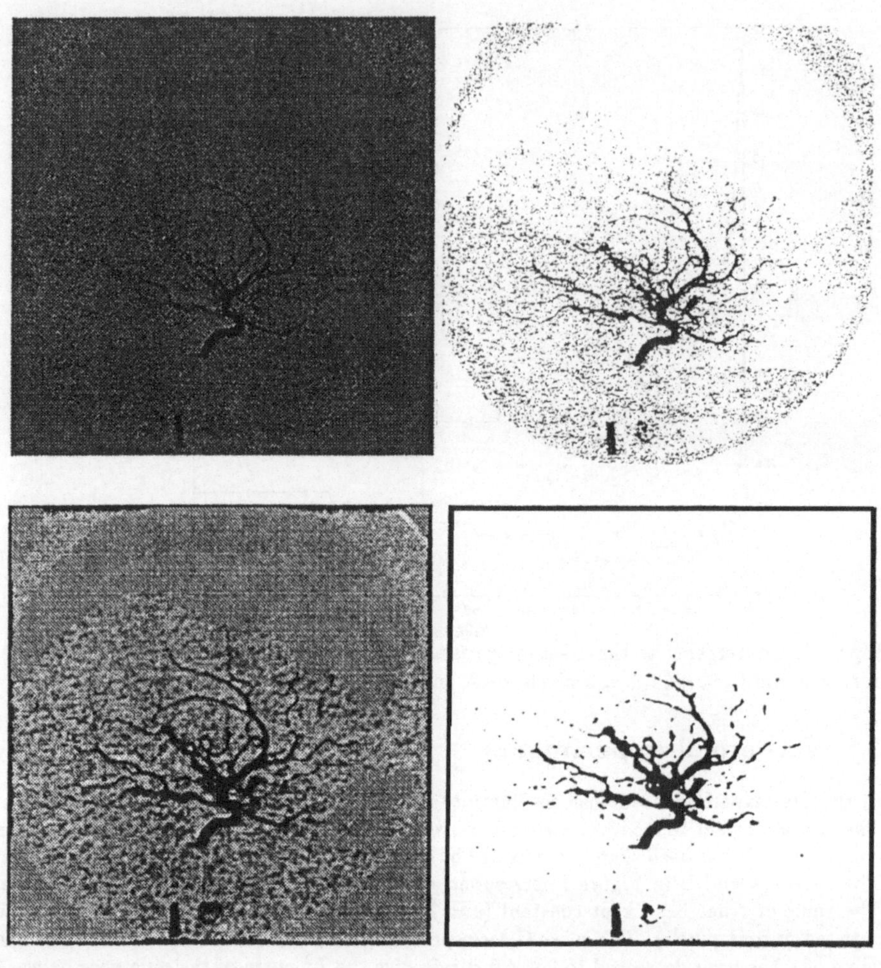

Figure 3: Image processing on a MRI image with one dimensional objects

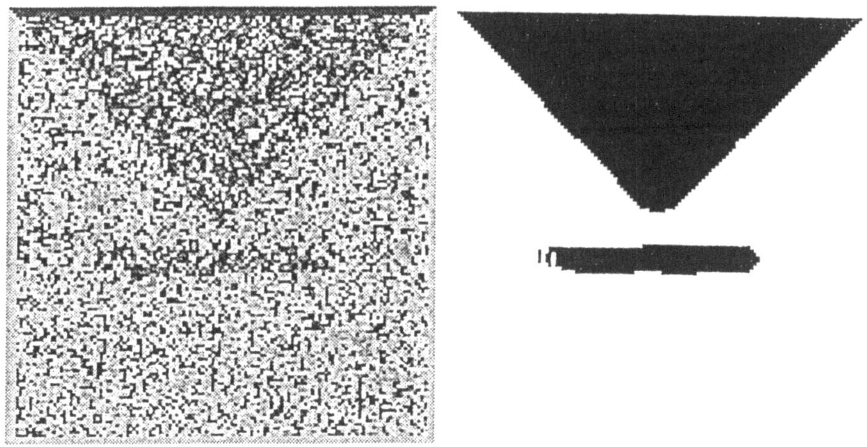

Figure 2: Image processing on a very noisy signal

In the top-left picture of Figure 3 we see the original image, and to show the necessity of processing this image, the top-right picture shows that a simple thresholding on this image does not allow to keep only the relevant strucutres. The bottom pictures correspond to a processed image where the regularization has been made on 4 pixels in each direction and the reaction function has been chosen with 4 stable zeros. Note also that in this example the diffusion matrix has only been computed at the first iteration. After thresholding (bottom right) some fine details appear in the core of the vessels. These results are promisisng and work is in progress to extend these techniques to 3D imaging and to combine them with segmentation tools that are more sophisitcated than simple thresholding.

References

[1] Berthomier, F. and al. 'Asymptotic behavior of neural networks and image processing', in A. Babloyantz (ed), Self-organization, Emerging properties and learning, Plenum Press, NATO Series (to appear).

[2] Cottet, G.-H. (1991) 'Modèles de reaction-diffusion pour des réseaux de neurones stochastiques et déterministes ', C. R. Acad. Sci. Paris, 312, 217-221.

[3] Cottet, G.-H. and Germain, L. (1992) 'Non-linear diffusion combined to reaction for image processing ', preprint.

[4] De Masi, A. and al. (1986) 'Reaction-diffusion equations for interacting particle systems', Journal of Stat. Phys., 44, 589-627.

[5] François, O. (1992) Thèse de l'Université Joseph Fourier, Grenoble

[6] Hopfield, J. J. (1984) 'Neurons with graded response have collective computational properties like those of two-states neurons', Proc. Natl. Acad. Sci., 81, 3088-3092.

[7] Alvarez, L. and al (1991) 'Image selective smoothing by nonlinear diffusion', preprint

A SURVEY ON SYSTEM IDENTIFICATION

M. Deistler
Institute for Econometrics, Operations Research
and Systems Theory
Technical University Vienna
Argentinierstraße 8
A-1040 Vienna, Austria

ABSTRACT: A short survey on linear system identification is given for an audience of applied mathematicians where little special knowledge of the subject is assumed. The main emphasis is put on the description of the basic features of the problem, rather than on giving an account of methods and theories.

1.Introduction

1.1 GENERAL REMARKS

The problem of identification is to obtain good models for (real or artificial) phenomena from data. This is a central issue in many branches of science. Such problems are often far from being trivial and in many cases they share common features. For these reasons systematic approaches to identification, in particular approaches that use mathematical tools for deriving or evaluating methods have been developed. From a formal point of view in identification one has to specify:

(i) The *model class*, i.e. the class of all a priori feasible candidate systems to be fitted to the data

(ii) The class of observations

(iii) The *identification procedure*, which is a rule (in the automatic case a function) attaching to the data for every sample size $T \in \mathbf{N}$ a system (or more general a set of systems) from the model class.

System identification has many different aspects and facets, depending in particular on the intended use for the identified model and on the amount of a priori information about the phenomenon to be modeled. Identification is performed for instance for the following purposes:

- for encoding data (i.e. giving a "short" description of data by system parameters)
- for analysis of systems and simulation
- for spectral estimation
- for prediction, filtering or interpolation
- for control
- for testing theories and for estimation of parameters (which have a "physical" meaning) in models obtained from theory.

System identification may be decomposed into the following steps:

(i) *Determination of the model class*: Here a "preliminary model" about the phenomenon under consideration is formulated using a-priori information and/or existing theories. This step in particular may include the selection of variables which are of potential interest, their

---

The author wants to thank Drs. W. Scherrer (Vienna) and J. van Schuppen (Amsterdam) for valuable discussions

classification into inputs and outputs, a specification of the functional form of the relation between the variables and some reasoning concerning the nature and the properties of the noise.

(ii) *Specification of the class of observations*:Here for instance sampling period, sample size and appropriate transformations of the data, such as seasonal adjustment are specified. In experimental situations, in addition, the experimental design as for instance the design of the inputs may be chosen.

(iii) *The identification procedure*: In most cases the identification procedures are obtained from optimizing a criterion function. In particular in a stochastic context a substantial body of theories has been developed for evaluation of estimation procedures and tests. In optimization a number of algorithmic problems arise.

(iv) *Model validation*: Here the identified model is evaluated according to other criteria; if the model turns out to be not satisfactory, identification may be redone in an "improved" setting.

It should be emphasized that in most cases identification is an art rather than a science in the sense that a number of subjective decisions has to be made. Only parts of the overall procedure are automatized. Recently in addition to "algorithmic" procedures also knowledge based approaches are used.

1.2 THE MAIN AREAS OF APPLICATION

The most important areas of application for system identification are:

1.2.1 *Signal processing*, in particular speech processing, sonar and radar applications. Examples for special problems here are: Channel equalization, blind deconvolution, echo chancellation, speech recognition or detection of sources from measurements at sensors. Here emphasis is laid on fast, real-time and on-line calculation and on tracking of time varying parameters.

1.2.2 *Control of technical systems*. In a number of applications in control, no sufficient a priori information (coming from "physical" theories) about the plant is available for the design of a controller. Then a model for the plant based on input-output measurements has to be built. This may be the case e.g. for the design of autopilots for airplanes or ships or for the control of chemical processes or steel mills. An important area is adaptive control, where at the same time the system is identified and the parameters of the identified system are used to automatically adapt the controller. This is of particular importance for time varying plants.

1.2.3 *Other technical applications* are e.g.: *monitoring,* where system identification together with jump point detection procedures are used (e.g. for monitoring of off shore oil platforms) or "data driven" *modeling for the purpose of simulation* experiments. Examples for the latter are modeling of jet- or car-engines where "real" experiments are more expensive.

1.2.4 *Economic and business applications:* Testing of economic theories, estimation of "deep" (i.e. theoretically meaningful) parameters. Forecasting and policy simulation models, e.g. for macro- or financial variables, forecasting models for inventories and sales.

1.2.5 *Geophysical applications* such as the analysis of seismic signals for exploration of oil or minerals, or the analysis of earthquakes.

1.2.6 *Biological and medical applications:* Analysis of EEG data or circumdianic rhythms. Monitoring for patients.

1.3 LINEAR SYSTEMS

In this paper we restrict ourselves to a special, however very important case, namely to identification of linear systems from discrete (and equally spaced) time series data (i.e. observations ordered in time). Both, with respect to the existing body of methods and theories and with resprect to the range of applications, linear system identification is a rich and extensive subject now; this is also documented by a number of recent books (Caines 1988, Hannan and Deistler 1988, Ljung 1987, Söderström and Stoica 1988; see also the book by Brockwell and Davis 1991 on time series analysis).

Even linear system identification has many different aspects. Nevertheless, a *"main stream theory"* of linear system identification has evolved during the last decades. This main stream theory now includes the multi-input, multi-output (MIMO) case. The basic framework for the main stream case may be described as follows:

(i) The model class consists of linear, finite dimensional, constant parameter, causal and stable systems only; the classification of the variables into inputs and outputs is given a priori

(ii) Noise is modeled by stochastic models, in particular by stationary, ergodic processes with rational spectral densities. These processes are modeled as the outputs of linear systems with unobserved white noise inputs. In most cases the noise characteristics are estimated too. For a number of cases, e.g. for spectral estimation or for unconditional forecasting systems without observed inputs, having only white noise inputs are used.

Clearly one advantage in using stochastic models is that we can evaluate the quality of identification procedures in the sense of statistical theory. It should be emphasized however, that in using stochastic models for noise, one always has to impose additional assumptions (such as stationarity and ergodicity) in order to make the problem statistically meaningful, i.e. in order to obtain "reasonable" estimators from one trajectory. On the other hand such assumptions in many cases are not justified by our a priori knowledge; thus the question arises whether such stochastic models for noise provide an important "test case" for the evaluation of our identification procedures for a given application.

(iii) The inputs are assumed to be observed free of noise. The noise, which is assumed to be orthogonal to the inputs, is added to the outputs or to the equations. This special (and in a sense unsymmetric) way of embedding a deterministic system in a stochastic environment is called the errors-in-equations approach in econometrics; it can be justified in a great number of applications, for instance in conditional (on the inputs) forecasting of the outputs.

(iv) The criteria for goodness of fit of a system to the data are of (Gaussian) maximum likelihood (ML) type. This includes criteria based on the prediction error variance.

(v) In many cases, the model class will be so large, that estimators obtained from optimizing criteria of goodness of fit only, will not give reasonable results, basically due to problems of overfitting. In these cases the model class is decomposed into (finite dimensional) subclasses. Each subclass is described by its socalled dynamic specification (e.g. by maximum lag lenghts) expressed by a vector of integers (a socalled multi-index). In general, the dynamic specification has to be determined from data too; this is done by optimizing a criterion taking into account both, the best fit within the subclass and the complexity of the subclass, described by its dimension. Once the subclass is fixed, we have a parametric estimation procedure, obtained by optimizing the fit. The overall approach (which includes estimation of the multiindex) thus could be called semiparametric.

(vi) As far as the desirable properties of the estimators are concerned, the main focus is on consistency and asymptotic efficiency.

Unless the contrary is stated explicitely we restrict ourselves to the main stream case. However, most of the basic ideas presented here also generalize to other cases.

There are several different representations for linear systems. Let z denote the complex variable as well as the backward shift on $\mathbf{Z}$, i.e. $z(y_t|t \in \mathbf{Z}) = (y_{t-1}|t \in \mathbf{Z})$. Then the *input-output representation* is of the form

$$y_t = k(z)u_t = k_1(z)z_t + k_2(z)\varepsilon_t \tag{1.1}$$

where y_t are the s-dimensional outputs, u_t are the m-dimensional inputs and

$$k(z) = \sum_{j=0}^{\infty} K_j z^j \quad ; \quad K_j \in \mathbf{R}^{sxm} \tag{1.2}$$

is the transfer function. In general $u_t' = \left(z_t', \varepsilon_t'\right)$ consists of observed inputs z_t and of an s-dimensional white noise component (ε_t) (i.e. $E\varepsilon_t = 0$; $E\varepsilon_s\varepsilon_t' = \delta_{st}.\Sigma$)which is not observed.

The ARMA(X) *representation* is of the form

$$a(z)y_t = b(z)u_t = c(z)z_t + d(z)\varepsilon_t \tag{1.3}$$

where

$$a(z) = \sum_{j=0}^{p} A_j z^j; \quad A_j \in \mathbf{R}^{sxs}; \quad b(z) = (c(z),d(z)) = \sum_{j=0}^{q} B_j z^j; \quad B_j \in \mathbf{R}^{sxm}$$

are polynomial matrices. In the main stream case we assume stability $\left(\text{i.e.} \quad \det a(z) \neq 0 \quad \text{for} |z| \leq 1\right)$ and we impose the miniphase assumption $\det \quad d(z) \neq 0$ for $|z| < 1$. The matrices A_j, B_j and Σ contain the realvalued parameters for the system and the noise; p and q are examples for integer parameters. The transfer function of (1.3) is given by

$$k(z) = a^{-1}(z).b(z) = (k_1(z), k_2(z)) \tag{1.4}$$

If (z_t) is stationary with spectral density f_z we have

$$f_y = k_1 f_z k_1^* + (2\pi)^{-1} k_2 \Sigma k_2^* \tag{1.5}$$

for the spectral density f_y of the outputs and

$$f_{yz} = k_1.f_z \tag{1.6}$$

for the cross spectrum f_{yz} between (y_t) and (z_t) Here $k^*(z) = k(z^{-1})$ and the spectral densities are defined on $\mathbf{C}$ rather than on $[-\pi, \pi]$.

The *state space representation* is of the form

$$x_{t+1} = Fx_t + Gu_t = Fx_t + Lz_t + K\varepsilon_t \tag{1.7}$$

$$y_t = Hx_t + \varepsilon_t \tag{1.8}$$

where x_t is the (n-dimensional) state at time t and $F \in \mathbf{R}^{n \times n}$, $G = (L, K) \in \mathbf{R}^{n \times m}$ and $H \in \mathbf{R}^{s \times n}$ are the parameter matrices (containing the real-valued parameters). The state dimension n is an example for an integer parameter. Again the stability and the miniphase assumption are imposed. The transfer function is given by

$$k(z) = H(I - Fz)^{-1} z \, G + (0, I) \qquad (1.9)$$

Of course f_y and f_{yz} are given by formulas (1.5) and (1.6) respectively.

From now on we will be mainly concerned with the mathematical aspects of identification, in particular with methods for identification and with the associated theory. From the point of view of the mathematical analysis it is convenient to decompose the problem into three "modules", namely structure theory, estimation (of the realvalued parameters) for given dynamic specification and estimation of the integers which characterize the dynamic specification.

2. Structure Theory

2.1 BASIC CONSIDERATIONS

Here we are concerned with the relation between the population second moments of the observations (and thus with f_z, f_y, f_{yz}) on the one side and the real and integer parameters of ARMA or state space systems and the parameters in Σ on the other side, in other words with the relation between "external" and "internal" system and noise characteristics.

The problem may be decomposed into two steps: The first problem is concerned with determining the transfer function k(z) and the variance matrix Σ from f_z, f_y, f_{yz} and is relatively simple: Under a socalled persistent excitation condition (e.g. if f_z is nonsingular except for a finite number of points) k_1 is uniquely determined from (1.6); determining k_2 from (1.5) leads to a spectral factorization problem which is unique under the additional assumptions $k_2(0) = I$ and $\Sigma > 0$ (see e.g. Hannan and Deistler 1988).

So we are left with the relation between the transfer functions k(z) and the internal characteristics. The main problem in this context is to find for given k(z) a corresponding ARMA or state space representation; this is the socalled *realization* problem. First note that every ARMA system (satisfying our assumptions) as well as every state space system has a causal, rational transfer function which is analytic in a disc containing the closed unit disk and where $\det k_2(z)$ has no zero for $|z| < 1$. In addition we impose the normalization k(0)=(0,I). Conversely, for every transfer function with the above mentioned properties there is a state space as well as an ARMA realization.

We have to distinguish between cases where additional a priori information, e.g. coming from physical theories, is available and cases where no such addtional a priori information is used. We restrict ourselves to the latter case here; also, for simplicity we consider ARMA realizations. For state space realizations most problems are analogous.

Let T_A denote the set of all (a,b) (for fixed s,m but for variable p,q) satisfying our assumptions, let U_A denote the set of all corresponding transfer functions $k = a^{-1}.b$ and let the mapping $\pi : T_A \to U_A$ be defined by $\pi(a,b) = a^{-1}.b$. The problem of *parametrization* is concerned with describing subsets of U_A by parameters. A set $T_\alpha \subset T_A$ is called *identifiable* if π restricted to T_α is injective. In this case there exists a mapping $\rho_\alpha : \pi(T_\alpha) = U_\alpha \to T_\alpha$ such that $\rho(\pi(a,b)) = (a,b)$ for all $(a,b) \in T_\alpha$ holds. Such a mapping is called an *ARMA parametrization* of U_α and T_α is a (preliminary) *parameter space*. The set U_A is endowed with the topology which corresponds to the relative topology in the product space $\left(\mathbf{R}^{s \times m} \right)^{\mathbf{Z}_+}$ for the power series coefficients $\left(K_j | j \in \mathbf{Z}_+ \right)$ of the transfer function $k(z)$. This topology for U_A is called the pointwise topology T_{pt}.

Now, neither is T_A a "good" parameter space nor is it reasonable to look for a parametrization for U_A. One justification for the first statement is that evidently T_A is not identifiable. A deeper result states that there does not exist a continuous parametrization for U_A (see e.g. Hazewinkel and Kalman 1976, Hazewinkel 1979). Thus T_A and U_A have to be broken into bits, T_α and U_α say, where α is a multiindex of integers ranging over an index set I, in order to allow for a "convenient" parametrization. Convenient parametrizations satisfy the following conditions:

(i) T_α is identifable

(ii) T_α is finite dimensional in the sense that the degrees of the $(a,b) \in T_\alpha$ are bounded. Furthermore $\bar{T}_\alpha$ is homeomorphic to a subset $\bar{T}_\tau$ of some Euclidian $\mathbf{R}^{d_\alpha}$ and $\bar{T}_\alpha$ contains an nonvoid open subset of $\mathbf{R}^{d_\alpha}$. The values of this homeomorphism are called the *free parameters* τ. We identify τ with $(a,b) \in T_\alpha$ and $\bar{T}_\alpha$ with T_α. In most cases the entries of τ can be chosen as specific entries of the A_j, B_j.

(iii) The parametrization $\rho_\alpha : U_\alpha \to T_\alpha$ is a $(T_{pt}-)$ homeomorphism. In addition it is required that U_α is open in its closure $\overline{U}_\alpha$.

(iv) $\underset{\alpha \in I}{\bigcup} U_\alpha = U_A$ (covering property)

2.2 ECHELON CANONICAL FORMS

For the single input, single output (SISO) case the "natural" integers for dynamic specification are the degrees in a and b or their maximum. For the multi-output case (s>1) a number of different possibilities for dynamic specification are available and the problems of realization and parametrization are significantly more complex. Here we only describe *Echelon (canonical) forms*. We commence from the (block) Hankel matrix $H = \left(K_{i+j-1} \right)$, $i,j = 1,2....$ of the transfer function k. If we define $\bar{k}(z) = k(z^{-1})$ and a corresponding ARMA system $\left(\bar{a},\bar{b}\right)$ by $\bar{a}^{-1}(z)\bar{b}(z) = \bar{k}(z)$ then, from a comparison of coefficients corres-ponding to the negative powers of z in the equation $\bar{a}\bar{k} = \bar{b}$ we obtain

$$\left(\bar{A}_0,\bar{A}_1,....\right)H = 0 \tag{2.1}$$

Now, since k is rational, H is of finite rank and this rank is equal to the degree of det $\bar{a}$ for every left coprime pair $\left(\bar{a},\bar{b}\right)$ such that $\bar{a}^{-1}.\bar{b} = \bar{k}$ This rank, n say, is called the *order* of k. Let h(i,j) denote the j-th row in the i-th block of rows of H. Due to the block Hankel structure of H, the first (w.r.t. natural ordering) rows of H which form a basis for the row space of H are of the form

$$h(1,1),...,h(n_1,1),h(1,2),...,h(n_2,2),\ ...,\ h(1,s),...,h(n_s,s)$$

for a suitably chosen multi-index $\alpha = (n_1,...,n_s)$; these $n_1,...,n_s$ are called the *Kronecker indices*. Expressing the $-h(n_i + 1, i)$, i=1,...,s as linear combinations of the preceding basis rows gives a prescription for uniquely determining $\bar{a}$ from (2.1). Clearly then $\bar{b}$ is also uniquely determined from $\bar{b} = \bar{a}\bar{k}$. The ARMA realization for k then is obtained from

$$(a(z),b(z)) = \text{diag}\left\{z^{n_i}\right\}\left((\bar{a}(z^{-1}),\bar{b}(z^{-1}))\right) \tag{2.2}$$

and this is called the Echelon form. Let U_α denote the set of all $k \in U_A$ with given Kronecker indices $\alpha = (n_1,...,n_s)$, let T_α be the set of all corresponding (a,b) in Echelon form and let $\rho_\alpha : U_\alpha \to T_\alpha$ denote the corresponding parametrization. Then for $I = \mathbf{Z}_+^s$ this gives convenient parametrizations in the sense defined above (see Hannan and Deistler 1988, chap 2). Let M(n) denote the set of all transfer functions of order n. M(n) is a real analytic manifold of dimension n(s+m), which for s>1, cannot be described by one coordinate system (Hazewinkel and Kalman 1976). The class of all U_α such that $n_1+...+n_s = n$ holds is a disjoint cover for M(n), where the U_α in general have different dimensions.

3. Estimation For Given Dynamic Specification

Here estimation of the real valued parameters, i.e. of the entries of the vector τ of free parameters of the systems and of the free parameters on Σ for given dynamic specification is considered. Since we do in general not impose additional restrictions of Σ its free parameters are its on- and above diagonal entries. Given dynamic specification means then the integer parameters in α and thus T_α and U_α are fixed. For simplicity of presentation we restrict ourselves to the case of no observed inputs, i.e. the case $u_t = \varepsilon_t$ for this and the next section.

In most cases - at least asymptotically - the only information exploited from the data in estimation is the information contained in the sample second moments, which are given by

18

$$\hat{\gamma}(s) = T^{-1} \cdot \sum_{t=1}^{T-s} y_{t+s} \cdot y_t' \qquad \text{for } s \geq 0 \quad \text{and} \quad \hat{\gamma}(s) = \hat{\gamma}(-s)' \quad \text{for } s < 0$$

where T denotes the sample size. Let Σ denote the set of all positive definite and symmetric sxs matrices and let V_α denote the set of all population second moments of (y_t) corresponding to U_α i.e.

$$V_\alpha = \left\{ \gamma : \mathbf{Z} \to \mathbf{R}^{sxs} \,\middle|\, \gamma(t) = \int_{-\pi}^{\pi} e^{i\lambda t} k(e^{-i\lambda}) \Sigma k(e^{-i\lambda})^* d\lambda ; k \in U_\alpha, \Sigma \in \mathbf{\Sigma} \right\}$$

In general the sample second moment function $\hat{\gamma}$ will not be contained in V_α , even if the true transfer function k_0 (i.e. the transfer function corresponding to the data generating process (y_t)) is contained in V_α. Thus in estimation we have to approximate a first estimate (e.g. $\hat{\gamma}$) within V_α . Such an approximation for instance is obtained by minimizing the socalled *Whittle likelihood*.

$$L_T^W(k(\tau), \Sigma) = \log \ \det \Sigma + (2\pi)^{-1} \int_{-\pi}^{\pi} \mathrm{tr}\left\{ \left[k(e^{-i\lambda}; \tau). \Sigma . k(e^{-i\lambda}; \tau)^* \right]^{-1} . I(\lambda) \right\} d\lambda \quad (3.1)$$

over $U_\alpha x \Sigma$ where $k(\tau) = k(e^{-i\lambda}; \tau)$ denotes the transfer function as a function of τ and where $I(\lambda)$ denotes the periodogram (i.e. the Fourier transform of $\hat{\gamma}$)

$$I(\lambda) = (2\pi)^{-1} \cdot \sum_{j=-T+1}^{T-1} \hat{\gamma}(j) e^{-i\lambda j}$$

The Whittle likelihood is an approximation to the *likelihood* function

$$L_T(k(\tau), \Sigma) = T^{-1} \log \det \ \Gamma_T(\tau, \Sigma) + T^{-1} y(T)' . \Gamma_T^{-1}(\tau, \Sigma) . y(T) \qquad (3.2)$$

where $y(T) = (y_1', ..., y_T')'$ is the stacked vector of the observations and where $\Gamma_T(\tau, \Sigma)$ denotes the s.T x s.T matrix of the second moments of a vector $(y_1', ..., y_T')'$ corresponding to an ARMA process with parameters τ and Σ More precisely L_T is equal to $-2^{-1}.T$ the Gaussian likelihood (up to a constant) and the estimators obtained from minimizing L_T (over $T_\alpha x \Sigma$ or over $U_\alpha x \Sigma$) are called the *maximum likelihood estimators* (MLE's) $\hat{\tau}_T, \hat{k}_T, \hat{\Sigma}_T$ of the true τ_0 , k_0 and Σ_0 respectively.

The MLE's and the estimators obtained from Whittle likelihood have the same properties in many respects; therefore we restrict ourselves to the MLE's.

In general there is no explicit expression available for the MLE's; thus numerical optimization prozedures have to be applied, each of which in a strict sense defines an estimator. There is still a number of open algorithmic problems, e.g. problems associated with multiple relative minima of L_T and with the appropriate choice of initial values.

Other problems associated with the MLE's are that in optimizing the likelihood function certain boundary points of $U_\alpha x\Sigma$ cannot be excluded and have to be taken into account (see e.g. Hannan and Deistler 1988, chapters 2 and 4). Also in general even the existence of the minimum of L_T cannot be guaranteed (Deistler and Pötscher 1984).

The asymptotic theory for the MLE's is rather complete now (see e.g. Dunsmuir and Hannan 1976 and Hannan and Deistler 1988, chap 4). Consistency of $\hat{k}_T$ and $\hat{\Sigma}_T$ can be shown under general conditions, in particular without imposing unnatural compactness assumptions on the parameter space. This even holds for overfitting, i.e. if e.g. the true Kronecker indices are smaller than the specified ones. For $k_0 \in U_\alpha$, by the continuity of the parametrization, $\rho_\alpha(\hat{k}_T) = \hat{\tau}_T$ clearly is consistent for $\tau_0 = \rho_\alpha(k_0)$. The asymptotic normality results for $\hat{\tau}_T$ and $\hat{\Sigma}_T$ can be obtained in the usual way.

An important generalization is concerend with the case where the true system is not contained in $\overline{U}_\alpha$, e.g. if it is infinite dimensional. Then the following *generalized consistency* result can be shown (Ljung 1978, Hannan and Deistler 1988, chap 4). Let D denote the set of all best approximations of the true transfer function k_0 within U_α (where boundary points of U_α may be included) in the sense that D is the set of all minimizers of the asymptotic likelihood function (i.e. the limit of L_T for $T \to \infty$)

$$L(k,\Sigma) = \log \det \Sigma + (2\pi)^{-1} \int_{-\pi}^{\pi} \text{tr}\left\{ \left(k(e^{-i\lambda})\Sigma k(e^{-i\lambda})^* \right)^{-1} . (k_0(e^{-i\lambda})\Sigma_0 k_0(e^{-i\lambda})^* \right\} d\lambda$$

D is not necessarily a singleton. Then the MLE's $\hat{k}_T$ converge a.e. to the set D. This is an important "robustness" property with respect to underfitting.

4. Determining the Dynamic Specification

In many applications, the dynamic specification is not known a priori and thus has to be determined from the data too. The development and evaluation of data based procedures for dynamic specification has constituted one of the most important contributions to the subject of system identification during the last three decades. This particularly applies for automatic procedures which now have replaced the non-automatic procedures such as the procedure proposed by Box and Jenkins (1970) to a certain extent.

Consider the problem of estimating the order first. Let $\overline{M(n)}$ denote the closure $\left(\text{in} \quad T_{pt} \right)$ of the set of all transfer functions of order n. Then $\overline{M(n_1)} \subset \overline{M(n_2)}$ holds for $n_1 < n_2$ and $M(n_1)$ has smaller dimension than $M(n_2)$ (see e.g. Hannan and Deistler 1988, chap 2).Thus in optimizing a criterion of goodness of fit only, we typically will obtain a transfer function of the largest order possible. One way to overcome this problem is to introduce a penalty term for complexity in the criterion function. An important class of such criteria is of the form

$$A(n) = \log \det \hat{\Sigma}_T(n) + 2ns\frac{c(T)}{T}; \quad 0 \le n \le N_T \tag{4.1}$$

Here $\hat{\Sigma}_T(n)$ is the MLE of Σ_0 over $M(n)x\Sigma$ (with sample size T), which is a measure of misfit, 2ns is the dimension of the manifold $M(n)$ and $c(T)$ and N_T are nonstochastic functions of T, which have to be prescribed. The estimator of n is defined as the minimizer of $A(n)$. N_T may be constant or increase with T, for instance at the rate $(\log\ T)^a$, $a < \infty$. The main question is the choice of $c(T)$ which defines the tradeoff between fit and complexity. Special cases of importance are the *AIC criterion*, defined by $c(T)=2$ (Akaike 1969, 1977) and the *BIC criterion*, defined by $c(T)=\log T$ (Rissanen 1983, Schwarz 1978).

These and related criteria have been derived from principles such as maximum entropy, from Bayesian arguments or from the idea of minimal code length for the data. For instance the idea behind AIC may be described as follows (see Findley 1985): Let $E_T(\theta)$, with $\theta = (k, \Sigma)$, denote the expectation of the likelihood function defined by (3.2) with respect to the true value $\theta_0 = (k_0, \Sigma_0)$, and let $\hat{\theta}_1, \hat{\theta}_2$ be two MLE's for two subclasses, i.e. for $M(n_1)x\Sigma$ and $M(n_2)x\Sigma$ respectively in our case. Then the suggestion is to prefer $\hat{\theta}_1$ over $\hat{\theta}_2$ if $E_T(\hat{\theta}_1) - E_T(\hat{\theta}_2) < 0$ holds. However, $E_T(\hat{\theta}_1) - E_T(\hat{\theta}_2)$ is unknown. It can be shown that

$$L_T(\hat{\theta}_2) - L_T(\hat{\theta}_1) + 2n_2s - 2n_1s = AIC(n_1) - AIC(n_2) + \eta_T$$

is an estimator for this quantitiy which is asymptotically unbiased and η_T is $o(1)$ under some additional assumptions..

The most important results concerning the asymptotic properties are:

(i) if the true transfer function is of order n_0, and $n_0 \le N_T$ (from a certain T onwards) holds, then (under some additional assumptions) BIC gives an a.e. consistent estimator for n_0. More generally, this is true for all criteria of the form (4.1) which satisfy

$$\lim_{T\to\infty} \inf(c(T) / \log T) > 0 \quad \text{and } \lim c(T)/T = 0$$

On the other hand AIC is not consistent for n_0 (Hannan 1980, 1981)

(ii) As has been shown by Shibata (1980, 1981) for the scalar autoregressive (i.e. $c(z)=0$, $d(z) = 1$) case, AIC has some optimality properties if the true system is of infinite order.

The Kronecker indices can be estimated by criteria analogous to (4.1), by just replacing 2ns by the dimensions d_α of U_α (see e.g. Akaike 1976, Hannan and Kavalieris 1984, Hannan and Deistler 1988). Again BIC is consistent, whereas AIC is not.

Alternative approaches for determining multi-indices such as the Kronecker indices, are based on an investigation of the linear dependence structure of a first estimate of the Hankel matrix H (Fuchs 1990).

5. Alternative Approaches in Linear System Identification

There is a relatively large number of alternative approaches and of extensions to the main stream approach. We only list some of the most important ones without going into

any detail
- Tracking of parameters for linear time varying systems by the use of recursive procedures
- Identification of unstable systems
- Identification if there is no a-priori distinction between inputs and outputs
- Identification under feedback; identification in the context of control
- Robust identification procedures
- The use of model reduction techniques (e.g. Hankel-norm approximation) in identification
- Identification if there is noise in the inputs too (errors-in-variables)

References

Akaike, H. (1969). Fitting autoregressive models for prediction. *Ann. Inst. Statist. Math.* **21**, 243-247.

Akaike, H. (1976). Canonical correlation analysis of time series and the use of an information criterion. in *System Identification: Advances and case studies* (eds. R.H. Mehra and D.G. Lainiotis), 27-96. New York, Academic Press.

Akaike, H. (1977). On entropy maximization principle. In *Applications of statistics* (ed. P.R. Krishnaiiah), 27-41. Amsterdam, North Holland.

Box, G.E.P. and G.M. Jenkins (1970). *Time Series Analysis, Forecasting and Control*, San Francisco, Holden Day.

Brockwell, P. and R. Davies (1991). *Time Series: Theory and Methods*. Second Ed., New York, Springer Verlag.

Caines, E.P. (1988). *Linear Stochastic Systems*, New York, John Wiley and Sons.

Deistler, M. and B.M. Pötscher (1984). The behavior of the likelihood function for ARMA models, *Adv. Appl. Probab.* **16**, 843-865.

Dunsmuir, W. and E.J. Hannan (1976). Vector linear time series models. *Adv. Appl. Probab.* **8**, 339-364.

Findely, D.F. (1985). On the unbiasedness property of AIC for exact or approximating linear stochastic time series models. *J. Time Series Anal.* **6**, 229-252.

Fuchs, J.J. (1990). Structure and order estimation of multivariable stochastic processes. *IEEE T I.AC* **35**, 1338-1341.

Hannan, E.J. (1980). The estimation of the order of an ARMA process. Ann. Statist. 8, 1071-1081.

Hannan, E.J. (1981). Estimating the dimension of a linear system. *J. Multivariate Anal.* **11**, 459-473.

Hannan, E.J. and M. Deistler (1988). *The Statistical Theory of Linear Systems*. New York, John Wiley.

Hannan, E.J. and L. Kavalieris (1984). Multivariate linear time series models. *Adv. Appl. Prob.* **16**, 492-561.

Hazewinkel, M. (1979). On identification and the geometry of the space of linear systems. *Lecture notes in Control and Information Sciences*, vol. 16, 401-415, Berlin, Springer Verlag.

Hazewinkel, M. and R.E. Kalman (1976). Invariants, canonical forms and moduli for linear, constant, finite dimensional, dynamical systems. In *Lecture notes in Economics and Mathematical Systems,* vol. 131 (eds. G. Marchesini and S.K. Mitter), 48-60. Berlin, Springer Verlag.

Ljung, L. (1978). Convergence analysis of parametric identification methods, *IEEE Trans. Autom. Control* **AC-23**, 770-783.

Ljung, L. (1987) *System Identification - Theory for the User*. Prentice Hall, Inc., Englewood Cliffs, New Jersey 07632.

Pötscher, B.M. (1991). Noninvertibility and pseudo-maximum likelihood estimation of misspecified ARMA models. *Econometric Theory* **7**, 435-449

Rissanen, J. (1983). Universal prior for parameters and estimation by minimum description length. *Ann. Statist.* **11**, 416-431.

Schwarz, G. (1978). Estimating the dimension of a model. *Ann. Statist.* **6**, 461-464.

Shibata, R. (1980). Asymptotically efficient selection of the order of the model for estimating parameters of a linear process. *Ann. Statist.* **8**, 147-164.

Shibata, R. (1981). An optimal autoregressive spectral estimate. *Ann. Statist.* **9**, 300-306.

Söderström, T. and P. Stoica (1988). *System identification*, Prentice Hall, Inc., Englewood Cliffs, New Jersey.

ON NUMERICAL SIMULATION IN THE MICROELECTRONICS INDUSTRY

ALBERT GILG
Siemens AG, ZFE BT SE 4
Corporate Research and Development
Otto-Hahn-Ring 6
D-8000 München 83
Germany

ABSTRACT. Microelectronics is an exploding industrial application area pushing progress in many areas, i.e. materials science, manufacturing technology, computer science, simulation technology and numerical mathematics. In the early 70s, circuit simulators were the first generally accepted software tools in microelectronic development. These applications initiated research acitivites e.g. in numerical treatment of differential-algebraic systems. Typical problems occur due to the uncertainties in the physical/chemical modelling leading to different and complicated sets of equations. These are a moving target for numerical analysis and the development of algorithms.

Today process, device and circuit simulation have found broad acceptance as they are integrated into TCAD systems (TCAD: Technology-Computer Aided-Design) which are easy to apply to engineering tasks.

Ongoing miniaturization poses new problems. Second order physical effects which have been neglected so far become dominant. Also simulations in three spatial dimensions are going to become standard engineering practice. To attack these size problems, new computer hardware, vector, parallel processores and specialized hardware are necessary and require new algorithmic approaches.

Again new application areas appear: Influence of manufacturing equipment has to be investigated. Therefore, special fluid dynamics problems have to be solved. Future applications also include integration of nonelectric devices on one chip. This coupling of electric with mechanical or optical devices into microsystems poses further challenges to mathematics for new algorithmic approaches.

1. Industrial Aspects

Microelectronics is a key technology for the next decades. Industrial competition in this field is driven by extremely short innovation cycles. The number of functions per unit area on a chip increases four times every three years (see Fig.). This speed has been pushing progress in many areas and poses exponentially increasing challenges to a number of technologies contributing to microelectronics development. This presentation focuses on silicon process technology since it is the leading edge of main stream industrial efforts. There are more demanding research acitivities in numerical simulation for microelectronics, but they are of less industrial importance.

Investment for a Fabrication Line

Structure Size Reduction and Investment for Fabrication of DRAM Chips

From an industrial point of view, a more and more dominating problem occurs in business topics such as capital expenditure and return on investment which affect long term planning decisions in a company. For example one of the main limitations in the development of DRAMs will not be a technological limitation, but costs associated with the construction of adequate processing lines to manufacture them. Cost of fabrication lines also grows exponentially: a typical line for production of 16M DRAM chips today is an investment of over 500 million dollars (see Fig.). A next generation plant necessary in the mid-nineties will cost far beyond one billion dollars. A simple but unavoidable answer are so-called strategic alliances. They combine efforts of two and more companies to share these tremendous costs and risks in development and manufacturing.

Two strategic upheavals for research activities are inevitable: First, a growing number of research challenges enforce concentration of widespread and multiple activities because of limited resources in industrial and academic research groups. And secondly, manufacturing science will become a very important if not dominating area for industrial progress in microelectronic technology.

2. Simulators in Semiconductor Technology

In the early 70s, circuit simulators were the first generally accepted software tools in microelectronic development. Pederson [1] gives a historical review on the beginnings of circuit simulation from an engineers view. Considerable effort was spent by electronics engineers during the 80s to catch up with the exploding complexity of circuit sizes. Most attempts focussed so-called waveform relaxation techniques. After several announcements of break throughs these activities turned down in late 80s. Consequently, simulators for large electronic circuits are now developed by company research groups with assistance of a few small academic research groups. Examples are Texas Instruments' simulator SUPPLE [2] and TITAN [3] by Siemens AG. It is noteworthy to have a look at these developments from a mathematicians viewpoint. The mathematical basis for circuit simulation are differential-algebraic systems, sometimes called descriptor systems. There is a well known theory for linear systems (Gantmacher [4]). One of the first numerical treatments was presented by Gear [5] in 1971 in an engineering journal. Mathematicians needed more than ten years to catch up with the ongoing engineering efforts and to contribute significantly in time to industrial developments. Todays research activities for circuit simulation are focussed on decomposition techniques like multi-rate integration and methods for non-linear circuits with highly oscillating devices.

In the late 70s, use of simulation tools was extended to semiconductor device characterization. Mathematically these tasks are governed by a set of nonlinear partial differential equations with very specific and tricky behaviour. These so-called semiconductor equations have been analyzed using analytical and numerical methods quite successful ([6]). From an industrial standpoint, satisfactory progress has been achieved almost comparable to the situation in circuit simulation.

The next step in covering process technology by simulation was to attack process simulation i.e. characterization of manufacturing steps on a microscopic level. Dominant effects are described by nonlinear equations of the diffusion type usually coupled with nonlinear Poisson-type equations. Typical problems arise due to uncertainty in physical/chemical modelling leading to different and very complicated systems of equations. Here, mathematicians are in a weak position. They usally need a considerable amount of time to investigate and analyze modified sets of nonlinear partial differential equations. But in practice targets move quickly. This is one of the main reasons why process simulation is a domain of industry inhouse developments. Short distances and strong collaboration of engineers, physicists and mathematicians are inevitable for successful progress with process simulators.

Nevertheless process, device and circuit simulation have found broad acceptance throughout the microelectronics industry. An essential step was to provide an engineering tool environment for these heterogenous simulator types. Today these tools are integrated into TCAD systems (TCAD: Technology Computer Aided Design) including optimization routines for predictive technology analysis [7]. Nobody can imagine further progress in semiconductor technology without use of such simulation systems.

3. New Challenges

Ongoing miniaturization poses new problems. Second order physical effects which have been neglected so far become dominant. We experience significant gaps in modelling in terms of microscopic, macroscopic and technological phenomena. Elaborate Monte Carlo codes supply extensive knowledge of atomic and molecular forces of crystal structure and electron properties. These simulators can handle implantation and diffusion from first principles but consume a tremendous amount of computing time. Model calibration to match experimental results causes severe problems because most parameters are not directly measurable. These simulators are an indispensible basis for further understanding of semiconductor physics but are of minor use for a device design engineer. He is overtaxed to deal with these basic phenomena. Therefore, macroscopic and technological models have to be abstracted from microscopic models and provided in fast simulators for engineering applications.

From an algorithmic view, previously homogeneous simulation areas split up into quite different problem types. Microscopic modelling is governed by Monte-Carlo or cellular automata-type algorithms. Macroscopic and

technological oriented models lead to differential equations with new nonlinear coupling terms and often to extended sets of partial differential equations demanding new analysis and algorithmic techniques.
In the near future, three-dimensional and transient simulators will become a general engineering tool. To attack these size problems new computer hardware, vector, parallel processors and specialized hardware is not only helpful but necessary. We are looking for new algorithmic approaches on parallel and distributed systems.

But also new application areas for numerical simulation appear: Influence of manufacturing equipment has to be investigated. There are three reasons. First, boundary conditions for microscopic process modelling are no longer available by thumb rules for deep submicron structures and have to be provided by a global calculation. Secondly, the number of properly functioning chips, called yield, depends on homogeneous conditions across the silicon wafers. Since size of wafers and sensitivity to disturbances increase every new generation a careful investigation becomes mandatory. Lastly, we already have mentioned the dramatically increasing costs of fabrication plants. Over 70% of this investment is due to equipment cost.
Equipment simulation is a broad subject, it can be separated again into three categories: technologic, macroscopic and microscopic, depending on geometrical dimensions and number of studied particles where the physical and chemical processes take place. At the meter level, the information that can be obtained from equipment simulation is the determination of pressure, temperature, velocity, concentration of chemical species for every location in the equipment, and uniformity across the wafer. The solution of these variables is the result of the macroscopic view of the equipment that is intrinsic to fluid-mechanics description. At the micron level, it provides structural information on the geometrical structure of the layers (typically of the same dimensions of transistor). In this level, the fluid dynamics description cannot acccount for the processes and a particle description is necessary because the dimensions of the structures in question are much smaller than the gas mean-free path. The models used in this regime are typically Monte Carlo and produce values related to step coverage, trench shape, and undercut. At the Angstrom level, equipment simulation focusses on the microscope nature of the chemical reaction at the wafer surface. For this regime the development of Monte Carlo techniques and quantum-mechanical computations is imperative and provide informtion about reaction-rate kinetics and their influence in deposition rates and film composition.

Very promising initial results are available for the fluid-dynamical approach [8]. Simulations are based on the solution of the mass-, chemical-species-, momentum-, and energy-continuity equations. These Navier-Stokes-type equations are very general and allow analysis and optimization of a variety of quite different equipment like diffusion furnaces, chemical vapor deposition (CVD) and etching reactors. Besides the impact of numerical methods, modelling of chemical phenomena is the key problem.

But not only molecular chemical and (fluid-)dynamical aspects gain importance. Future applications also include integration of nonelectric devices on one chip. This coupling of electric with mechanical or optical devices into microsystems is not only a simple coupling of existing codes but enforces new algorithmic approaches on a powerful computing evironment.

4. Conclusions

Industrial microelectronics is an interdisciplinary effort: It is commonly accepted now that it is imperative to include mathematics as a basis and backbone for microelectronics progress especially in the field of numerical simulation. Research results in applied mathematics have been some steps behind industrial applications for a long time but have now succeeded in significantly contributing in time. One must be aware of the many problems caused by an interdisciplinary effort. One critical topic is a significant lack in analysis and measurement techniques necessary to calibrate physical models. Simply fitted parameters may cause useless results and a waste of expensive resources. Strong interaction of engineers, physicists and mathematicians at research institutes and in industry are required for the usefulness of numerical simulation.
New challenges show up in microelectronics development. Research for control of manufacturing by software techniques will be a key issue and a continuing source of incentives for mathematics in industry.

References

[1] D.O. Pederson, A Historical Review of Circuit Simulation. IEEE Trans. Circuits and Systems CAS-31(1), 103-111, 1984

[2] P. Cox et al., SUPPLE: Simulator utilizing parallel processing and latency exploitation. Proc. ICCAD, 368-371, Santa Clara, CA, 1987

[3] U. Feldmann, R. Schultz, TITAN: A universal circuit simulator with event control for latency exploitation. Proc. ESSCIRC, 183 - 185, Manchester, U.K., 1988

[4] F. R. Gantmacher, The Theory of Matrices, vol. 2, Chelsea, New York, 1964

[5] C. W. Gear, Simultaneous Numerical Solution of Differential-Algebraic Equations. IEEE Trans. Circuit Theory CT-18, 89-95, 1971

[6] S. Selberherr, Analysis and Simulation of Semiconductor Devices. Springer, Wien 1984

[7] A. Gilg, Silicon Technology Development and Optimization by Integrated Process, Device and Circuit Simulation System SATURN. ECMI 7, 235 - 238, Teubner, Stuttgart 1991

[8] J. I. Ulacia, C. Werner, Equipment Simulation: Part I, Solid State Technology, 107-113, Nov. 1990

INSTABILITIES IN ROLLER CONVEYORS USED IN FLAT GLASS MANUFACTURE

D.GELDER
Pilkington Technology Centre
Hall Lane, Lathom
Ormskirk,Lancashire, L40 5UF
England

ABSTRACT.   Mathematics in industry may not often involve working at the
limits of what is possible, but it can still be challenging in requiring
a variety of techniques to be brought together to provide a theoretical
basis for a manufacturing process. Such is the case in considering the
cooling of a ribbon of glass from a hot viscous condition to the cold
elastic state as it is carried on a train of steel rollers. These
rollers sometimes become eccentric, and a thermal instability which can
lead to this is described. Further when a roller in the viscous regime
becomes eccentric it is capable of putting a permanent wave into the
glass - but it is by no means obvious how the magnitude of this wave is
related to the temperature, diameter and pitch at the offending roller.
Algebraic and numerical studies of the problem are shown to produce some
surprising results.

1. MATHEMATICS IN INDUSTRY

Mathematics degrees are not generally thought of as vocational by any of
the parties involved. This is good in that employment using the specific
skills developed is not generally available. However industry does need
a chosen few, and the work to be done is conveniently thought of in
three categories. The first is obviously the exploitation of computers
in the optimisation of industrial products and processes. The adage
"The shorter you are of ideas the bigger the computer you need"
is a reminder that computer studies can easily grow beyond the data on
which they are based and away from the problem they are intended to
address. However industry clearly needs real specialists in numerical
methods and all the related computing skills. Some will be working with
supercomputers, some with PCs, but on the right application they can all
make a tremendous impact on product and process development. Perhaps
less often, industrial pressures lead to the development of mathematical
techniques which are essentially new. Quite rightly, such cases can
attract special interest in universities. The present topic fits neither
the specialist computing nor the novel mathematics category. Rather it
is an illustration of how material found in a reasonably applied degree

course can find direct application. It will be seen that a surprising amount of what is taught is included in the core of material which a working mathematician calls on day after day.

2. GLASS ON ROLLERS

When first formed flat glass has to be cooled from temperatures where it behaves as a very viscous liquid down to temperatures where it is an elastic solid. This cooling occurs as a continuous ribbon is carried along a train of rollers, which will typically be at a few hundred mm pitch. The choice of roller material involves several issues - among others it should not damage the glass, and it must be robust enough not to be damaged by the glass when the inevitable breakages occur. Steel is sometimes the preferred material, and it is hard to imagine that anything much can go wrong in carrying the glass over the roller train. It turns out that this is far from being the case. One problem is that various metallurgical inhomogeneities may result in rollers becoming bowed as they heat up to and run at operating temperatures of several hundred °C. However when a severe bow develops a thermal origin seems more likely, and various questions related to the initiation, persistence and decay of thermal bow are amenable to mathematical investigation. The particular case of initiation due to a thermal instability is examined below.

Bowed rollers are undesirable in that they increase the risk of surface damage to and breakage of the glass. If they occur where the glass is still viscous, the wave with pitch equal to roller circumference which is imparted to the glass is more of a problem. It is important to know the temperature range in which the glass is particularly prone to develop a wave, and to determine whether there are combinations of roller diameter and pitch which are to be preferred, or avoided. A locally isothermal theory is developed algebraically below and a numerical treatment is also discussed which is more general but as it stands suffers from stiffness at the higher temperatures of interest.

3. BASIS OF ROLLER STABILITY THEORY

The worst of the observed eccentricities correspond to a temperature difference across the roller of some tens of °C - which is the sort of difference seen between rollers running in contact with the glass and rollers which, for some operational reason, have been dropped out of contact. Further badly eccentric rollers run in contact with the glass for only part of their rotation. This seems to be the basis of how an eccentricity can be maintained, although it is surprising how effective the very limited contact zone must be in transferring heat from the glass to the roller. The possibility of an inherent instability due to contact heat transfer variations amplifying any initial eccentricity is studied below and illustrates the sort of analysis needed for a range of related problems.

33

3.1 Elastic Aspects

For the present purpose the glass is assumed to be in the elastic regime, and the following notation is used:

$$
\begin{aligned}
h &= \text{glass thickness} \\
E &= \text{Young's modulus of glass} \\
v &= \text{Poisson's ratio of glass} \\
\rho &= \text{density of glass} \\
E_s &= \text{Young's modulus of steel} \\
v_s &= \text{Poisson's ratio of steel} \\
X &= \text{roller pitch} \\
r &= \text{roller radius} \\
g &= \text{acceleration due to gravity} \\
F &= \text{roller reaction per unit width} \\
l &= \text{glass to roller contact length.}
\end{aligned}
$$

Suppose one roller has a small eccentricity e, small enough for the glass to remain in contact with the roller throughout a revolution. Elastic plate bending theory is fairly easily applied to the case of a plate with periodic line support to give when the roller is arched up (+) or bent down (-):

$$
F = \rho g h X \pm \frac{eEh^2}{0.84\rho g X^4}
$$

The main result required of elastic theory is the contact length l. It turns out that in the situations of interest it is self-consistent to work on the basis that l is small compared with h, when:

$$
l = 2\sqrt{\frac{4rF}{\pi}\left(\frac{1-v^2}{E} + \frac{1-v_s^2}{E_s}\right)}
$$

3.2 Thermal Aspects

The combination of radiation and air flow will lead to a cooling regime which may be regarded as giving an overall heat transfer from the roller surface at mean temperature T to an ambient temperature T_A of

$$
H(T-T_A)
$$

The radiative contribution to the coefficient H will be larger at higher temperatures, and T_A may be close to T in an enclosed section or much lower in an open section or where cold air is being blown. The radiative transfer from the glass plays a large part in controlling the roller temperature, and can be estimated relatively reliably as

$$
H_R(T_G-T)
$$

where H_R is a heat transfer coefficient and T_G is the glass temperature.

Because the thermal conductivity of steel is much larger than that of glass, and assuming a high interface heat transfer coefficient, the average heat transfer coefficient over the contact time l/u - u being the glass velocity - may be approximated by

$$\sqrt{\frac{k\rho c u}{\pi l}}$$

where k,c = thermal conductivity and specific heat of glass. Averaging over a revolution, the effect of contact is to provide a mean heat transfer coefficient to a part of the roller which experiences contact length l of

$$H_C = \frac{1}{2\pi r}\sqrt{\frac{k\rho c u l}{\pi}}$$

Applying this formula to the situation of rollers running entirely in or out of contact gives a result in remarkably good agreement with the observed temperature difference. There is though a complication: the heat transfer through the film of air on each side of the contact zone is comparable with the above surface heat transfer rate over a distance considerably greater than the contact zone! However the separations involved are only of the order of the surface roughness, so the idealised theory overestimates conduction through the air, while also being suspect in assuming effectively perfect contact. For all its faults, since it gives a good answer the above formula can perhaps be accepted as a working approximation.

3.3 Overall Stability

Summarising the situation, as a roller of small eccentricity e rotates:

$$H_C \text{ changes periodically with amplitude} \propto \frac{eh}{rX^5}\sqrt{u}\sqrt[4]{rhX}$$

Although the rollers have high thermal conductivity, they are thin walled, and conduction round the walls can be shown to be relatively small. Radiation across the bore is a bit more important but a reasonable approximation to the peripheral temperature variation can often be made neglecting this to give a range across the roller ΔT where

$$\Delta T = \frac{2H\,H_C(T_G - T_A)}{(H_R + H)^2}$$

This will induce a curvature along the roller and a corresponding eccentricity e'. The situation is unstable if e' > e.

3.4 Conclusions

It is clear that there is a potential thermal instability in running steel rollers - for various reasons (which can be seen from the above theory) other types are not at risk in the same way. Inserting realistic data, situations can be envisaged which appear to be at risk. Often in industrial situations "good" and "bad" operating regimes can be identified from past experience, and the role of mathematics is to provide a general understanding of which good features must be retained as a process is developed. In this case desirable features are:

> slower moving and thinner glass
> larger rollers with smaller thermal expansion
> lower glass to ambient temperature differentials
> more than anything, larger pitch.

While this analysis gives a degree of understanding of steel roller instability, and indeed provides a rather neat if somewhat idealised result, it also illustrates the somewhat messy nature of much industrial mathematics. All sorts of factors have to be checked quantitatively to see whether or not they are relevant to the main course of the argument. The other aspect of glass on rollers examined below is rather different, involving more mathematical and less physical intuition.

4. ROLLER WAVE

This analysis is concerned with the waves frequently seen in flat glass whose wavelength is the circumference of the lehr rollers. The presumed mechanism of formation is that, when the inevitable eccentricity in the rollers becomes too large, a waviness is introduced which persists sufficiently to have an unacceptable amplitude in the final product. It is sufficient to study just one eccentric roller in various positions. Two essentially different mathematical approaches to solving this problem have been identified. The first is a direct numerical simulation of the bending of the glass along the roller train. This is a very flexible approach, but it can be time consuming to use, as it is necessary to check that discretisation errors are not masking the true situation. The second is an algebraic treatment. This is fast and reliable, but limited to idealised cases with constant roller pitch and temperature. It is particularly useful in providing exact solutions to check the accuracy of the numerical approach, and in understanding the general nature of the problem.

In both cases it is assumed here that the glass remains in contact with the rollers - this is certainly true for small eccentricities. Another somewhat less satisfactory approximation is needed to make the calculation tractable - a Maxwell model is used, taking viscoelastic stress decay to be exponential. Since the temperature range believed to be most important for the introduction of roller wave is well above the annealing zone this should be reasonably reliable.

4.1 The Basic equations

Neglecting any variation across the width, the displacement is a function of x, the distance along the lehr, and t, time. The displacement can be split into a component y, the shape the ribbon would have if elastic gravitational displacements did not keep it in contact with the rollers, and a component w, the elastic displacement. With the following additional notation:

$$M = \text{bending moment}$$
$$\mu = \text{viscosity}$$

between the rolls w satisfies the normal plate deflection equation:

$$\frac{Eh^3}{12(1-v^2)}\frac{\partial^2 w}{\partial x^2} = M \tag{1}$$

y satisfies its viscoelastic analogy ($E=3\mu$, $v=1/2$) - taking into account the forward velocity u:

$$\frac{\mu h^3}{3}\left(\frac{\partial^3 y}{\partial x^2 \partial t} + u\,\frac{\partial^3 y}{\partial x^3}\right) = M, \tag{2}$$

while M satisfies the normal moment equation:

$$\frac{\partial^2 M}{\partial x^2} = \rho gh. \tag{3}$$

At the rollers:

$$w+y = 0 \text{ (or the displacement at the eccentric roller),} \tag{4}$$

$$w,\ \frac{\partial w}{\partial x},\ y,\ \frac{\partial y}{\partial x},\ \frac{\partial^2 y}{\partial x^2} \text{ and M are continuous.} \tag{5}$$

Taking w+y to be zero at the eccentric roller too, and with constant μ and roller spacing X, there is clearly a steady periodic solution. The interest is in the residual shape left as the ribbon passes into the cooler elastic regime, and the steady component is of no significance. Instead of (3) then, it is only necessary to consider

$$\frac{\partial^2 M}{\partial x^2} = 0 \tag{6}$$

w and y are now defined as their departures from the steady solution. If there are displaced rollers or top to bottom temperature differences the equations can be used to predict (steady) residual curvature, and this has been considered to some extent in [1], [2] and [3]. However the interest here is in the case of an eccentric roller, when at every point the solution can be expected to be periodic with period $2\pi/\omega$ where:

$$\omega = u/(\text{radius of eccentric roller})$$

Somewhat intricate arguments are needed to define appropriate boundary conditions for (1), (2) and (6) at the first and last rollers considered. These are not developed here, but M can safely be taken to be zero at both. In principle one could also control the inlet curvature, and this can be taken as zero to provide the final condition. Two further conditions are apparently needed to solve the equations. However it is always possible to add a linear function of x to w and subtract it from y, so nothing is lost by constraining the result to have (for example) zero y upstream and downstream.

4.2 Numerical Solution

Various options can be envisaged for discretisation of (1), (2) and (6), but a simple explicit scheme requires a minimum of programming effort and does not take significant computer time - except when the eccentric roller is in the top end of the temperature range or it is necessary to ensure the discretisation errors are much smaller than is generally acceptable. A successful program has been written based on a uniform mesh in the x direction, with rollers spaced at integer numbers of mesh intervals (but not necessarily uniformly spaced). A time dependent calculation is used, starting from an initial condition with y defined (in fact as zero) at each mesh point.

At each step w is known from the contact condition at each roller. (1), (6) and the boundary conditions on M are satisfied taking for w a natural cubic spline. w can then be evaluated at each mesh point, and M is conveniently taken (away from the end points where it is known) as:

$$M_i = \frac{Eh^3}{12(1-v^2)} \cdot \frac{w_{i+1} + w_{i-1} - 2w_i}{s^2} \tag{7}$$

where the subscript indicates the mesh point considered, and s is the length of the mesh interval.

(2) can be solved taking an explicit increment in time τ: this can be expected to be stable and reasonably reliable so long as the following two conditions are satisfied:

$$\frac{u\tau}{s} \leq 1 \quad \text{and} \quad \frac{E\tau}{4(1-v^2)\mu_i} \leq 1 \tag{8}$$

However since a train of many rollers is considered in practice it is highly desirable to have

$$\frac{u\tau}{s} = 1 \quad \text{and hence} \quad \mu_i \geq \frac{Es}{4(1-v^2)\,u} \tag{9}$$

Otherwise there is numerical damping of the travelling wave which is difficult to allow for in interpreting the results. As a result of this constraint the explicit method is somewhat unsatisfactory above 630°C.

4.3 Algebraic Solution

As previously noted, equations (1), (2) and (6) with the conditions (4), (5) and the imposed additional upstream and downstream conditions are amenable to algebraic solution. This solution can be developed writing y and w as:

$$y = A(x)\ \sin \omega t + B(x)\ \cos \omega t \tag{10}$$
$$w = C(x)\ \sin \omega t + D(x)\ \cos \omega t \tag{11}$$

In any interval between two rollers, introducing the parameters

$$\alpha = \frac{XE}{4\mu u(1-v^2)} \qquad\qquad \beta = \frac{\omega X}{u}$$

A, B, C and D may be written as as:

$$A(x) = \alpha\,(a_1(\cos(\omega x/u)-1)+b_1\sin(\omega x/u)+c_1 x^3/X^3+d_1 x^2/X^2+e_1 x/X)+f_1 \tag{12}$$

$$B(x) = \alpha\,(-a_1\sin(\omega x/u)+b_1(\cos(\omega x/u)-1)+c_2 x^3/X^3+d_2 x^2/X^2+e_2 x/X)+f_2 \tag{13}$$

$$C(x) = -c_2\beta x^3/X^3 + (3c_1 - d_2\beta)x^2/X^2 + g_1 x/X + h_1 \tag{14}$$

$$D(x) = c_1\beta x^3/X^3 + (3c_2 + d_1\beta)x^2/X^2 + g_2 x/X + h_2 \tag{15}$$

It is necessary to relate the coefficients in one interval to those in the next so as to satisfy all the boundary and continuity conditions. Away from the eccentric roller, this is most conveniently done expressing the coefficients as a combination of individual eigenfunctions, each of which are such that

$$w,\ \frac{\partial w}{\partial x},\ \frac{\partial^2 w}{\partial x^2} \ (\text{i.e. M}),\ y,\ \frac{\partial y}{\partial x} \text{ and } \frac{\partial^2 y}{\partial x^2}$$

at the downstream end of an interval are λ times their values at the upstream end. h_1 and h_2 may be readily eliminated in terms of f_1 and f_2, leaving a set of equations which may be written:

$$Pv = \lambda\ Qv$$

where P and Q are 12 by 12 matrices and v is the vector consisting of
elements ordered as say

$$f_1, \; f_2, \; c_1, \; c_2, \; d_1, \; d_2, \; e_1, \; e_2, \; g_1, \; g_2, \; a_1, \; b_1$$

P and Q involve only the two dimensionless parameters α and β. Q has 2
zero rows: there are four unit eigenvalues, and six important values of
λ which turn out to consist of three complex pairs, two of modulus less
than 1 and one of modulus greater than 1. There is an exceptional case
when $\beta=2\pi$ when one pair has modulus 1 - this corresponds physically to
the circumference and hence wavelength being equal to the pitch so that
a deformed ribbon can sit on the rollers without any bending and ride
downstream unchanged.

In the usual situation, it is clear now why four upstream and two
downstream boundary conditions - two and one for each phase - are needed
to give a well-posed problem. Suppose the eigenvalues of modulus
greater than one are λ_1, λ_2 (complex conjugates) with eigenvectors

$$E_1(1) \; \dots \; E_1(12) \quad E_2(1) \; \dots \; E_2(12) \quad \text{for } f_1 \dots b_1$$

Moving on to the case of an isolated eccentric roller, at the roller
upstream of it and beyond the solution must be a linear combination of
E_1 and E_2. Similarly suppose the eigenvalues of modulus less than 1 are
λ_3, λ_4, λ_5, λ_6 with eigenvectors E3, E4, E5 and E6. At and beyond one
roller downstream of the eccentric roller the solution must be a linear
combination of E_3 to E_6. Taking the eccentric roller to have a
displacement $r \sin \omega t$, the full solution can be obtained considering one
interval upstream and downstream in detail and matching to the
eigenfunctions at one roller upstream and downstream of the eccentric
one. Allowing for the upstream and downstream linear components
corresponding to the unit eigenvalues gives a system of 38 equations in
38 variables which is conveniently solved using the Harwell Library.

It was initially assumed that the algebraic solution would give one
quickly decaying and one slowly decaying downstream eigenfunction pair,
and that the latter would be much the more important since its remains
would become "frozen in" as the glass cooled. Certainly there clearly is
a slowly decaying pair, but the glass in fact cools quickly enough for
it not to be good enough to seek an approximate solution to the
practical situations of interest solely in terms of this pair being
frozen in. The direct numerical solution is thus an essential tool.
However, the general features of the algebraic results are instructive.
A check using realistic data confirms that the two approaches give
consistent results so long as extrapolation as $1/n$ is applied to the
numerical results with the largest n at least 30 (n being the number of
mesh points between rollers).

4.4 Algebraic Theory Results

While it has become clear that the algebraic theory cannot be adapted readily to approximate the non-isothermal situations which occur in practice, the results do help to provide some understanding of the behaviour of the system. Attention is here concentrated on the amplitude (relative to the roller eccentricity at the eccentric roller) and the roller to roller decay factor of the slowly decaying wave. A typical speed of .11 m/s and spacing of 300mm is considered at various values of roller circumference and temperature with a viscosity variation characteristic of flat glass.

Looking first at the decay factors, these approach 1 at lower temperatures, as is to be expected with the higher viscosity and the glass moving into an essentially elastic regime. More surprisingly, the decay factors also approach 1 at higher temperatures. As regards the variation with roller circumference, the calculation reproduces the obvious result that for a value of 300 mm equal to the roller pitch the wave can "ride" the rollers with no need for elastic deformation and hence no decay. Surprisingly, the decay factor stays close to 1 until the wavelength is considerably greater than the 300 mm pitch. The minimum factors (fastest decay) occur around 590°C, being .97, .84 and .42 at wavelengths of 400mm, 500mm and 600mm respectively.

Apart from the expectation that the amplitude of the generated wave should become small at low temperatures, it is not clear what results to expect. It turns out that the equivalent magnitude is remarkably close to (but not identical with) 1 minus the decay factor. In particular for 300 mm circumference the magnitude is zero - what appears to be the worst situation when the generated wave could pass down the lehr without any decay is (in this idealised isothermal case) the best! Only the eigenfunctions which decay rapidly away from the eccentric roller are involved. In fact there is a rather limited range of temperatures where substantial magnitudes can occur - this analysis suggests that roller wave does not come from temperatures above 640°C because
a) the magnitude of the generated wave is small
b) what there is decays in its passage over subsequent rollers.
However overall these results fail to explain the plant observation that at 400 mm circumference the final relative magnitude (not just the value at the eccentric roller) can reach .2. In terms of the algebraic theory, the cooling rate is presumably fast enough to lock in a proportion of the faster decaying eigenfunction as well.

References

[1]   Bourne, D.E. and Marshall, E.A. (1986) 'Roll-out mechanics of glass sheet manufacture I', Int. J. Engng. Sci. **24**, 119-134.
[2]   Marshall, E.A. (1986) 'Roll-out mechanics of glass sheet manufacture II', Int. J. Engng. Sci. **24**, 135-159
[3]   Bourne, D.E. and Marshall, E.A. (1989) 'Residual curvature in glass sheet roll-out', Glass Tech. **80**, 128-134

Acknowledgement: This paper is published by permission of the Directors of Pilkington plc and of Dr. A. Ledwith, Director of Group Research.

INVERSE CASCADES IN TURBULENCE

ALAN C. NEWELL
Arizona Center for Mathematical Sciences
University of Arizona
Tucson, Arizona 85721

ABSTRACT. In this article, I report on some joint work with Vladimir E. Zakharov. We suggest that inverse cascades of finite flux motion constants can play a central role in the production of intermittency in both optical and hydrodynamic turbulence and that they may also be responsible for building the dipole component of planetary and solar dynamos.

1. Introduction

Kolmogorov (1941) proposed a universal theory for small-scale eddies in high Reynolds number turbulent flows. It rests on several assumptions, that the transfer of spectral energy to large wavenumbers is local over a window of transparency (the inertial range) in wavenumber space, that statistical information is lost in the cascade so that average flow quantities are scale invariant and determined by the mean rate of energy flux ϵ which is constant in a statistically steady state. If $E(k)$ is the spectral energy ($\overline{u^2} = \int E(k)\, dk$) then, for isotropic turbulence, the equation for $E(k)$ is

$$\frac{\partial E(k)}{\partial t} = T(k) - 2\nu k^2 E(k) + f(k) \tag{1}$$

where $T(k)$ is the energy transfer integral given by a linear functional of the third-order moment in velocities. It is called a transfer integral because it neither produces nor dissipates energy and it has the property that $\int T(k)\, dk = 0$. The terms $f(k)$ and $-2\nu k^2 E(k)$ represent the production and dissipation of energy. The former could be proportional to $E(k)$ if energy is introduced by an instability process. Locality means that there is a transparency window (called the inertial range) in the wavenumber spectrum (between the scale k_0^{-1} at which energy is introduced and the dissipation scale $k_d^{-1} = (\frac{\nu^3}{\epsilon})^{\frac{1}{4}})$ where $T(k)$ exists. In this window, the production and dissipation terms $f(k)$ and $-2\nu k^2 E(k)$ are unimportant. In that case, the balance in (1) has conservation law form $E_t = T(k) = -P'(k)$, where $P(k)$ is the energy flux, positive when the flow of energy is to small scales and large wavenumbers. We call $\int_a^b E(k)\, dk$ $(0 \le a < b \le \infty)$ a true constant of the motion if $P(k) = 0$ at both $k = a$ and $k = b$, so that the total energy is trapped in the interval (a, b) for all time because of zero flux through the boundaries. However, the presence of viscosity makes these thermodynamic equilibria uninteresting because there is a constant leakage of energy through to the dissipation scales. Therefore, in any interval (a, b) of the transparency window where $a > k_0$, $b < k_d$, $\frac{\partial}{\partial t} \int_a^b E(k)\, dk$ is zero by virtue of the fact that the fluxes $P(k)$ at $k = a$ and $k = b$ are not zero but the same. In this case, we call $\int_a^b E(k)\, dk$ a <u>finite flux constant of the motion</u>, and it is finite flux constants which are important to us here. Moreover, within the window of transparency, the interval (a, b) is arbitrary, so that then $P(k)$ is a constant and equal to the mean dissipation rate ϵ throughout the inertial range. From dimensional consideration $E(k)$ $(\ell^3\, t^{-2})$, $k(\ell^{-1})$ and $\epsilon(\ell^2 t^{-3})$ are related by the well-known Kolmogorov law $E(k) = c_2\, \epsilon^{\frac{2}{3}}\, k^{-\frac{5}{3}}$, where c_2 is a universal constant. Suitably normalized higher order moments of velocity differences $u_\ell = u(\vec{x} + \vec{\ell}) - u(\vec{x})$ (velocity gradients in the limit $\ell \to 0$) are also universal constants. The relevance of such solutions to turbulence rests on the Kolmogorov assumption that the energy dissipation rate $\epsilon = -\frac{d}{dt}\, \overline{u^2} = 2\nu \int k^2 E(k)\, dk$ does indeed settle down to a steady state value in which the energy production rate $\int f(k)dk$ is balanced by the dissipation rate.

We suggest that the primary cascade of spectral energy to small scales is necessarily accompanied by an inverse cascade of the density of another finite flux motion invariant to large scales and that the accumulation of this second density there leads to fast instabilities which result in a secondary cascade of energy towards small scales. It is this secondary cascade that gives rise to intermittency. In many respects our ideas are consistent with the works of Kraichnan (1990) and She (1991). Kraichnan shows how the deviation from Gaussian behavior in the probability density function (pdf) for velocity gradients (which can be taken as a definition of intermittency) can be explained by following the dynamical and nonlinear evolution of the pdf due to the combined influences of straining and viscous relaxation. She follows Kraichnan but is more specific in attributing the non-Gaussian behavior to local structures with high amplitude fluctuations in the velocity gradient field. By contrast we suggest a physical mechanism for intermittency by identifying a source (the inverse cascade of squared angular momentum) for building the large scale structures whose instabilities lead to intermittent events, a source which is present even when these structures are not directly forced or even when the external forcing has been switched off. The details are given in two recent articles, Kuznetsov, Newell, Zakharov (1991), and Dyachenko, Newell, Pushkarev, Zakharov (1992).

We also suggest that the inverse cascade of magnetic helicity $H_m = \int \vec{B} \cdot \vec{A} \, d\vec{x}$ $(\vec{B} = \nabla \times \vec{A})$ may have an important role to play in building the dipole component of planetary and solar dynamos. Bayly and Childress (1988) have shown from the magnetic induction equation

$$\frac{\partial \vec{H}}{\partial t} = \nabla \times (\vec{u} \times \vec{H}) + R_M^{-1} \nabla^2 H \tag{2}$$

that a chaotic velocity field $\vec{u}$ (not necessarily consistent with the actual velocity field compatible with $\vec{H}$) can deposit significant amounts of magnetic energy (fast dynamo action) at small scales. Unfortunately, in the limit of infinite magnetic Reynolds number R_M and zero mean field $\langle H \rangle$, no helicity can be generated. However, the presence of a finite, albeit large, magnetic Reynolds number and/or the presence of a weak external field $\langle H \rangle \ll \langle H^2 \rangle^{1/2}$ could produce helicity at large wavenumbers. But the presence of helicity and magnetic field energy at small scales k_0^{-1} will not produce a significant far field component. It is in this stage of dynamo building that the full self-consistent velocity field is crucial. We suggest that the flux of total energy to larger wavenumbers $k \gg k_0$ will be accompanied by an inverse flux of helicity to large scales and that the far field component of these large scale components will be significant.

At this time, however, we want to stress that the aforementioned ideas are just that, ideas. We do not claim to have constructed a watertight theory in which quantitative predictions are supported by experimental evidence or by the results of numerical simulations. Nevertheless, we believe that there is merit in our claims and present a brief outline of the relevance of an inverse cascade in optical turbulence to support our contentions.

2. Optical Turbulence

We illustrate the ideas in the context of the optical turbulence connected with the regularized nonlinear Schrödinger (NLS) equation

$$i\psi_t + \nabla^2\psi + \alpha\psi^2\psi^* = i\gamma \cdot \psi \tag{3}$$

where $\gamma \cdot \psi$ means $\int \gamma(k)\, A(\vec{k})\, e^{i\vec{k}\cdot\vec{x}}\, d\vec{k}$ where $A(\vec{k})$ is the Fourier transform of $\psi(\vec{x}, t)$. The damping function $\gamma(k)$ is positive near $k = 0$ and $k = k_d \gg k_0$ and is negative, that is, amplifying, in a narrow window $(k_0 - \Delta k, k_0 + \Delta k)$. In a spatially homogeneous random field we take $i \int (\psi\nabla\psi^* - \psi^*\nabla\psi)d\vec{x}$ to be zero and then the two "invariants" are number density $N = \int \psi\psi^*\, d\vec{x}$ and energy $H = \int (|\nabla\psi|^2 - \frac{\alpha}{2}|\psi|^4)d\vec{x}$. When the amplification rate is sufficiently small, the primary fluxes of number and energy density in wavenumber space are described by weak turbulence theory, and a kinetic equation $\dfrac{\partial N_\omega}{\partial t} + 2\gamma(\omega)N_\omega = T(n)$ can be written for the number density averaged over angle, $N = \int n_k\, d\vec{k} = \int N_\omega\, d\omega$ where $n_k\delta(\vec{k} - \vec{k'}) = \langle A(\vec{k})\, A^*(\vec{k'})\rangle$ and $N_\omega = \Omega_0\, k^{d-1}\dfrac{dk}{d\omega}\, n_k$ with $\omega = k^2$ and Ω_0 is the solid angle in d dimensions. The kinetic equation is closed because $T(n)$ can be approximated by an integral involving triple products of n_k whose form makes clear that the principal transfer mechanism is a four wave resonant interaction. Moreover, one can write $T = \dfrac{\partial^2 R}{\partial\omega^2}$, $R = \int_0^\omega (\omega - \omega')\, T(n)d\omega'$, $Q = \dfrac{\partial R}{\partial\omega}$, $P = R - \omega\dfrac{\partial R}{\partial\omega}$ and then it is easy to see that $Q(P)$ represents the flux of number density N_ω (energy density $E_\omega = \omega N_\omega$) towards low (high) wavenumbers. Equilibrium solutions are of (a) thermodynamic type, $n_k = \tau(s + \omega)^{-1}$ where τ is temperature and s chemical potential for which both fluxes Q and P are zero, (b) pure Kolmogorov type, $n_k = c_1\, Q^{1/3}\omega^{-d/2-1/3}$, $n_k = c_2 P^{1/3}\omega^{-d/2}$ which correspond respectively to a constant finite flux $P(Q)$ towards low (high) wavenumbers and are valid $d \geq 3$ and (c) a combination of (a) and (b), $n_k = \tau(s + \omega + aQ\tau^{-3}\omega^2\ln^2\frac{\omega}{\omega_C})^{-1}$, which describes the equilibrium state for $0 < \omega < \omega_0$ $(\omega_0 = k_0^2)$ for $d = 2$, and has constant finite flux Q. Further, all the usual Kolmogorov assumptions obtain for the primary fluxes of number and energy densities. Interactions are local ($T(n)$ converges for solutions n_k in the neighborhood of the finite flux equilibrium spectra), statistical information is lost (all non-resonant interactions are ignored and the kinetic equation is irreversible) and averages flow quantities in the windows $(0, \omega_0)$ and (ω_0, ω_d) are scale invariant and depend only on Q and P respectively.

However, weak turbulence theory eventually fails. The reason is that the flux of energy density $E_\omega = \omega N_\omega$ towards high wavenumbers is necessarily accompanied by a flux of particle number density towards $\omega = 0$. This occurs because of the conservation laws. A particle born at $\omega = \omega_0$ carries with it an energy ω_0. When it dies at the damping frequency $\omega_d \gg \omega_0$ it has significantly greater energy. Therefore very few particles born at ω_0 get to $\omega = \omega_d$. Instead most end up with energies $\omega < \omega_0$. The redistribution of energy is achieved by nonlinear interactions in which a small number of particles N_d ($N_d\omega_d \sim N_0\omega_0$) pick up energy but most, i.e. $N_0 - N_d$, lose energy. What is the fate of the particles that drift to low frequencies? Near $k = \omega = 0$, the low amplitude theory fails because the quadratic term in H no longer dominates the quartic and one must resort to a fully nonlinear theory. In the defocusing case, $\alpha = -1$, condensates $\psi = \psi_0 \exp(i\alpha|\psi_0|^2 t)$ are built and their growth can only be controlled by damping at $k = 0$. Further, the weak turbulence theory of fluctuations about condensates has a different character (see ref. [2]). In the focusing case $\alpha = +1$, the condensate state is unstable, a saddle point in the phase space of the system. While the condensate itself is never attained, its unstable manifold, which in physical space consists of collapsing filaments (see Kosmatov, Shvets and Zakharov (1991)), is reached, and a secondary flux of number density reverses the direction of the inverse cascade and sends particle number density (but not energy density because H = 0 for a collapsing filament) back towards high frequencies. No damping at $k = 0$ is required. The nature of the secondary flow is entirely different to that of the primary flows. It is simply the manifestation of a collapsing filament in physical space in which number density is squeezed from large to small scales in a highly organized and coherent manner. No statistical information is lost in each event. Statistical considerations are introduced, however, by the intermittent nature of events, the uncertainty in time and space as to when and where they occur. The process is probably governed by Poisson statistics whose parameters depend on the primary flux of particle numbers towards the origin. Because these events involve large amplitude fluctuations, their effect is experienced principally by the tails of the probability density function for $\psi(\vec{x}, t)$. Their manifestation in the particle number dissipation rate is seen as an intermittent sequence of spikes superposed on a background arising from those particles which reach large frequencies through four wave mixing. Further, the inverse flux appears to be enhanced when intermittency is present because the incomplete burnout of collapsing filaments leads to the production of new wavetrain particles, some of which gain in frequency due to four wave mixing but most of which lose.

This work was supported by AFOSR contract FQ8671-900589 and NSF grants DMS8922179 and DMS9021253.

REFERENCES

1. Bayly, B. and Childress, S., (1988) Geophys. Astrophys. Fluid Dynamics 44, 211-240.

2. Dyachenko, S., Newell, A. C., Pushkarev, A. and Zakharov, V. E., Physica D, in press.

3. Kolmogorov, A. N. (1941) C. R. Acad. Sci. URSS 30 (4), 301-305.

4. Kosmatov, N., Shvets, V. and Zakharov, V. E. (1991) Physica D 52 (1), 16-35.

5. Kraichnan, R. H. (1990) Phys. Rev. Lett. 65, 575.

6. Kuznetsov, E., Newell, A. C. and Zakharov, V. E. (1991) Phys. Rev. Lett. 67 (23), 3243-3246.

7. She, Z.-S. (1991) Phys. Rev. Lett. 66, 600-603.

NUMERICAL COMPUTATION
OF ELECTROMAGNETIC FIELDS

Jukka Sarvas, Rolf Nevanlinna Institute,
Teollisuuskatu 23, 00510 Helsinki, Finland

1. INTRODUCTION

Computational electromagnetics has become an important field both for academic research and for the engineering design of electrical devices. It is an involved region of applied mathematics where advanced mathematical methods are applied to paritcular questions of electrical engineering. In this talk, some basic computational methods for time harmonic fields in scattering problems are discussed. The emphasis will be on wave function series and integral representations of the fields which, alone with the finite element and difference methods, still play an important role in the practical field computation.

We start by treating the scattering problem of a homogeneous body in a 3-dimensional space. First, it is solved using the spherical vector wave functions and a straightforward point matching method. Next, we apply the fundamental surface integral field representations to treat the same problem as a surface integral equation.

The finite element method (FEM) provides one of the most versatile field computation methods, although in three dimensions, it is numerically a rather heavy procedure. The FEM is normally applied to boundary value problems rather than scattering problems. However, it can be extended to scattering problems, for instance, by an appropriate mesh termination method, like the unimoment method, which is based on wave function series, and a discussion on that method finishes the talk.

The theoretical and practical approach to scattering problems in this talk has arisen from a joint project with a Finnish Company, Instrumentarium Imaging Co., on the antenna design problems of the magnetic resonance imaging device.

2. MAXWELL'S EQUATIONS

In an isotropic, linear and homogeneous media the Maxwell's equations are as follows:

$$(2.1) \qquad \begin{aligned} \operatorname{curl} \mathbf{E} &= -\frac{\partial \mathbf{B}}{\partial t} & ; & \quad \operatorname{div} \mathbf{E} = \frac{1}{\varepsilon}\varrho \\ \operatorname{curl} \mathbf{H} &= \mathbf{J} + \varepsilon\frac{\partial \mathbf{E}}{\partial t} & ; & \quad \operatorname{div} \mathbf{B} = 0, \end{aligned}$$

where $\mathbf{E}$ is the electric field, $\mathbf{H}$ is the intensity of the magnetic field, $\mathbf{B}$ is the density of the magnetic flux, $\mathbf{J}$ is the total current density (of free charges) and ε is the electric permittivity of the media. In addition, $\mathbf{B} = \mu\mathbf{H}$ where μ is the magnetic permeability of the media.

We assume that all our fields are time-harmonic, which means that the time factor is $e^{-i\omega t}$ and the fields are of the form

$$(2.2) \qquad \mathbf{E}(\mathbf{r}, t) = \mathbf{E}(\mathbf{r})e^{-i\omega t}, \quad \text{etc...},$$

where $\omega = 2\pi f$ and f is the frequency. The interpretation of this notation is as follows: the physical field $\mathbf{E}_{\mathrm{phys}}(\mathbf{r}, t)$ is the real part of $\mathbf{E}(\mathbf{r})e^{-i\omega t}$, and accordingly,

$$(2.3) \qquad \mathbf{E}_{\mathrm{phys}}(\mathbf{r}, t) = \mathrm{real}\,(\mathbf{E}(\mathbf{r}))\cos(\omega t) + \mathrm{imag}\,(\mathbf{E}(\mathbf{r}))\sin(\omega t),$$

and, similarly, with $\mathbf{H}$ and $\mathbf{B}$.

If we insert the presentation (2.2) into the equations (2.1), the time factor cancels and we get the time-independent Maxwell's equations:

$$(2.4) \qquad \begin{aligned} \mathrm{curl}\,\mathbf{E} &= i\omega\mathbf{B} \\ \mathrm{curl}\,\mathbf{H} &= \mathbf{J} - i\omega\varepsilon\mathbf{E}. \end{aligned}$$

The solutions of (2.3) are 3-dimensional complex fields $\mathbf{E}$, $\mathbf{H}$, and $\mathbf{B}$, and the physical fields are obtained by (2.3).

3. SPHERICAL VECTOR WAVE FUNCTIONS

To evaluate the fields numerically we need an explicit presentation of them and the spherical wave functions provide a convenient way. In a homogeneous sphere with center at the origin, with no sources inside the sphere, the field, say $\mathbf{E}$, can be presented in incoming wave functions, $\mathbf{F}_j^{\mathrm{in}}$, $j = 1, 2, ...$, as a series expansion

$$\mathbf{E}(\mathbf{r}) = \sum_{j=1}^{\infty} a_j \mathbf{F}_j^{\mathrm{in}}(\mathbf{r})$$

Similarly, if all sources are inside the same sphere, and the outside is homogeneous, then the field can be presented there as a series expansion in the outgoing wave functions $\mathbf{F}_j^{\mathrm{out}}$, $j = 1, 2, ...$. The wave functions $\mathbf{F}_j^{\mathrm{in}}$ and $\mathbf{F}_j^{\mathrm{out}}$, $j = 1, 2, ...$, are solutions of the homogeneous Maxwell's equations in a homogeneous space. Their explicit definition looks somewhat complicated due to the six-folded indexing system: the whole function family consists of $\mathbf{M}$- and $\mathbf{N}$-type functions as follows: for $\mathbf{r} = (r, \theta, \varphi)$ in spherical coordinates

$$(3.1) \qquad \begin{aligned} \mathbf{M}_{\sigma mn}^{\{\substack{\mathrm{in}\\\mathrm{out}}\}}(\mathbf{r}) &= \mathrm{curl}\left(Y_{\sigma mn}(\theta, \varphi)\left\{\begin{matrix} j_n(kr) \\ h_n(kr) \end{matrix}\right\}\mathbf{r}\right), \\ \mathbf{N}_{\sigma mn}^{\{\substack{\mathrm{in}\\\mathrm{out}}\}} &= \tfrac{1}{k}\,\mathrm{curl}\,\mathbf{M}_{\sigma mn}^{\{\substack{\mathrm{in}\\\mathrm{out}}\}}, \end{aligned}$$

where σ = even or odd, $m = 0, 1, ..., n$, $n = 1, 2, ...$, and $k = \sqrt{\omega\mu(\varepsilon\mu + i\sigma)}$ is the wave number of the space, and

$$Y_{\{\,{}^{\text{even}}_{\text{odd}}\,\}mn} = \left\{\begin{array}{c} \cos(m\varphi) \\ \sin(m\varphi) \end{array}\right\} P_n^m(\cos\theta),$$

where P_n^m is the associated Legendre function. For a more explicit formulas, e.g., see [M-F, part II, p.1898]. After having written programs for these wave functions, they are easily handled with a computer.

4. HOMOGENEOUS BODY; SOLUTION BY WAVE FUNCTIONS

Let D be a finite homogeneous, simply connected body, or scatterer, with a smooth boundary S lying in a homogeneous 3-dimensional space. Let the electromagnetic parameters be ε_1, μ_1 and $\sigma_1 < \infty$ outside D and ε_2, μ_2 and $\sigma_2 < \infty$ inside D. Let a time-harmonic transmitter, or source, lie outside D with a known primary field $\mathbf{E}_p$; i.e. the field of the source in the homogeneous entire space without the scatterer. Our field problem is to find the total electric and magnetic fields inside and outside D.

We solve this problem by using wave functions. We devide $\mathbf{E}$ as follows:

$$(4.1) \qquad \mathbf{E} = \left\{\begin{array}{ll} \mathbf{E}_p + \mathbf{E}_s & \text{outside } D, \\ \mathbf{E}_t & \text{inside } D, \end{array}\right.$$

where $\mathbf{E}_s$ is called the scattered field and $\mathbf{E}_t$ the transferred, or penetrated, field. The idea of the solution is to find $\mathbf{E}_s$ and $\mathbf{E}_t$ which satisfy the boundary conditions on S. For that end, we approximate $\mathbf{E}_s$ and $\mathbf{E}_t$ with a finite linear combination of the outgoing and incoming wave functions, respectively:

$$\mathbf{E}_s \simeq \sum_{j=1}^{n} \left(a_j \mathbf{M}_j^{1,\text{out}} + b_j \mathbf{N}_j^{1,\text{out}}\right),$$

$$(4.2)$$

$$\mathbf{E}_t \simeq \sum_{j=1}^{n} \left(\alpha_j \mathbf{M}_j^{2,\text{in}} + \beta_j \mathbf{N}_j^{2,\text{in}}\right),$$

where superindices 1 and 2 refer to the wave numbers k_1 and k_2, $k_j = \sqrt{\omega\mu_j(\omega\varepsilon_j + i\sigma_j)}$, $j = 1, 2$. These approximations are possible, because the tangential components of $\mathbf{E}_s$ and $\mathbf{E}_t$ on S determine $\mathbf{E}_s$ and $\mathbf{E}_t$ uniquely (provided that ω is not a resonance frequency of D), and further more, the tangential components of both $\mathbf{F}_j^{\text{in}}$ and $\mathbf{F}_j^{\text{out}}$, $j = 1, 2, ...$, on S form dense sets in the space of all tangential fields on S (more precisely, in the L^2-sense).

In (4.2) we have written the wave functions in terms of **M**- and **N**-functions because of the following property:

$$(4.3) \qquad \operatorname{curl} \mathbf{M} = k\mathbf{N} \quad \text{and} \quad \operatorname{curl} \mathbf{N} = k\mathbf{M},$$

where **M** and **N** are of the same type. Because $\mathbf{H} = (i\omega\mu)^{-1} \operatorname{curl} \mathbf{E}$ due to (2.4), the property (4.3) is very helpful: from the presentations (4.2) we immediately get (approximate) presentations for corresponding **H**-fields. Next we state the boundary conditions: $\mathbf{n} \times \mathbf{E}$ and $\mathbf{n} \times \mathbf{H}$ are continuous across the interface S, where $\mathbf{n}$ is the unit outer normal of D, or in other words: on S

$$(4.4) \qquad \begin{aligned} \mathbf{n} \times (\mathbf{E}_p + \mathbf{E}_s) &= \mathbf{n} \times \mathbf{E}_t \\ \mathbf{n} \times (\mathbf{H}_p + \mathbf{H}_s) &= \mathbf{n} \times \mathbf{H}_t. \end{aligned}$$

Inserting the corresponding approximations into (4.4), we get the equations:

$$\begin{aligned}
\mathbf{n} \times \mathbf{E}_p \ + \ \sum_{j=1}^{n} \left(a_j \mathbf{n} \times \mathbf{M}_j^{1,\text{out}} + b_j \mathbf{n} \times \mathbf{N}_j^{1,\text{out}} \right) &= \\
\sum_{j=1}^{n} \left(\alpha_j \mathbf{n} \times \mathbf{M}_j^{2,\text{in}} + \beta_j \mathbf{n} \times \mathbf{N}_j^{2,\text{in}} \right) &, \\
\mathbf{n} \times \mathbf{H}_p \ + \ \frac{k}{i\omega\mu} \sum_{j=1}^{n} \left(b_j \mathbf{n} \times \mathbf{M}_j^{1,\text{out}} + a_j \mathbf{n} \times \mathbf{N}_j^{1,\text{out}} \right) &= \\
\frac{k}{i\omega\mu} \sum_{j=1}^{n} \left(\beta_j \mathbf{n} \times \mathbf{M}_j^{2,\text{in}} + \alpha_j \mathbf{n} \times \mathbf{N}_j^{2,\text{out}} \right) &.
\end{aligned}$$

We can solve these equations for the unknown coefficients a_j, b_j, α_j and β_j, $j = 1, ..., n$, for instance, by using point-matching: we force the equations to be valid at points $p_1, ..., p_N$ on S, with $N \geq n$. This leads to an overdetermined matrix equation

$$(4.5) \qquad Wd = c,$$

where the vector d contains the unknown coefficients and c contains the values of $\mathbf{n} \times \mathbf{E}_p$ and $\mathbf{n} \times \mathbf{H}_p$ at points $p_1, ..., p_N$.

For the practical solution of (4.5) it is good to take $N \geq 2n$ and solve (4.5) in the regularized sense of the least squares to avoid numerical instabilities; this means that we compute d from the formula:

$$(4.6) \qquad d = (W^*W + \delta I)^{-1} W^* c,$$

where W^* is the Hermitian adjungate of W, I is the unit matrix and $\delta > 0$ is the regularization parameter. An appropriate $\delta > 0$ is easily found by experimenting; if n is large (say ≥ 30), δ is often 100-1000 times larger than the machine-epsilon of the computer.

In Figure 1, a numerical example of the method is presented. The scatterer is an ellipsoid and the transmitter is an electric dipole at point $(0,2,1)$ with moment $\mathbf{p} = (1, -\frac{1}{2}, 1)$. In addition, $f = 100$ MHz, $\varepsilon_1 = \varepsilon_0$, $\mu_1 = \mu_0$, $\sigma_1 = 0$, $\varepsilon_2 = 10\varepsilon_0$, $\mu_2 = \mu_0$, $\sigma_2 = 0.1\Omega^{-1}$ and $n = 24$.

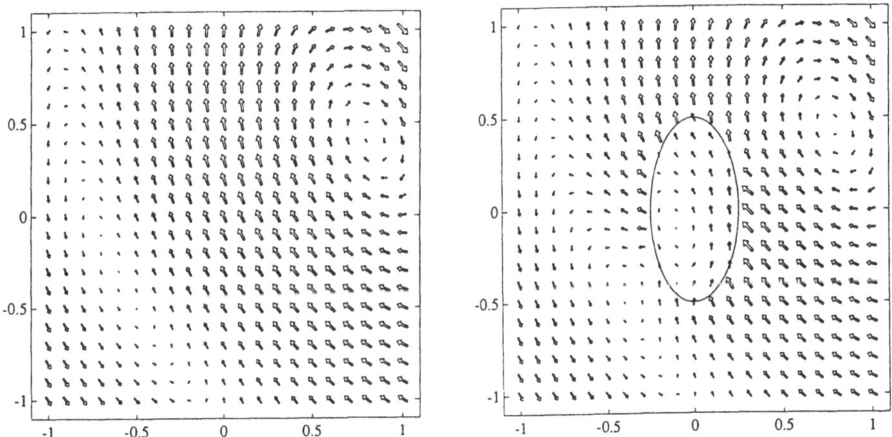

Figure 1. *The real parts of the yz-component of the primary field* $\mathbf{E}_p$, *on the left, and the total field* $\mathbf{E}$ *on the right with the scatter.*

The above field computing method with wave functions works quite well for scatterers whose shape is close to that of a sphere. Especially, for a sphere, the method yields the classical results which can be derived directly using the orthogonality of spherical wave functions on a spherical surface.

If the scatterer is strongly non-spherical we need a more powerful and complicated method; for a homogeneous scatterer, perhaps, the best choice in that case is a surface integral equation.

5. SURFACE INTEGRAL EQUATION

We again consider the same field problem as in the previous section with scatterer D, the source outside D with primary field $\mathbf{E}_p$, the scattered field $\mathbf{E}_s$ and the penetrated field $\mathbf{E}_t$.

For deriving the surface integral equations for this problem we need surface integral presentations for $\mathbf{E}$ and $\mathbf{H}$ called Huygen's principle, e.g., see [C-K, p.110]: $\mathbf{E}$ and $\mathbf{H}$ are solutions of the homogeneous Maxwell's equations (i.e. their source is outside

D), then

$$(5.1) \qquad -(L\tilde{E})(x) - \frac{i\omega\mu}{k^2}(M\widetilde{H})(x) = \begin{cases} E(x), & \text{if } x \in D, \\ 0, & \text{if } x \notin \bar{D}, \end{cases}$$

$$(5.2) \qquad -(L\widetilde{H})(x) - \frac{1}{i\omega\mu}(M\widetilde{H})(x) = \begin{cases} \mathbf{H}(x), & \text{if } x \in D, \\ 0, & \text{if } x \notin \bar{D}, \end{cases}$$

where

$$(5.3) \qquad \tilde{E} = \mathbf{n} \times \mathbf{E} \quad \text{and} \quad \widetilde{H} = \mathbf{n} \times \mathbf{H}$$

on S, and we have denoted by L and M the surface integral operators defined by

$$\left(L\tilde{F}\right)(x) = \operatorname{curl}_x \int_S \Phi(x,y)\tilde{F}(y)\mathrm{d}y,$$

$$(5.4)$$

$$\left(M\tilde{F}\right)(x) = \operatorname{curl}_x^2 \int_S \Phi(x,y)\tilde{F}(y)\mathrm{d}y,$$

with

$$(5.5) \qquad \Phi(x,y) = \frac{1}{4\pi}\frac{e^{ik|x-y|}}{|x-y|},$$

for any smooth tangential vector field $\tilde{F}$ on S.

We choose any x in the outside of D and not on S. We apply (5.1) to $\mathbf{E}_p$ and D with ε_1, μ_1 and σ_1 in the entire space and get

$$(5.6) \qquad 0 = L_1\tilde{E}_p(x) + \frac{i\omega\mu_1}{k_1^2}\left(M_1\widetilde{H}_p\right)(x).$$

The subindex 1 in L_1 and M_1 refers to using $k = k_1$ in (5.4). Next, we apply (5.1) to $\mathbf{E}_s$ and the domain D' consisting the exterior of D with ε_1, μ_1 and σ_1 and $\mathbf{n}$ replaced by $-\mathbf{n}$. Because $\mathbf{E}_s$ has no sources in D' we get:

$$(5.7) \qquad \mathbf{E}_s(x) = \left(L_1\tilde{E}_s\right)(x) + \frac{i\omega\mu_1}{k_1^2}\left(M_1\widetilde{H}_s\right)(x).$$

Next apply (5.1) to $\mathbf{E}$ in D with ε_2, μ_2 and σ_2. Because $\mathbf{E}$ has no sources in D, we get

$$(5.8) \qquad 0 = \left(L_2\tilde{E}\right)(x) + \frac{i\omega\mu_2}{k_2^2}\left(M_2\widetilde{H}\right)(x).$$

Now, add (5.6) and (5.7) together, add $\mathbf{E}_p(x)$ to the both sides and multiply the resulting equation with k_1^2/μ_1 and add this equation to (5.8) multiplied by $^-k_2^2/\mu_2$. Finally, multiply the resulting equation with μ_1/k_1^2 and get

$$
\text{(5.9)} \qquad
\begin{aligned}
\mathbf{E}(x) = \mathbf{E}_p(x) + & \left((L_1 - \tfrac{\mu_1 k_2^2}{\mu_2 k_1^2} L_2)\tilde{E} \right)(x) + \\
& \frac{i\omega\mu_1}{k_1^2} \left((M_1 - M_2)\widetilde{H} \right)(x).
\end{aligned}
$$

Next multiply the both sides of (5.8) by $\mathbf{n}$ using the cross product. Let $x \to z \in S$. Obviously, $\mathbf{n} \times \mathbf{E}(x) \to \tilde{E}(z)$ and $\mathbf{n} \times \mathbf{E}_p(x) \to \tilde{E}_p(z)$. The limits involving the surface integrals are more involved. By extending $\Phi(x,y)$ in Taylor series with respect to $|x - y|$ and taking curl_x^2, we notice that the singularity in $((M_1 - M_2)\widetilde{H})(x)$ as $x \to y$ is of order $\mathcal{O}(1/|x - y|)$ which implies that
$((M_1 - M_2)\widetilde{H})(x) \to ((M_1 - M_2)\widetilde{H})(z)$ as $x \to z$.

The limiting behavior of $(L_j\tilde{E})(x)$, $j = 1,2$, is more complicated. However, the behavior resembles that of a double layer potential, and we, fortunately, know that

$$
\text{(5.10)} \qquad
\lim_{x \to z} \mathbf{n}(x) \times (L_j\tilde{E})(x) = \mathbf{n}(z) \times (L_j\tilde{E})(z) + \frac{1}{2}\tilde{E}(z),
$$

for $j = 1,2$ see [C-K, p.47]. Taking all these limit results into account, and treating $\mathbf{H}$ in a similar way, we get the final surface integral equations written out in the explicit way: for all $x \in S$,

$$
\begin{aligned}
c_1\tilde{E}(x) \;=\; \tilde{E}_p(x) \;&+ \int_S \mathbf{n}(x) \times \left[\left(\nabla_x\Phi_1(x,y) - \frac{\mu_1 k_2^2}{\mu_2 k_1^2}\nabla_x\Phi_2(x,y) \right) \times \tilde{E}(y) \right] dy + \\[2mm]
&\frac{i\omega\mu_1}{k_1^2} \int_S \mathbf{n}(x) \times \mathrm{curl}_x^2 \left[(\Phi_1(x,y) - \Phi_2(x,y))\widetilde{H}(y) \right] dy,
\end{aligned}
$$

$$\text{(5.11)}$$

$$
\begin{aligned}
c_2\widetilde{H}(x) \;=\; \widetilde{H}_p(x) \;&+ \int_S \mathbf{n}(x) \times \left[\left(\nabla_x\Phi_1(x,y) - \frac{\mu_2}{\mu_1}\nabla_x\Phi_2(x,y) \right) \times \widetilde{H}(y) \right] dy + \\[2mm]
&\frac{1}{i\omega\mu_1} \int_S \mathbf{n}(x) \times \mathrm{curl}_x^2 \left[(\Phi_1(x,y) - \Phi_2(x,y))\tilde{E}(y) \right] dy,
\end{aligned}
$$

$$
\text{with } c_1 = \tfrac{1}{2}\left(1 + \frac{\mu_1 k_2^2}{\mu_2 k_1^2} \right), \qquad c_2 = \tfrac{1}{2}\left(1 + \frac{\mu_2}{\mu_1} \right).
$$

The integrals above have singularities as $x = y$, but it is not difficult to see that they all are of order $\mathcal{O}(1/|x - y|)$, and therefore, integrable in a surface integral.

Next, we discuss the numerical solution of the equations (5.11). We use the boundary element method and divide the surface S into a mesh of triangles Δ_i, $i = 1,...,n$.

We can approximate the fields $\tilde{E}$ and $\widetilde{H}$ by either a piecewise linear or piecewise constant approximation. For brevity, the piecewise constant case is explained here; the linear one is essentially similar.

Let p_i be the center point of Δ_i, $i = 1, ..., n$. We take $\mathbf{U}_i = \mathbf{E}(p_i)$ and $\mathbf{V}_i = \mathbf{H}(p_i)$, $i = 1, ..., n$, be the unknowns. Then

$$(5.12) \qquad \tilde{E} \simeq \sum_{i=1}^{n} \mathbf{n}_i \times \mathbf{U}_i \varphi_i, \quad \widetilde{H} \simeq \sum_{i=1}^{n} \mathbf{n}_i \times \mathbf{V}_i \varphi_i,$$

where $\mathbf{n}_i = \mathbf{n}(p_i)$ and φ_i is the caracteristic function of Δ_i, i.e. $\varphi_i(x) = 1$ if $x \in \Delta_i$ and $\varphi_i(x) = 0$ if $x \notin \Delta_i$, $i = 1, ..., n$. We insert the representations (5.12) into the equations (5.11) and demand that the resulting equations are valid for $x = p_i$, $i = 1, ..., n$ (point-matching). This yields a matrix equation for the unknowns $\mathbf{U}_i$, $\mathbf{V}_i$, $i = 1, ..., n$. It is easy to see that the equation only determines tangential components of $\mathbf{U}_i$ and $\mathbf{V}_i$, or $\mathbf{n} \times \mathbf{U}_i$ and $\mathbf{n} \times \mathbf{V}_i$. Therefore, if we solve the matrix equation using the Moore-Penrose pseudoinverse, i.e. the solution will be the minimum norm least squares solution, the resulting solutions $\mathbf{U}_i$ and $\mathbf{V}_i$ will just be tangential components of $\mathbf{E}(p_i)$ and $\mathbf{H}(p_i)$, $i = 1, ..., n$, (approximatively). For a large number of unknowns one can use an alternative method. Insert $\mathbf{U}_i$ and $\mathbf{V}_i$ instead of $\mathbf{n} \times \mathbf{U}_i$ and $\mathbf{n} \times \mathbf{V}_i$ into the equations and solve them, say, by Gauss-Seidel iteration. If the method converges, as it did in our experimenting, the obtained solutions $\mathbf{U}_i$, $\mathbf{V}_i$ have no normal components, as you can see from the equations (5.11) because their right hand sides automatically have no normal components. Therefore, the obtained solutions are the wanted fields $\tilde{E}(p_i)$ and $\widetilde{H}(p_i)$ (approximatively).

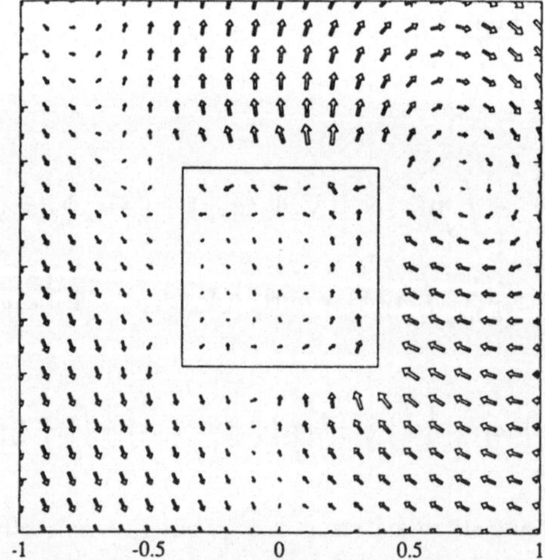

Figure 2: *The real part of the yz-component of the total field* **E** *with a cube as the scatterer.*

After solving $\tilde{E}$ and $\widetilde{H}$, we get the fields both inside and outside D using the integral presentations (5.1) and (5.2) for $\mathbf{E}$ and $\mathbf{H}$ inside D and for $\mathbf{E}_s$ and $\mathbf{H}_s$ outside D.

In Figure 2, a numerical example of the above method is presented with a cube as a scatterer. The primary field and the electromagnetic parameters are as in the example at the end of Section 4.

6. SCATTERING PROBLEM AND FEM

For field problems with a non-homogeneous scatterer the finite element method (FEM) provides one of the most versatile field computation methods, although, in 3 dimensions it is numerically quite elaborate. Furthermore, the FEM is normally applied to boundary value problems of finite domains, and the scattering problem is not of this type because the problem domain is the entire space. However, the FEM can be also applied to scattering problems, for instance, by an appropriate mesh termination method like the unimoment method, e.g., see [Mo]. This method is based on the spherical wave functions, and in this section we present a 3-dimensional version of it.

Assume that we have a FEM-solver for time harmonic boundary value problems using the boundary values of the vector potential $\mathbf{A}$ and scalar potential φ on the boundary S of the problem domain.

We consider the same field problem as before concerning a scatterer D, with ε_2, μ_2 and $0 < \sigma_2 < \infty$, lying in a homogeneous space with ε_1, μ_1 and $\sigma_1 < \infty$. We encircle D by a spherical surface S_1 not meeting the boundary of D and leaving the source outside S_1. The problem domain for the FEM-solver is the inside of S. As boundary values on S_1 we have: $\mathbf{A} = (i\omega)^{-1}\mathbf{E}$ and $\varphi = 0$, where $\mathbf{E}$ is the still unknown total electric field.

We want to show how we can, using the FEM-solver, find the restriction $\mathbf{E}|S_1$ of $\mathbf{E}$ on S_1 which is the needed boundary value.

We denote by L the FEM-solver. Accordingly, L is a linear operator mapping any boundary value $\mathbf{E} \mid S_1$ to $L(\mathbf{E} \mid S)$ which presents the total electric field inside S_1 uniquely determined by the boundary value $\mathbf{E}|S_1$. Again, we divide the total field $\mathbf{E}$ as follows:

$$\mathbf{E} = \begin{cases} \mathbf{E}_p + \mathbf{E}_s & \text{outside of } D, \\ \mathbf{E}_t & \text{inside of } D, \end{cases}$$

where $\mathbf{E}_p$ is the given primary field. Let S_2 be another spherical surface, concentric with S_1, and between S_1 and D. Then we can expand $\mathbf{E}_s$ in outgoing wave functions $\mathbf{F}_j$ outside S_2:

$$(6.2) \qquad \mathbf{E}_s \cong \sum_{j=1}^{n} c_j \mathbf{F}_j,$$

where we have used a truncated series. Now, outside S_2 we also have:

$$(6.3) \qquad \mathbf{E} = \mathbf{E}_p + \mathbf{E}_s,$$

and, especially, $\mathbf{E}\,|\,S_1 = \mathbf{E}_p\,|\,S_1 + \mathbf{E}_s\,|\,S_1$. Applying the operator L to $\mathbf{E}\,|\,S_1$ yields:

$$(6.4) \qquad \mathbf{E} = L(\mathbf{E}\,|\,S_1) \simeq L(\mathbf{E}_p\,|\,S_1) + \sum_{j=1}^{n} c_j L(\mathbf{F}_j\,|\,S_1)$$

inside S_1. Equate (6.3) and (6.4) on S_2 and obtain

$$\mathbf{E}_p + \sum_{j=1}^{n} c_j \mathbf{F}_j = L(\mathbf{E}_p\,|\,S_1) + \sum_{j=1}^{n} c_j L(\mathbf{F}_j\,|\,S_1)$$

on S_2, or by rearranging the terms,

$$(6.5) \qquad \sum_{j=1}^{n} c_j \left[\mathbf{F}_j(x) - L(\mathbf{F}_j\,|\,S_1)(x)\right] = L(\mathbf{E}_p\,|\,S_1)(x) - \mathbf{E}_p(x),$$

for all $x \in S_2$. In this equation, only the coefficients c_j are unknown; we can solve for them, for instance, by using point-matching on S_2. After having obtained c_j, $j = 1, ..., n$, we know $\mathbf{E}_s$, approximatively, and we get $\mathbf{E} = \mathbf{E}_p + \mathbf{E}_s$ outside S_2. Applying the FEM-solver once more to $\mathbf{E}\,|\,S_1$, we get $\mathbf{E}$ inside S_1, and the entire scattering problem is solved.

REFERENCES

[C-K] C. Colton and R. Kress: *Integral equation methods in scattering theory.* - John Wiley & Sons, Inc., New York, 1983.

[Mo] M. A. Morgan (Ed.): *Finite element and finite difference methods in Electro-magnetic Scattering.* - Elsevier Science Publishing Company, New York, 1990.

[M-F] P. Morse and H. Feshbach: *Methods of Theoretical Physics.* - McGraw-Hill, New York, 1953.

NONLINEAR HYPERBOLIC SYSTEMS IN CHROMATOGRAPHY

H. P. URBACH
Philips Research Laboratories
P.O. Box 80000
5600 JA Eindhoven
The Netherlands

ABSTRACT. The mathematical theory of nonlinear hyperbolic systems is applied to the modelling of chromatography. The elegant ω or h transformation which can be applied to the ideal Langmuir model of chromatography, is explained. Furthermore, a computational method for a more advanced model is discussed and results of a comparison of the two models are presented.

1. Introduction

In liquid chromatography a mixture of dissolved species flows through a column containing a stationary medium that partially sorbs the species of the mixture. The species that are more strongly sorbed lag behind. Under appropriate conditions at the inlet this can lead to complete separation of the mixture into pure zones that move with equal velocity through the column.

The separation process for a mixture of two species is illustrated in Fig. 1. Before the mixture is introduced, a pure state called the carrier flows through the column. The carrier is chosen such that it is less sorbed than all species in the mixture. After the mixture has been introduced, another pure zone called the displacer is led in. The displacer is more strongly sorbed than all species in the mixture. As shown in Fig. 1, the mixed zone splits off two pure zones, the lengths of which increase downstream. When the column is sufficiently long the mixture eventually disappears.

In the modelling of separation processes the flow is often assumed to be one-dimensional and uniform with velocity U in the direction of the axis of the column. Furthermore, the concentrations c_i and n_i of species i in the liquid and the stationary phase, respectively, are assumed to be functions of only the time t and the axial coordinate z. By neglecting diffusion, the conservation law for the ith species reads

$$(1 - \epsilon)\frac{\partial n_i}{\partial t} + \epsilon\frac{\partial c_i}{\partial t} + U\epsilon\frac{\partial c_i}{\partial z} = 0, \tag{1}$$

where ϵ is the void fraction. The ion exchange between the liquid and the stationary phase

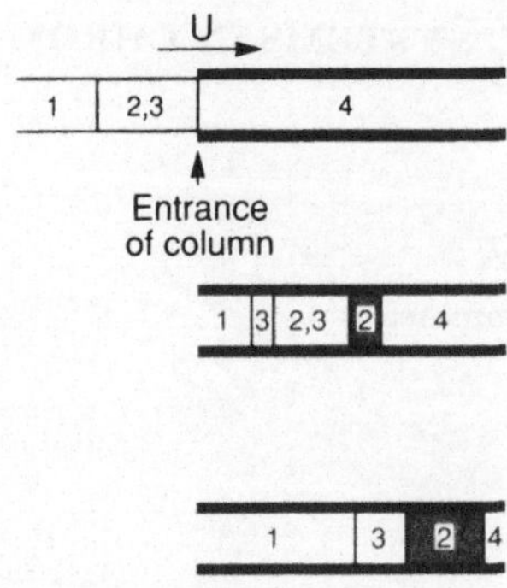

Figure 1: A separation process for a mixture consisting of two species. Species 2 is more strongly sorbed than 3. Species 1 and 4 are the displacer and the carrier, respectively. The top figure shows the situation just before the mixture enters the column. In the bottom figure, complete separation is attained.

is described by an isotherm which expresses the c_i in terms of all n_j. By substitution of the isotherm into (1), a coupled system of first-order nonlinear partial differential equations results. One would therefore expect discontinuities to occur in the solutions and, in fact, the boundaries between the zones shown in Fig. 1 are discontinuities in concentrations.

The aim of this paper is to describe the mathematical properties of system (1) for two isotherms, namely the classical Langmuir isotherm and the more strongly nonlinear BB isotherm. In Section 2, a convenient dimensionless formulation of the problem is given. In Section 3 the systems are classified as hyperbolic and a fascinating property of the Langmuir model is described. This concerns the existence of a transformation of dependent variables, called ω (or h) transformation (Helfferich et al. [7]), by which the system is decoupled and the solution can be calculated very efficiently (Rhee et al. [9],[10]). In Section 4 conditions are formulated which are to be satisfied by the isotherm in order that the ω transform exists. The BB isotherms do not satisfy these conditions and this not only complicates the computation of the solutions of separation processes, but also means that the number of shocks that occur according to the two models differs for mixtures of at least 4 species. Finally, in Section 5 examples of calculated solutions of the two models are shown.

2. Mathematical Formulation

Let there be m species and let the numbering be in order of decreasing sorption so that $i = 1$ is the displacer, $i = 2, \ldots, i = m - 1$ are the species in the mixture and $i = m$ is the carrier. The species we are considering are homovalent ions. Electroneutrality in the stationary phase then implies that the total concentration in the stationary phase, $\mathcal{N} \overset{\text{def}}{=} \sum_{i=1}^{m} n_i$ is constant. For the sake of simplicity we assume that the total concentration in the liquid phase $C \stackrel{\text{def}}{=} \sum_{i=1}^{m} c_i$ is kept constant at the inlet. Then system (1) implies that C is constant everywhere in the column.

We define dimensionless concentrations $x_i = c_i/C$ and $y_i = n_i/\mathcal{N}$ and dimensionless time and distance τ and ζ by $\tau = (Ut - z)/(Ut_e)$ and $\zeta = (1 - \epsilon)\mathcal{N}z/(\epsilon C U t_e)$, where t_e is the inlet time of the mixture. Then system (1) simplifies to

$$\frac{\partial y_i}{\partial \tau} + \frac{\partial x_i}{\partial \zeta} = 0, \tag{2}$$

for $i = 1, \ldots, m$.

The ion-exchange process is described by an isotherm which expresses the liquid-phase concentrations x_i in terms of the concentrations y_j in the stationary phase ([3]):

$$x_i = \frac{\alpha_i(y)y_i}{V(y)}, \tag{3}$$

with $y \stackrel{\text{def}}{=} (y_1, \ldots, y_{m-1})$ and

$$V(y) = 1 + \sum_{j=1}^{m-1} [\alpha_j(y) - 1])y_j. \tag{4}$$

A priori, the functions α_j depend on all stationary-phase concentrations $y_1, \ldots, y_m$, but y_m has been eliminated by using $\sum_{i=1}^{m} y_i = 1$. The functions α_i satisfy $0 < \alpha_i < \alpha_{i+1} < 1$ and depend on the model used. We shall consider the following two cases.

1. <u>Langmuir isotherms</u>. In this case the α_i are constants. This means that interactions between sorbed species are neglected.

2. <u>BB isotherms</u>. In this case interactions between sorbed species are taken into account. The α_i are given by

$$\alpha_i(y) = \alpha(y)^{p_i}, \tag{5}$$

where $0 < p_{i+1} < p_i < 1$ are constants and

$$\alpha(y) = \exp\left[\varrho\left(\sum_{k=1}^{m-1} p_k y_k - 1\right)\right]. \tag{6}$$

The parameter $\varrho > 0$ depends on the temperature T and on the class of ions considered. As explained in [12], (5) follows from a special case of the model derived by Boots & De Bokx [2], [4], [5].

After substitution of (3) into (2) for $i = 1, \ldots, m-1$, a system of $m-1$ coupled nonlinear differential equations for the $y_1, \ldots, y_{m-1}$ is obtained. The boundary conditions follow from the conditions at the inlet $\zeta = z = 0$. Initially, for $\tau < 0$ say, the carrier flows through the column, hence $y_m = 1$. Then for $0 < \tau < 1$ the mixture with given constant concentrations

$$60$$

$y_2^F, \ldots y_{m-1}^F$ is introduced. For $\tau > 1$ only the displacer is present, hence $y_1 = 1$ for $\tau > 1$. This results in the following values for $y(\tau, 0) = (y_1(\tau, 0), \ldots, y_{m-1}(\tau, 0))$:

$$y(\tau, 0) = \begin{cases} (0, \ldots, 0) & \text{for } \tau < 0, \\ (0, y_2^F, \ldots, y_{m-1}^F) & \text{for } 0 < \tau < 1, \\ (1, 0, \ldots, 0) & \text{for } \tau > 1. \end{cases} \tag{7}$$

We thus see that for sufficiently small ζ the solution is obtained by solving two Riemann problems. For larger ζ, i.e. further downstream the column, these Riemann problems will interact.

3. Riemann Problems

In the Langmuir case the eigenvalues λ_ν of the Jacobian matrix $(\partial x_i / \partial y_i)$ are the roots of the equation

$$\sum_{i=1}^{m-1} \frac{(1 - \alpha_i)\alpha_i y_i}{\lambda_\nu V - \alpha_i} - V = 0. \tag{8}$$

If $y_j = 0$, then $\lambda = \alpha_j / V$ is an eigenvalue. Using the fact that the denumerators $(1 - \alpha_i)\alpha_i y_i$ are all nonnegative, it is easy to see that all roots of (8) are real and in general distinct. In particular, if $y_j > 0$ and $y_{j+1} > 0$, then there exists an eigenvalue in the interval $(\alpha_j / V, \alpha_{j+1}/V)$. Hence in the Langmuir case system (2) is hyperbolic. The right and left eigenvectors r^ν and l^ν corresponding to λ_ν are given by

$$r^\nu(y) = \left(\frac{\alpha_1 y_1}{\lambda_\nu V - \alpha_1}, \ldots, \frac{\alpha_{m-1} y_{m-1}}{\lambda_\nu V - \alpha_{m-1}} \right), \tag{9}$$

and

$$l^\nu(y) = \left(\frac{1 - \alpha_1}{\lambda_\nu V - \alpha_1}, \ldots, \frac{1 - \alpha_{m-1}}{\lambda_\nu V - \alpha_{m-1}} \right), \tag{10}$$

The spectral properties of the Jacobian matrix corresponding to the BB isoterm are more difficult to analyze. The characteristic equation for the λ_ν is now ([12]):

$$\sum_{i=1}^{m-1} \frac{(1 - \alpha_i)\alpha_i y_i}{\lambda_\nu V - \alpha_i} + \varrho \sum_{i=1}^{m-1} \frac{\left[p_i V - \sum_{k=1}^{m-1} \alpha_k y_k p_k \right] \alpha_i y_i p_i}{\lambda_\nu V - \alpha_i} + $$
$$+ \varrho \sum_{i=1}^{m-1} \sum_{\substack{j=1 \\ j \neq i}}^{m-1} \frac{\alpha_i y_i p_i (1 - \alpha_j)\alpha_j y_j (p_j - p_i)}{(\lambda_\nu V - \alpha_i)(\lambda_\nu V - \alpha_j)} - V = 0, \tag{11}$$

and the corresponding left eigenvector has components given by

$$l_i^\nu(y) = \frac{(1 - \alpha_i)}{\lambda_\nu V - \alpha_i} + $$
$$+ \varrho \sum_{\substack{j=1 \\ j \neq i}}^{m-1} \left(p_j - \sum_{s=1}^{m-1} \alpha_s y_s p_s / V \right) \frac{[(1 - \alpha_i)p_j - (1 - \alpha_j)p_i]\alpha_j y_j}{(\lambda_\nu V - \alpha_i)(\lambda_\nu V - \alpha_j)} \tag{12}$$

We shall not write down the formulae for the right eigenvectors because we shall not need them in this paper. It can be shown ([12]) that the roots of (11) are real and in general distinct, so that system (2) is again hyperbolic. Note that by substituting $\varrho = 0$ at all places in (11) and (12) where ϱ occurs explicitly, the formulae (8) and (10) of the Langmuir case are obtained.

Since for both isotherms the system of conservation laws is hyperbolic, the general mathematical theory of systems of this type can be applied. Because of the importance of the Riemann problem in the chromatographic model, we shall first recall some general properties of this problem. For more details, see e.g. Lax [8] or Smoller [11].

Consider a hyperbolic system of $m - 1$ equations

$$\frac{\partial y_i}{\partial \tau} + \frac{\partial x_i(y)}{\partial \zeta} = 0, \tag{13}$$

for $y = (y_1, \ldots, y_{m-1})$, with boundary value

$$y(\tau, 0) = \begin{cases} y^L & \text{for } \tau > 0, \\ y^R & \text{for } \tau < 0, \end{cases} \tag{14}$$

where y^L and y^R are constant states. Let $\lambda_1(y) < \lambda_2(y) < \ldots < \lambda_{m-1}(y)$ be the eigenvalues of the Jacobian matrix. A solution of (13), (14) is a function of $\epsilon = \zeta/\tau$. For $\nu = 1, \ldots, m-1$, a ν−simple wave (also called ν−rarefaction wave) is a smooth solution $y(\epsilon)$ of (13) such that

$$\frac{d\,y(\epsilon)}{d\epsilon} \propto r^\nu(y(\epsilon)). \tag{15}$$

Discontinuous solutions are obtained by interpreting (13) in the weak sense. If $\zeta = s\tau$ is a shock separating two constant states, y and $\tilde{y}$ say, then

$$s(\tilde{y}_i - y_i) = \tilde{x}_i - x_i, \tag{16}$$

for $i = 1, \ldots, m - 1$, where s is the shock speed. Shocks are only acceptable when smooth solutions do not exist. This happens when characteristics of the same type intersect. If $\tilde{y}$ is to the left of y in the (τ, ζ) plane, then the ν−characteristics of these states intersect if $\lambda_\nu(\tilde{y}) > \lambda_\nu(y)$. When in addition to this inequality there holds

$$\lambda_\nu(\tilde{y}) > s > \lambda_\nu(y) \quad \text{and} \quad \lambda_{\nu+1}(\tilde{y}) > s > \lambda_{\nu-1}(y), \tag{17}$$

then the shock is called a ν−shock.

Provided the eigenvectors are genuinely nonlinear in the sense of Lax [8], which they are for both isotherms considered in this paper, the following existence theorem applies.

Theorem 1 (Lax [8]) *If y^L and y^R are sufficiently close, then problem (13), (14) has a unique solution that consists of m constant states y^ν, $\nu = 0, \ldots, m - 1$ with $y^0 = y^L$ and $y^{m-1} = y^R$, and such that $y^{\nu-1}$ and y^ν are linked by either a ν−simple wave or a ν−shock.*

An example of such a solution of a Riemann problem is given in Fig. 2. Computing such a solution can be rather involved because in general it is not known beforehand how many simple waves and shock there will be.

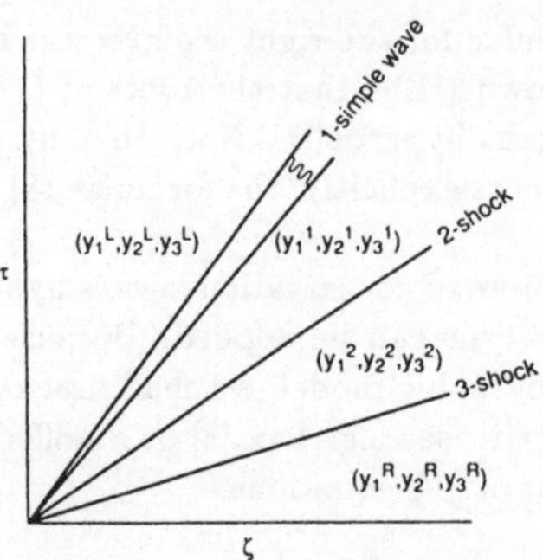

Figure 2: The solution of a Riemann problem for a system with $m - 1 = 3$. The states y^1 and y^2 are constant and the transitions between the states are a 1-simple wave, a 2-shock and a 3-shock.

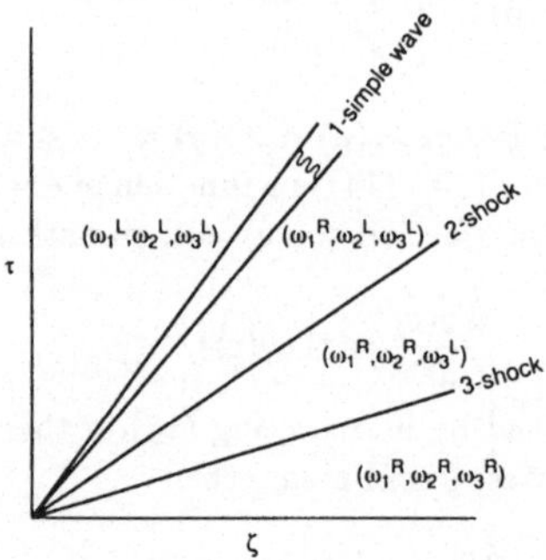

Figure 3: The solution of the Riemann problem of Fig. 2 in terms of the ω_i. ω_2 and ω_3 remain constant across the 1-simple wave, ω_1 and ω_3 remain constant across the 2-shock and ω_1 and ω_2 do not change across the 3-shock.

However, in the Langmuir case the solving of a Riemann problem is greatly simplified by the ω (or h) transformation which was first introduced in the context of chromatography by Helfferich et al. [7]. This transformation is a mapping $\mathbf{R}^{m-1} \ni y \mapsto \omega \in \mathbf{R}^{m-1}$ such that across a $\nu-$simple wave and a $\nu-$shock only ω_ν changes, while all other ω_i remain constant. Furthermore, the transition between two states is a shock when ω_ν decreases from the left to the right in the (τ, ζ) plane (i.e. downstream) and is a simple wave when ω_ν increases downstream. Finally, the shock speeds can be expressed in terms of the ω_i of the left and right states. Hence, the solution of a Riemann problem can immediately be constructed in ω space. An example is given in Fig. 3.

4. The ω transformation

The remarkable simplification of the Riemann problem obtained by applying the ω transformation raises the question of when this transformation exists for general hyperbolic systems. In order to gain insight in this problem we first remark that according to (15), the set of states that can be linked to a given state $\tilde{y}$ by means of a ν–simple wave is an integral curve of the vector field r^ν. Now a function ω_ν is constant across all μ–simple waves with $\mu \neq \nu$ if and only if

$$\nabla \omega_\nu(y) . r^\mu(y) = 0, \quad \text{for } \mu \neq \nu. \tag{18}$$

Because the sets of right and left eigenvectors are biorthogonal, (18) is equivalent to

$$\nabla \omega_\nu(y) \propto l^\nu(y), \tag{19}$$

hence, the left eigenvector l^ν is proportional to a gradient field. According to the theory of Pfaffian systems ([1], [6]), this is equivalent to

$$\left(\frac{\partial l_i^\nu}{\partial y_j} - \frac{\partial l_j^\nu}{\partial y_i} \right) l_k^\nu + \left(\frac{\partial l_k^\nu}{\partial y_j} - \frac{\partial l_i^\nu}{\partial y_k} \right) l_j^\nu + \left(\frac{\partial l_j^\nu}{\partial y_k} - \frac{\partial l_k^\nu}{\partial y_j} \right) l_i^\nu = 0, \quad \text{for all } i,j,k. \tag{20}$$

We thus conclude that functions ω_ν with the property that they remain constant across all μ–simple waves for $\mu \neq \nu$, exist if and only if all left eigenvectors l^ν satisfy (20).

Using (8) and (10), it is easy to verify that in the Langmuir case the left eigenvectors satisfy (20) and that the functions ω_ν are given by

$$\omega_\nu(y) = \lambda_\nu(y) V(y). \tag{21}$$

However, the left eigenvectors (12) corresponding to the BB isotherm do not satisfy (20) in all states y and hence the ω transformation does not exist in this case. This means that solving a Riemann problem is more difficult for the BB isotherm than for the Langmuir isotherm.

The connection between the ω_ν and the Riemann invariants is as follows. Recall that a μ–Riemann invariant is a function ψ which satisfies $\nabla \psi . r^\mu = 0$. Since the dimension of the space is $m - 1$, there exist $m - 2$ Riemann invariants for every μ, so that there are $(m-1)(m-2)$ in all. The existence of functions ω_ν satisfying (18) for every ν means that ω_ν is a μ–Riemann invariant for all $\mu \neq \nu$, hence only $m - 1$ of the Riemann invariants are different.

Notice that in three dimensions, i.e. for $m-1 = 3$, (20) can be written as l^ν . curl $l^\nu = 0$. Furthermore, if $m-1 = 2$, (20) is always satisfied so that the ω transformation always exists when $m = 3$.

It is important to note that the conditions (20) are *not* sufficient in order that the ω_ν are also constant across all μ–shocks for $\mu \neq \nu$. The change of μ–Riemann invariants across μ–shocks is of third order in the shock strength ([11]) and therefore is often small, but it is in general not zero. Nevertheless, in the case of the Langmuir isotherm it can be shown that the ω_ν remain constant across all μ–shocks with $\mu \neq \nu$ ([9]). This is a very nice property because most transitions that occcur in chromatography are shocks.

64

5. Simulation of Separation Processes

We consider a system of 5 cations. The displacer is Rb, the carrier Li and the mixture consists of K, Na and H in order of decreasing sorption. Hence, $m = 5$ and the indices 1,2,3,4 and 5 correspond to Rb, K, Na, H and Li, respectively. The values of the α_i in the case of the Langmuir isotherm and of the p_i for the BB isotherm, are listed in [12].

Fig. 4 shows results of the simulation of a separation process using the Langmuir isotherm. The dimensionless liquid-phase concentrations in the feed are $x_2^F = 0.25$, $x_3^F = 0.50$ and $x_4^F = 0.25$. At the inlet ($\zeta = 0$) the carrier state (superscript C) enters first, then the feed mixture (superscript F) and finally the displacer (superscript D) are introduced. The values of the ω_i of these three states follow by substituting the solutions λ_ν of (8) into (21). One finds ($i = 5$ has been eliminated):

$$
\begin{aligned}
\omega^C &= (\alpha_1, \alpha_2, \alpha_3, \alpha_4), \\
\omega^F &= (\alpha_1, \omega_2^F, \omega_3^F, 1), \\
\omega^D &= (\alpha_2, \alpha_3, \alpha_4, 1).
\end{aligned}
\tag{22}
$$

The remark below (8) in Section 3 about the position of the eigenvalues implies that $\alpha_2 < \omega_2^F < \alpha_3 < \omega_3^F < \alpha_4$. For sufficiently small ζ the states as functions of time are determined by the two solutions of the Riemann problems at $\tau = 0$ and at $\tau = 1$. Since the feed mixture and the carrier state differ in the concentrations of only three species (after eliminating Li of course), Theorem 1 implies that the solution of the Riemann problem at $\tau = 0$ will consist of two intermediate states y^1, y^2 and three transitions. Since for every i we have $\omega_i^F \geq \omega_i^C$, it follows that all transitions are shocks. In a similar manner one infers that the solution of the Riemann problem at $\tau = 1$ also consists of three shocks and two intermediate states.

When the fastest shock of the Riemann problem at $\tau = 1$ catches up with the slowest shock of the Riemann problem at $\tau = 0$, we have to solve a third Riemann problem (at the point of intersection analogous to P_1 in Fig. 5). Because the left and right states differ in only two of the ω_i, there are only one intermediate state and two transitions. Both transitions turn out to be shocks. Further downstream two other intersections of shocks occur (analogous to point P_2 and P_3 in Fig. 5), which lead to two additional Riemann problems, which again can be solved conveniently by using the ω_i. From the calculated ω_i of the different states, the concentrations of the species can be determined numerically. The compositions of the states are indicated in Fig. 4. It is seen that new states that occur as a result of the intersection of shocks contain less different species further downstream the column until finally complete separation is attained.

The preceding can be generalized to mixtures of an arbitrary number of species. It can be proved that the Langmuir model implies that only shocks occur and that eventually complete separation always occurs. The property that only shocks occur is a consequence of the fact that the displacer is more strongly sorbed than all species in the mixture, while the carrier is less sorbed than all species in the mixture. If the carrier and the displacer were interchanged, simple waves instead of shocks would occur and complete separation would not happen.

Fig. 5 shows the solution of the same separation process as in Fig. 4 but now simulated using the BB isotherm. Since the ω_i do not exist in this case, solving the Riemann problem is now more involved. For example, in order to verify that the solution of a given Riemann

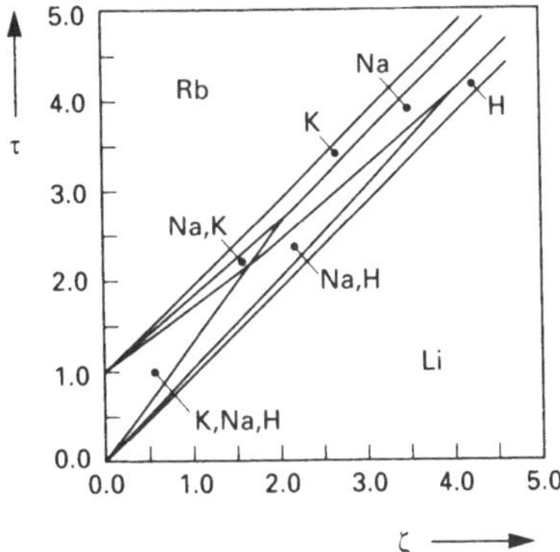

Figure 4: Calculated solution of the separation process using the Langmuir isotherm. The displacer is Rb, the carrier is Li and the mixture consists of K, Na and H in order of decreasing sorption with dimensionless liquid-phase concentrations given by 0.25, 0.50 and 0.25, respectively. The straight lines are shocks between states with compositions as indicated. To the right of the point of intersection of shocks, analogous to P_3 in Fig. 5, complete separation is attained.

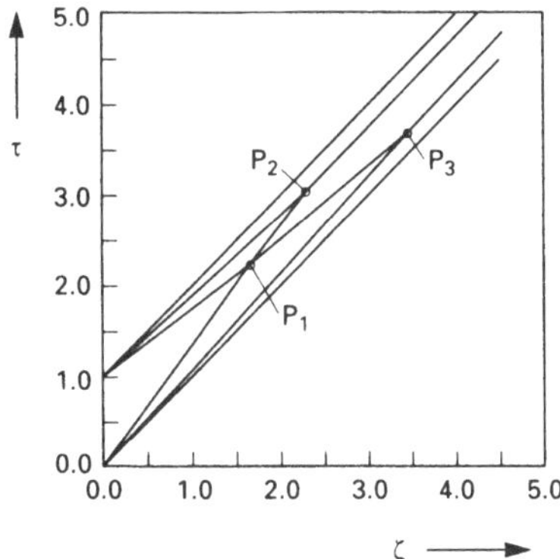

Figure 5: Calculated solution of the separation process using the BB isotherm for the same system as in Fig. 4. The shock speeds and the points of intersection P_1, P_2 and P_3 differ considerably from those in Fig. 4.

problem consists of only shocks, one solves the $(m-1)^2$ jump equations

$$s^{\nu}(y_i^{\nu} - y_i^{\nu-1}) = x_i^{\nu} - x_i^{\nu-1}, \quad \nu = 1,\ldots,m-1, \ i = 1,\ldots,m-1, \tag{23}$$

for the $(m-1)^2$ unknowns $s^1,\ldots,s^{m-1}$, and $y_i^1,\ldots,y_i^{m-2}$ for $i = 1,\ldots,m-1$ (recall that y^0 and y^{m-1} are the known left and right state of the Riemann problem). Once system (23) has been solved, the so called Lax conditions (17) have to be checked in order to decide whether the solution with only shocks and no simple waves is indeed physically acceptable. If not, one or more simple waves will occur. These are determined by integrating (15).

In all simulations based on the BB isotherm it was found that, just as in the case of the Langmuir isotherm, only shocks occur and that complete separation is obtained. The compositions of the states are identical for the two cases. However, the concentrations of the species, the shock speeds and hence also the point where complete separation is attained, differ considerably. Comparison with experiment showed that the BB isotherm yields more accurate results ([3]).

But the differences between the predictions of the two models are not always merely of a quantitative nature. Qualitative differences also occur in the development graphs. The Langmuir model implies that always two shocks emanate from point P_1 (i.e. the first point of intersection of shocks), whatever the number of species. The reason is that the two states that meet at point P_1 differ in only two of the ω_i. But these states differ in general in a much larger number of concentrations y_i and hence, when the functions ω_i do not exist, such as in the BB model, one would expect from Theorem 1 the number of shocks that emanate from point P_1 will increase with the number of species. Experiments performed to verify this conjecture showed that the predictions of the BB isotherms are correct, whereas the Langmuir model predicts too few shocks when $m \geq 6$. For more details see [13].

References

[1] Abraham, R., Marsden, J.E. & Ratiu, T. (1983) Manifolds, Tensor Analysis and Applications, pp. 376-379, Addison-Wesley Publishing Company Reading MA.

[2] Boots, H.M.J. & De Bokx, P.K. (1989) 'Theory of Enthalpy-Entropy Compensation', J. Phys. Chem., 93, 8240-82443.

[3] De Bokx, P.C., Baarslag, P.C. & Urbach, H.P. 'The modelling of displacement chromatography using non-ideal isotherms', accepted by J. Chromatogr.

[4] De Bokx, P.K. & Boots, H.M.J. (1989) 'The Ion-Exchange Equilibrium', J. Phys. Chem., 93, 8243-8248.

[5] De Bokx, P.K. & Boots, H.M.J. (1990) 'The Compensating-Mixture Model for Multicomponent Systems and Its Application to Ion Exchange', J. Phys. Chem., 94, 6489-6495.

[6] Haack, W. & Wendland, W., (1972) Lectures on Partial and Pfafian Differential Equations, Pergamon Press, Oxford.

[7] Helfferich F. & Klein, G. (1970) Multicomponent chromatography, Marcel Dekker, New York.

[8] Lax, P.D. (1973) 'Hyperbolic Systems of Conservation Laws and the Mathematical Theory of Shock Waves', in SIAM Regional Conference Series on Applied Mathematics, Philadelphia, PA.

[9] Rhee, H-K, R. Aris, R. & Amundson, N.R. (1970) 'On the Theory of Chromatography', Phil. Trans., Roy. Soc. Lond. A., 267, 419-455.

[10] Rhee, H-K & Amundson, N.R. (1982) 'Analysis of Multicomponent Separation by Displacement Development', AIChE Journal, 28, No. 3, 423-433.

[11] Smoller, J. (1983) Shock Waves and Reaction-Diffusion Equations, Springer-Verlag, New York.

[12] Urbach, H.P., 'Comparison of two Hyperbolic Systems in Chromatography', accepted by Proc. Roy. Soc. Lond., Series A.

[13] Urbach, H.P. & De Bokx, P.K., 'On the Number of Shocks in Chromatography', submitted to Proc. Roy. Soc. Lond., Series A.

Contributed Papers

ON THE MATHEMATIC MODELING OF KINEMATICALLY RELATED
TRANSFORMING ROTATIONS TOOTH SURFACES

V.ABADJIEV,   D.PETROVA
Institute of Mechanics and Biomechanics
Bulgarian Academy of Sciences
Acad.G.Bonchev str., bl. 4
1113 Sofia
Bulgaria

ABSTRACT. Two approaches to mathematic modeling of skew-axes gears whose
tooth surfaces are generated by the Olivier's second principle are discussed. Which of them will be used depends on whether a definite quality
in the whole region of mesh or in a vicinity of some point is required.

1.Introduction

The tooth surfaces that transform rotations between skewed axes are characterized with a complicated geometry and expensive manufacture technology. Their synthesis must guarantee optimal gears exploitation qualities. The quality control in the whole region of mesh or in a vicinity
of some point only determines the approaches to the design process.

2. Pitch Point Approach

When applying this approach, the mathematic model is based on the assumption that the transformation of rotations between skewed axes is
realized using one point of the space. It is also assumed that at every
mo- ment two tooth surfaces contact at the same point. The character of
the kinematic contact in this point is such that one may define uniquely
the dimensions of the pitch spirals of the contacting tooth surfaces,
namely the tangent to the pitch spiral in this point is collinear with
the sliding velocity between the tooth surfaces. This assumption gives
the name of the common point which is called a pitch point. We require
this point to be the common point of two circles whose centers are situated on the skewed axes and whose planes are perpendicular to the
skewed axes. The authors have called these circles primary circles [1].
   If these circles are circles belonging to surfaces of revolution whose
contact is linear and whose axes of rotation are the skewed axes, then
the pitch point is the common point of primary surfaces. The concept of
primary surfaces is applied for example by Nelson [2].
   It is evident that this approach gives preconditions for synthesizing

gears with "guaranteed" qualities in the pitch point vicinity.

The illustrated in Fig.1 mathematic model is extremely convenient when synthesizing skew-axes gears is generated by the Olivier's second principle, i.e. the shape of one member is chosen, and after that the shape of the other is defined as conjugate to the first. The described model has served as a basis of creating algorithms and program system for spiroid gears synthesis. Angular velocity vectors ω_1 and ω_2 correspond to tooth surfaces operating with their low sides when the gear with a tooth surface Σ_1 is the driving element.

Let two skewed axes 1-1 (first gear axis) and 2-2 (second gear axis) be given. Their mutual position is completely determined by the angle δ and by the offset a_w. Let a pitch point P be given too. The primary circle $_2H_G$ lies in a plane which is perpendicular to the axis $i-i$ and H_G and H_G are externally tangent at P. Let T be the plane including the tangents at P to these circles. Eight parameters determine the diameters and the mutual position of H_G and H_G: r_i, a_i, θ_i – semi-polar coordinates of P in the system S_i (i=1,2) and two angles δ_1 and δ_2 which define the orientations of the circles planes with respect to the straight-line $m-m$ which is perpendicular to T (in Fig. 1 a_1 $_1$ is the distance between the offset line and the plane of the circle H_G). Expressing the vector O_1P (see Fig.1) and the unit vector m of $m-m$ both by r_1, a_1, θ_1, δ_1 and by r_2, a_2, θ_2, δ_2 one obtains the following set of equations:

$$r_1\cos\theta_1 = r_2\cos\theta_2\cos\delta + a_2\sin\delta$$

$$r_1\sin\theta_1 = a_w - r_2\sin\theta_2$$

$$a_1 = r_2\cos\theta_2\sin\delta - a_2\cos\delta \qquad\qquad (1)$$

$$\cos\delta_1\sin\theta_1 = \cos\delta_2\sin\theta_2$$

$$\sin\delta_1 = \sin\delta_2\cos\delta + \cos\delta_2\cos\theta_2\sin\delta$$

To solve (1) we consider the unknowns a_w, δ, δ_1, r_1 and a_1 as free parameters. The analytic solution of system (1) and the conditions which the free parameters should satisfy so that (1) to be solved have been discussed in details in [1].

Let us define tooth surfaces Σ_1 and Σ_2 which have a common tangent at the point P at every moment and which rotate with angular velocities ω_1 and ω_2, respectively. It means that the speed ratio $i_{12}=\omega_1/\omega_2$ is known. It is also known the direction of the pitch helix passing through the pitch point which is determined by the angle [3]
$\beta_1 = arctg((i_{12}r_1 - r_2\cos\mu)/r_2/\sin\mu)$ where $\mu = arcsin(\sin\theta_2\cos\delta/\cos\delta_1)$.

Using the primary circles synthesis results and just defined worm helix angle β_1 the authors have created algorithm for evaluating spiroid gear-set teeth dimensions and spiroid worm and gear widths [4].

It is essential to mention that the program system realizes the optimal synthesis of this type of gears by the help of criteria which control the mesh quality in the pitch point vicinity.

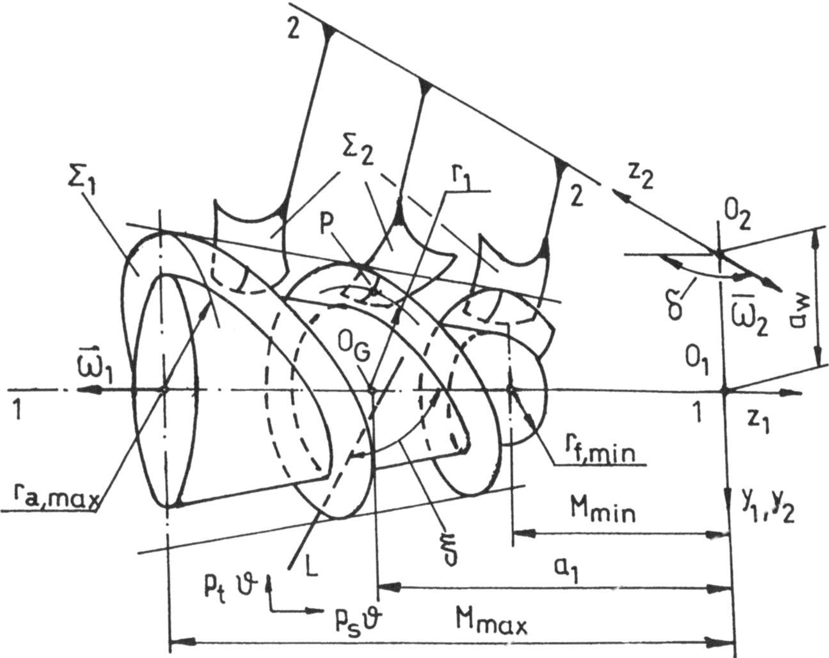

Figure 1. Kinematic scheme and parameters of skew-axes gears generated by Olivier's second principle

3. Region of Mesh Approach

The second approach to mechano-mathematical modeling of the discussed gears is based on a situation which is close to the real conditions when transforming rotations. In this case we have in mind that the tooth surfaces contact at every moment in a given region of mesh. It is not obligatory that the contact point of the conjugate tooth surfaces to be a pitch point. The aim is to look for such orientation of the tooth surfaces in a direction of tooth and in a normal direction and for such gears dimensions that definite kinematic characteristics or quality characteristics related to the forces in the whole region of mesh to be fulfilled

This approach is used when modelling the working and tool meshing of double-enveloping wormgear drive with non-orthogonally skewed axes and with plane tooth surfaces of the straight-line teeth of the wormgear.

The authors have created criteria that control definite quality characteristics which refer to the whole region of mesh of spiroid gears. Using these criteria one may control and eliminate from the region of mesh: a)points of undercutting; b) ordinary nodes that cause increased specific friction, wedging of the tooth surfaces and as a result a decrease of the gears hydrodynamic loading capacity. The last criterion, i.e. the criterion for non-existence of ordinary nodes in the whole region of mesh, in the case of archimedean spiroid gears has the analytic expression [5]:

$$M_{max} < r_{f,min}\, a_w cotg\delta\, tg(\xi-\pi/2)/p_1 \,,$$

$$cos\delta + a_w cos(\xi-\delta)/p_1/sin\xi + M_{max} sin\delta/r_{f,min} < i_{12} < \qquad (2)$$

$$cos\delta - a_w cos(\xi+\delta)/p_1/sin\xi - M_{max} sin\delta/r_{f,min}$$

and

$$M_{min} > r_{a,max}\, a_w cotg\delta\, tg(\xi-\pi/2)/p_1 \,,$$

$$cos\delta - a_w cos(\xi+\delta)/p_1/sin\xi - M_{min} sin\delta/r_{a,max} < i_{12} < \qquad (3)$$

$$cos\delta + a_w cos(\xi-\delta)/p_1/sin\xi + M_{min} sin\delta/r_{a,max}$$

Here $r_{f,min}$ and $r_{a,max}$ are the minimum internal and maximum external radii of   the   spiroid worm, M_{min} and M_{max} are the minimum and maximum mounting displacements starting   from   the$_j$ offset of the gears, ξ^j is the obtuse angle between the generatrix $_jL$ $_j$of the spiroid worm$_j$ thread surface and the worm axis z_1, $p_1 = p_s \pm p_t cotg\xi$, p_s and p_t are the helical parameters$_j$ with   which is performed   the helical motion of L ($j=1$ corresponds to low-side driving and $j=2$ refers to high-side driving).

In a number of cases, however, when designing skew-axes type gears we combine both approaches for mathematic modeling. Thus, two synthesis stages are fixed. The first one, that can be called *extended local*, represents an optimal synthesis in the pitch point and in its vicinity. The second stage, called *global*, ensures definite quality characteristics in the whole region of mesh.

The authors consider that the suitable combination of the two discussed approaches involves considerable increase of skew-axes gears design effectiveness.

References
[1]. Abadjiev, V. and Petrova, D.(1989) 'Synthesis of geometric primary surfaces of externally meshed skew-axes gears', Automation Design' 89 Conference of ASME, Vol. H0509C.
[2]. Nelson, M.(1961) 'Spiroid gearing. Part 1-Basic design practices', Machine design, Vol. 33, No 4.
[3]. Abadjiev, V.(1984) 'On the synthesis and analysis of spiroid gears' Ph.D. Thesis, Bulgarian Academy of Sciences.
[4]. Abadjiev, V. and Petrova, D.(1991) 'Analytical and computer design of spiroid gears', in M.Heiliö (ed.), Proceedings of the fifth European Conference on Mathematics in Industry, Kluwer Academic Publishers, pp. 121-126.
[5]. Petrova, D. and Abadjiev, V. (1991) 'Some aspects of the analysis of high reduction gears', Proceedings of Eight World Congress on the Theory of Machines and Mechanisms, Vol. 5, 1335-1338.

MATHEMATICAL MODELLING OF THERMOSYPHONS IN CRYOGENIC AIR SEPARATION PLANTS

C.J. ALDRIDGE AND A.C. FOWLER
Mathematical Institute
University of Oxford
24-29 St. Giles'
Oxford OX1 3LB
United Kingdom

ABSTRACT: A simple mathematical model for the flow in a thermosyphon is described, and some preliminary analysis is reported.

1. Introduction

A thermosyphon is a circulating fluid system in which the flow is driven by thermal buoyancy forces. Thermosyphons are used in cryogenic air separation processes. Specifically, a fluid loop of liquid nitrogen is heated by the external condensation of oxygen at cryogenic temperatures. The nitrogen boils, and the production of vapour leads to a flow around the loop.

The system under consideration is an experimental facility designed to test the flow, and is illustrated in Figure 1. A plate-fin heat exchanger is employed, which basically consists of a series of parallel channels through which the nitrogen flows as it is heated.

The aim of the work reported here is to model the flow in the thermosyphon, particularly the heated two-phase flow, incorporating a single-channel description of the heat exchanger. A complete description of the model is given by Aldridge and Fowler (1991).

2. Model Summary

We model the system in terms of its separate components: reservoir, feeder pipe, heat exchanger and exit pipe. The heat exchanger is modelled using one-dimensional conservation equations, while in the other sections algebraic balances are used.

The reservoir of liquid nitrogen is assumed to be so large that its state is effectively decoupled from the flow in the heat exchanger. We assume that the level of the reservoir is constant, so providing a constant pressure head for the flow. The enthalpy in the reservoir is also assumed to be constant, with thermodynamic equilibrium at the surface and the liquid nitrogen at saturation enthalpy there. As the pressure in the fluid increases towards

$$[\,\delta(1-\alpha)(u_t + uu_z) + \,]\,p_z = -(1-\alpha) - \tau_{mw},$$

$$H_t + (uH)_z = q_{mw} - \{\alpha V_j/(1-\alpha)\}_z,$$

$$H = a_1(1-\alpha)p + \alpha, \tag{2}$$

where α is the void fraction, H is the enthalpy per unit volume, τ_{mw} is the two-phase friction, q_{mw} is the heat input, and V_j is the drift velocity, which must be constituted as a function of α. The $O(\delta)$ acceleration terms in $(2)_2$, where $\delta = \rho_g/\rho_l \ll 1$, are neglected, but are included for the purposes of computing characteristics (below).

Appropriate boundary conditions are

$$p = a_2 - a_7 u_0^{1.6}, \; h = 0 \text{ at } z = 0;$$

$$H = a_1 p, \; \alpha = 0, \; u = u_0, \; [p]_-^+ = 0 \text{ at } z = r;$$

$$p = 0 \text{ at } z = 1. \tag{3}$$

Values $a_1 \sim 3$, $a_2 \sim .75$, $a_7 \sim 3.7$ are typical.

For prescribed heat flux q_{lw}, we can solve (1) to obtain

$$r = \int_{t-a_1 p_r/q_{lw}}^{t} u_0(s)ds, \tag{4}$$

where $p_r = p|_{z=r}$, and if f_{lw} is a constant which depends on u_0, then

$$\begin{aligned} p_r &= a_2 + a_7 u_0^{1.6} - [1 + f_{lw} u_0^2]r \\ &= p_0(u_0) - f_l(u_0)r. \end{aligned} \tag{5}$$

4. Analysis

If we write $\alpha = 1 - \rho$, $\nu = 1/\rho$, $h = a_1 p + \nu - 1$, $W(\rho) = \alpha V_j(\alpha)/\rho$, and $q(\nu) = -W'(\rho)/\nu$, then (2) can be written as

$$A\underline{\psi}_t + B\underline{\psi}_z = \underline{c}, \tag{6}$$

where

$$A = \begin{pmatrix} -1 & 0 & 0 \\ 0 & a_1\delta & 0 \\ 0 & 0 & 1 \end{pmatrix}, \; B = \begin{pmatrix} -u & \nu & 0 \\ -\nu & a_1\delta u & \nu \\ q & 0 & u \end{pmatrix}, \tag{7}$$

and $\underline{\psi} = (\nu, u, h)^T$. For $\delta \ll 1$, the characteristics are $dz/dt = \lambda$, where $\lambda \approx u + q$, $\pm(\nu^2/a_1\delta)^{1/2}$, and are all real. By putting $\delta = 0$, we neglect rapidly propagating 'sound' or density waves. The system is hyperbolic, and thus well-posed.

If u_0, r vary slowly, then a quasi-static solution in the two-phase region is appropriate. For example, if q_{mw} is constant, we have

$$u = u_0/(1-\alpha),$$

$$1 - \alpha = \frac{u_0 + V_j}{u_0 a_1(p_r - p) + (u_0 + V_j) + q_{mw}(z - r)}, \tag{8}$$

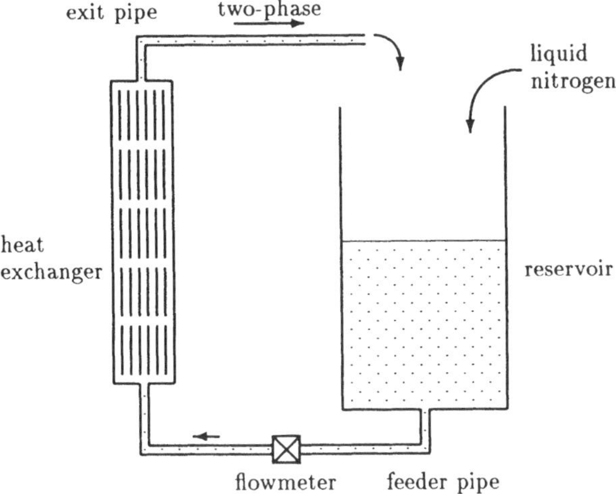

Figure 1: Schematic of thermosyphon system

the base, the saturation temperature increases due to the Clapeyron effect, and hence at the reservoir base the fluid is subcooled.

The flow in the feeder pipe is of single-phase unheated nitrogen. Bernoulli's law is applied at each end, and we use an appropriate friction correlation to give the pressure drop in the feeder pipe. In the heat exchanger, subcooled liquid nitrogen enters and is heated, causing boiling to occur. The heat exchanger contains regions of subcooled single-phase and evaporating two-phase flow. We describe the model in this region below. The exit pipe contains an unheated two-phase mixture; again an appropriate pressure drop correlation is used, with Bernoulli's law applied at the heat exchanger exit.

3. Heat Exchanger Model

In a channel, suitable dimensionless reduced equations for the subcooled region $0 < z < r(t)$ are those of mass, momentum and enthalpy:

$$u = u_0(t),$$

$$p_z = -1 - f_{lw} u_0^2,$$

$$h_t + u h_z = q_{lw}, \tag{1}$$

where $u_0(t)$ is the inlet velocity, f_{lw} is a turbulent friction factor, and q_{lw} is the dimensionless heat input. The boiling boundary is at $z = r(t)$, and in the two-phase region $r < z < 1$, a drift-flux model (Zuber 1967, Ishii and Zuber 1979) can be reduced to

$$-\alpha_t + [(1 - \alpha)u]_z = 0,$$

thus $\alpha = \alpha(p, z; p_r, u_0, r)$, and a quadrature gives

$$p_r = \int_r^1 (1 - \alpha + \tau_{mw})dz. \tag{9}$$

In principle, then, p_r can be determined as the solution of an equation $p_r = F(u_0, p_r, r)$, whence $p_r = p_r(u_0, r)$, and the model reduces to the delay-integral equation

$$r(u_0) = \int_{t-\tau(u_0)}^t u_0(s)ds, \tag{10}$$

where $r(u_0)$ is determined from solving (5) with $p_r = p_r(u_0, r)$.

It appears that the dependence of r on u_0 may admit the existence of multiple steady states, with more than one value of u_0 corresponding to a given value of r over a certain range. Further investigations are currently being made in this area.

References

C.J. Aldridge and A.C. Fowler. A single channel two-phase flow model for a plate-fin heat exchanger. *OCIAM Technical Report Number 3*, Oxford Centre for Industrial and Applied Mathematics, 1991.

M. Ishii and N. Zuber. Drag coefficient and relative velocity in bubble, droplet or particulate flows. *AIChE Journal*, 25(5):843–855, 1979.

N. Zuber. Flow excursions and oscillations in boiling, two-phase flow systems with heat addition. In *Proc. Symp. Two-Phase Flow Dynamics, Eindhoven, EUR 4288e*, pages 1071–1089, 1967.

STABILITY ESTIMATE FOR THE NONLINEAR SEAKEEPING PROBLEM

P. BASSANINI
Dept. of Mathematics, Univ. of Rome "La Sapienza"
Città Universitaria, 00185 ROMA, Italy

U. BULGARELLI
INSEAN, National Ship Model Basin
Via di Vallerano 139, 00128 ROMA, Italy

ABSTRACT.   The paper deals with the nonlinear I-BV problem describing
the motion of a 3D surface-piercing heavy rigid body in an irrotational
and incompressible fluid .   An energy estimate is proven and several
conclusions concerning uniqueness, continuous dependence and blow-up of
solutions are inferred.

1. Introduction

The motion of a heavy rigid body (a ship) under the action of the sea
has received considerable attention in the literature [1,2].   Usually
some simplifying assumptions are made, namely, viscous effects, surface
tension and wave breaking are neglected, and attention is restricted to
incompressible and irrotational flow past a rigid 3D body, with free
surface extending to infinity in all directions, in an atmosphere at
rest.   Even so, one still has to solve a highly nonlinear I-BV problem
for a time-dependent harmonic function (the velocity potential) in an
unbounded 3D domain with free and moving boundaries (see F. John [3]).
This is a formidable problem, so much that no existence and uniqueness
results are available except in the linearized case (see e.g. [4]) and
most of the work deals with the time-harmonic case (see e.g. [5-8]).
   In this paper we derive an energy estimate for general nonlinear ship
motions in the context of John's approach [3].   The proof is based on
the Gronwall Lemma, by enforcing the conditions of static equilibrium
of the hull in its unperturbed state [2]. As a consequence the solution
depends continuously on initial data and forcing terms (in a suitable
sense) and cannot "blow-up" in finite time.

2. The Mathematical Problem

We denote by m the mass of the rigid body (the ship), by G its center
of mass , by ω its angular velocity , and by K its angular momentum

(with respect to G). The position of the body may be specified by giving G and the three "naval Euler angles" [2] θ (roll), ϕ (pitch), ψ (yaw) as functions of time, t. It is well-known that K can be expressed as a linear function of ω through the inertia tensor, so that

$$\dot{K}.\omega = \frac{d}{dt}(\tfrac{1}{2}K.\omega) \quad , \quad K.\omega \geq A|\omega|^2 \qquad (1)$$

where A is a positive constant. We denote by B=B(t) the wet part of the hull at time t , which must be determined as part of the solution, by ρ the fluid density and by $\phi = \phi(P,t)$ the flow potential.

The equations of motion of the hull are:

$$m\dot{v} = R + R' - mg \quad , \quad \dot{K} = M + M' \qquad (2)$$

where $v = \dot{G}$, g is the acceleration of gravity, R'(t) and M'(t) are given (continuous) forcing terms, R and M are the resultant and moment due to the fluid forces on the wetted hull:

$$R = \int_B p\, n\, dS \quad , \quad M = \int_B (P-G)\char94 n\, p\, dS \qquad (3)$$

(n is the inner normal to the body, $\char94$ the cross product), and, by the Bernouilli theorem,

$$p + \rho[\frac{\partial\phi}{\partial t} + \tfrac{1}{2}\,grad^2\phi + g\,z] = 0 \qquad (4)$$

(z is a vertical coordinate oriented upwards). The interface condition of no fluid crossing the wetted hull (unknown moving boundary) is

$$\partial\phi / \partial n = n.[v + \omega\char94(P-G)] \qquad P\,\epsilon\,B(t)\ ,\ t>0 \qquad (5)$$

The free surface elevation $\zeta(x,y,t)$ satisfies the kinematical condition on the unknown free boundary $F \equiv z - \zeta(x,y,t) = 0$:

$$\partial\phi / \partial n = -|grad\,F|^{-1}\,\partial F/\partial t \qquad (6)$$

and the dynamical equation

$$\partial\phi/\partial t + \tfrac{1}{2}\,grad^2\phi + g\,z = 0 \qquad (7)$$

The velocity potential is harmonic and vanishes uniformly at infinity:

$$\Delta\phi = 0 \quad in\ \Omega(t)\ ,\quad \phi \longrightarrow 0 \quad P \longrightarrow \infty \quad (t>0) \qquad (8)$$

($\Omega(t)$ is the unbounded flow domain with the wetted hull and the free surface as finite boundary). Finally the initial conditions

$$\phi(P,0) = \phi_o(P) \quad \text{on} \quad z = \zeta(x,y,0) = \zeta_o(x,y)$$

$$G = G_o \ , \quad v = v_o \ , \quad \theta = \theta_o \ , \quad \phi = \phi_o \ , \quad \psi = \psi_o \ , \quad \omega = \omega_o$$

(9)

are satisfied for t=0. Note that, if at any given time t $\phi(P,t)$ is zero on the free surface and $v = \omega = 0$, then, by a property of harmonic functions, $\phi(P,t)$ vanishes everywhere.

Equations (2)-(9) will be referred to as the <u>seakeeping problem.</u>

3. Stability Estimate

Suppose that a sufficiently smooth solution of the seakeeping problem exists, that ϕ and ζ vanish at infinity sufficiently fast, and that the hull is sufficiently smooth, so that the Gauss Lemma holds in $\Omega(t)$ and all integrals considered below converge (explicit conditions can be easily inferred from the context). Then we can define the total energy of the system (body+waves), $E(t) = T' + T'' + V' + V''$, where

$$T' = \frac{\rho}{2} \int_{\Omega(t)} \text{grad}^2 \phi \ dV \quad , \quad T'' = \tfrac{1}{2}(m \, |v|^2 + K.\omega)$$

are the kinetic energies of the fluid and the body, and

$$V' = \frac{\rho}{2} g \int_{\partial\Omega(t)} z^2 n_z \ dS \quad , \quad V'' = mgz'' \qquad (z''=z_G)$$

are the potential energies of the fluid and the body. By means of the transport theorem, the divergence theorem and manipulations, from (2)-(8) it follows that any solution of the seakeeping problem satisfies the energy theorem

$$\dot{E} = v.R' + \omega.M' \tag{10}$$

Thus energy is conserved, $E(t)=E(0)$, when the forcing terms are zero. When $R' = M' = 0$ there exists an equilibrium solution (static solution)

$$\phi = \zeta = \omega = \theta = \phi = \psi = 0 \ , \ G = \bar{G}$$

provided Archimedes' principle and the static balance of moments hold:

$$m = \rho \, vol(\bar{V}) \ , \qquad (C - \bar{G})^\wedge g = 0 \tag{11}$$

where $\bar{V}$ is the static displacement volume and C its center of mass (center of buoyancy) [2]. Let us denote by C' the point corresponding to C through the rigid motion of the body at time t and by $z'(t)$ its elevation. We then define the modified (mathematical) energy

$$W(t) = E(t) + mg[z'(t) - z''(t)] \qquad (12)$$

(From now on we shall suppose all equations reduced to non-dimensional form we shall denote by c, c', c'',... some fixed positive constants which can be explicitly evaluated). By means of (1), (11), (12), the formula of rigid motions, and manipulations it is not difficult to see that $W(t) \geq T'' \geq 0$, and $W(t) = 0$ iff the solution is static (hence $W(t)$ is a "norm" for hull motions), and that

$$W(t) \leq cW(t)+c'\Delta +c''N(t) \ , \ N(t)=|R'|^2 +|M'|^2 \ , \ \Delta =|[C'(0)-G(0)]\char94 g|^2$$

Hence by the Gronwall Lemma we finally find that $W(t)$ satisfies the stability estimate

$$W(t) \leq \exp(ct) \{W(0) + \int_0^t [c'\Delta + c''N(\tau)]\exp(-c\tau)d\tau \} \qquad (SE)$$

for $t \geq 0$, where, by virtue of (11), Δ vanishes in static conditions. Hence the following conclusions can be immediately drawn:
(I)   The solutions (the hull motions) depend continuously on initial data and forcing terms, and cannot "blow-up" in finite time
(II)   For static initial data ($\Delta = W(0) = 0$) and zero forcing terms ($N(t) = 0$) the solution is unique and is static ($W(t) = 0$) for every t
(III)   If the initial data are "nearly static", the solutions remain nearly static in finite time intervals.
The same results hold when the fluid depth is finite and Φ satisfies the boundary condition $\partial \Phi / \partial n = 0$ at the bottom.
Note that we have not required (Ljapunov) stability of hydrostatic equilibrium. The stability estimate (SE) implies global uniqueness in the linearized case, even if the equilibrium is unstable (cf [3]).

ACKNOWLEDGMENT. This research was partially supported by CNR under contract n. 91.01319.CT01, and by MURST (Fondi 40%).

[1]   J.V. Wehausen and E.V. Laitone, "Surface Waves". In: Handbuch der Physik Bd. IX, Springer (1960), 446-778
[2]   J.N. Newman, "Marine Hydrodynamics", MIT Press, 1978
[3]   F. John, "On the motion of floating bodies. I and II", Comm. Pure Appl. Math. II (1949), 13-57; III (1950), 45-101
[4]   J. T. Beale, "Eigenfunction expansions for objects floating in an open sea", Comm. Pure Appl. Math. XXX (1977), 283-313
[5]   M.J. Simon and F. Ursell, "Uniqueness in linearized two-dimensional water-wave problems", J. Fluid Mech. 148 (1984), 137-154
[6]   F. Ursell , "Irregular frequencies and the motion of floating bodies", J. Fluid Mech. 105 (1981), 143-156
[7]   T.S. Angell, G.C. Hsiao, and R.E. Kleinman, "Recent Developments in Floating Body Problems". In: Math. Approaches in Hydrodynamics, Ed. by Touvia Miloh, SIAM, Philadelphia, 1991
[8]   T.S. Angell, G.C. Hsiao, and R.E. Kleinman, "An integral equation for the floating body problem", J. Fluid Mech. 166 (1986), 161-171

Modelling optical channel-waveguides using a domain-integral equation technique

H.J.M. Bastiaansen
PTT Research
P.O. Box 421, 2260 AK Leidschendam
The Netherlands

ABSTRACT. A domain-integral equation method is presented to determine propagation constants and electromagnetic field distributions of guided modes in integrated optical waveguides. Both the waveguide and its multi-layered embedding may be anisotropic. A scattering-matrix formalism is used to construct the Green's tensor. A numerical example for an anisotropic waveguide configuration is given.

1 Introduction

Straight open dielectric channel waveguides in a multi-layered planar embedding (figure 1) are basic building blocks of integrated optical devices. For tight design criteria, accurate full-vectorial mathematical models describing its electromagnetic field are necessary. The infinite transversal cross-section poses the major problem in modelling the electromagnetic field: the field is nowhere equal to zero. Therefore the development of rigorous mathematical methods, allowing the computational domain to be restricted to the finite cross-section of the waveguide D^w, is desirable. The domain-integral method is such a method. In [1] the method was proposed for straight waveguides in a homogeneous isotropic embedding. In this paper, an extension to multi-layered anisotropic configurations is given. All permittivity tensors $\underline{\varepsilon}$ are diagonal with respect to the coordinate system $Oxyz$ (figure 1). For isotropic materials the permittivity tensor transforms into the scalar ε.

2 The Domain-Integral Equation

Propagation of electromagnetic fields in straight waveguides is described using the concept of guided modes. Guided modes are separable solutions of Maxwell's equations having harmonic time- and z-dependence $\{\underline{E}, \underline{H}\}(x, y, z) = \{\underline{\tilde{E}}, \underline{\tilde{H}}\}(x, y; k_z) \exp[-j(\omega t - k_z z)]$, in which $\underline{E}$ and $\underline{H}$ are the electric and magnetic field strength, ω is the angular frequency, t is the time coordinate and k_z the axial wavenumber. Only for a finite number of k_z-values a non-trivial solution exists. Each k_z-value is the propagation constant of a guided mode, the solution is its electromagnetic field distribution.

In the domain-integral equation method, the channel waveguide D^w is regarded as a perturbation of its embedding. Lorentz's reciprocity theorem yields an integral equation for the electric field. In Fourier representation, the domain-integral equation [2] is

$$\underline{\tilde{E}}(x, y) = j\omega \int_{-\infty}^{\infty} \iint_{D^w} \underline{\underline{G}}(x, x'; k_y, k_z) \{\underline{\varepsilon}^w - \underline{\varepsilon}^k\} \underline{\tilde{E}}(x', y') \exp[-jk_y(y - y')] dx'dy'dk_y, \quad (1)$$

in which $\underline{\underline{G}}$ is the Green's tensor. The i-th ($i = 1, 2, 3$) column of $\underline{\underline{G}}$ is denoted as $\underline{E}^G$. It is the electric field distribution for an electric point-source, situated in the multi-layered embedding (permittivity $\underline{\epsilon}^b(x)$)

$$-\underline{\nabla}_t \times \underline{H}^G(x;x';k_y,k_z) + j\omega\underline{\epsilon}^b(x)\underline{E}^G(x;x';k_y,k_z) = -\underline{e}_i.\delta(x-x'),$$
$$\underline{\nabla}_t \times \underline{E}^G(x;x';k_y,k_z) + j\omega\mu_0\underline{H}^G(x;x';k_y,k_z) = \underline{0}, \qquad (2)$$

with $\underline{\nabla}_t = (\partial/\partial x, -jk_y, -jk_z)$, e_i the i-th unit vector, and $\delta(x)$ the Dirac delta function. The equations (2) also yield the magnetic field $\underline{H}^G$ of the point-source problem. For the solution of (1) however, $\underline{H}^G$ is not used.

The integral equation (1) yields non-trivial solutions for k_z-values equal to the propagation constants. The solution $\tilde{\underline{E}}$ equals the electric field distribution of the guided mode. Before solving the integral equation, the point-source problem (2) has to be solved to obtain the Green's tensor. This is done in the next section. In figure 2, the configuration and notation are shown.

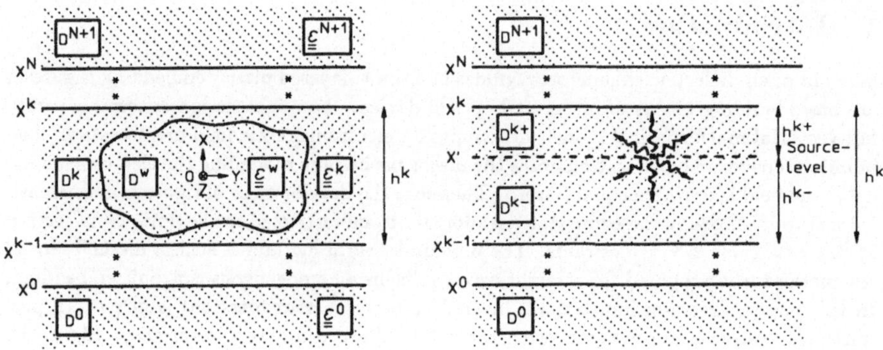

Figure 1: Dielectric channel waveguide.　　　　Figure 2: Point-source problem.

3　The Green's Tensor

Maxwell's equations (2) consist of 4 first-order ordinary differential equations and two algebraic equations. The latter are used to eliminate the components E_1^G and H_1^G. The resulting set of ordinary differential equations is represented in matrix-notation;

$$\frac{\partial \underline{F}^G(x;x')}{\partial x} = A(x).\underline{F}^G(x;x') + \underline{J}.\delta(x-x'), \qquad \underline{F}^G = (E_2^G, E_3^G, H_3^G, H_2^G)^T. \quad (3)$$

Within subdomain D^l, the system is homogeneous and the matrix $A(x)$ is constant, $A(x) \equiv A^l$. The diagonal blocks of A^l are equal to zero. The field vector $\underline{F}^G$ is continuous accross the interfaces $x = x^l$.

The solution of (3) is constructed using the general solution of the homogeneous system:

$$\underline{F}^G(x;x') = T^l.\exp[\bar{\Lambda}^l(x-x^{ref})].\underline{f}^l(x'), \qquad x^{l-1} \le x \le x^l. \quad (4)$$

T^l and $\bar{\Lambda}^l$ are respectively the decomposition matrix and the diagonal matrix in the diagonal decomposition of A^l. The reference vector $\underline{f}^l(z')$ equals $(T^l)^{-1} \cdot \underline{F}^G(z^{ref}; z')$ in which the reference level z^{ref} of D^l equals the lower interface $z = z^{l-1}$ for subdomains above the source-point and the upper interface $z = z^l$ for subdomains underneath the source-point. Because the permittivity tensor in D^l is diagonal, the eigenvalues and eigenvectors of A^l occur in pairs:

$$\bar{\Lambda}^l \;=\; \mathrm{diag}(-\Lambda^l, \Lambda^l), \quad \Lambda^l = \mathrm{diag}(\lambda_1^l, \lambda_2^l), \quad \Re e\{\lambda_1^l, \lambda_2^l\} \geq 0, \quad T^l = \begin{bmatrix} T_{11}^l & -T_{11}^l \\ T_{21}^l & T_{21}^l \end{bmatrix} \quad (5)$$

With the scattering matrix formalism, the unknown reference vectors $\underline{f}^l$ can be determined from the boundary conditions:

    1) Boundedness of the field vector in cover and substrate,

    2) Continuity of the field vector at the interfaces $z = z^l$, $\; l\epsilon\{0, 1, ..., N\}$,

    3) Connection condition at $z = z'$: $\underline{f}^{k+}(z') - \underline{f}^{k-}(z') = (T^k)^{-1} \cdot \underline{\mathcal{J}}$.

For subdomains above the source-level, a downward recursive scheme is used; for subdomains underneath the source-point, an upward recursive scheme is used. Both schemes are connected at $z = z'$.

Notation: Let $\underline{v} = (v_1, v_2, v_3, v_4)^T$, then $\underline{v}_{12} = (v_1, v_2)^T$ and $\underline{v}_{34} = (v_3, v_4)^T$.

Downward Recursion: The transmission matrix T^l and the reflection matrix $\mathcal{R}^l$ for subdomain D^l, $l \in \{k+, ..., N+1\}$, are defined through

$$\underline{f}_{12}^{N+1}(z') \;=\; T^l \cdot \underline{f}_{12}^l(z'), \qquad \underline{f}_{34}^l(z') = \mathcal{R}^l \cdot \underline{f}_{12}^l(z'), \qquad l \in \{k+, ..., N+1\}. \quad (6)$$

The continuity of the field vector yields the recursive scheme for T^l and $\mathcal{R}^l$:

$$T^l \;=\; T^{l+1} \cdot \{B_+^l + B_-^l \cdot \mathcal{R}^{l+1}\}^{-1} \cdot \exp[-\Lambda^l h^l],$$

$$\mathcal{R}^l \;=\; \exp[-\Lambda^l h^l] \cdot \{B_-^l + B_+^l \cdot \mathcal{R}^{l+1}\} \cdot \{B_+^l + B_-^l \cdot \mathcal{R}^{l+1}\}^{-1} \cdot \exp[-\Lambda^l h^l], \quad (7)$$

with $\qquad B_+^l = \dfrac{1}{2} \cdot \{C_{21}^l + C_{11}^l\}, \qquad B_-^l = \dfrac{1}{2} \cdot \{C_{21}^l - C_{11}^l\}, \qquad C_{ij}^l = (T_{ij}^l)^{-1} \cdot T_{ij}^{l+1}. \quad (8)$

Boundedness of $\underline{F}^G$ in the cover implies the initialization: $T^{N+1} = I$, $\mathcal{R}^{N+1} = 0$.

Upward Recursion: The transmission matrix T^l and the reflection $\mathcal{R}^l$ for subdomain D^l, $l \in \{0, ..., k-\}$, are defined through

$$\underline{f}_{34}^0(z') \;=\; T^l \cdot \underline{f}_{34}^l(z'), \qquad \underline{f}_{12}^l(z') = \mathcal{R}^l \cdot \underline{f}_{34}^l(z'), \qquad l \in \{0, ..., k-\}. \quad (9)$$

Its recursive scheme is equal to (7) with $l+1$ replaced by $l-1$, and is initialized through $T^0 = I$, $\mathcal{R}^0 = 0$.

Interconnecting the Reference Vectors: With the recursive schemes the reference vectors of all subdomains are solved as function of $\underline{f}_{12}^{k+}(z')$ and $\underline{f}_{34}^{k-}(z')$. These last vectors are determined with the connection condition at $z = z'$. Subsequently, the recursive schemes and the general solution of each subdomain generate the solution of (2) and therefore the Green's tensor in the entire configuration.

	r=0.5	r=1.0	r=1.5	r=2.0	r=2.5
DIM	1.59114	1.59011	1.58947	1.58915	1.58903
EIM	1.59119	1.59028	1.58965	1.58922	1.58909

Table 1: Effective refractive index N_{eff} versus r.

4 Numerical Implementation and Results

To find the nontrivial solutions of the domain-integral equation, the method of moments is applied [3]. A rectangular mesh for D^w is used. Pulse functions are used as expansion functions. Dirac functions are used for the weighting functions (point-matching). Spatial integrations are performed analytically. A homogeneous system of linear algebraic equations for the expansion coefficients of the electric field results. The values $k_z = \beta^m$ for which a nontrivial solution exists are calculated. Each β^m is a propagation constant of a guided mode, the solution is its electric field distribution.

The theory described in this paper is illustrated for a polymeric single-rib waveguide (figure 3) developed within the EC-project RACE 1019: on a glass substrate three polymeric layers are deposited; two isotropic polyurethane buffer layers and a central guiding layer of an uniformly poled uniaxially electro-optical polymer. In the electro-optical polymer, a rib is photochemically induced, causing the refractive indices of the anisotropic poled polymers to decrease to the isotropic value 1.556. The waveguide is operated at the free-space wavelength $\lambda_0 = 1.335\mu m$. Table 1 gives the effective refractive indices $N_{eff} = \beta^m/k_0$ as a function of r for the TM_{00} mode. The results of the Domain-Integral equation Method (DIM) described in this paper are compared with the frequently used approximative Effective Index Method (EIM). Clearly, the effective index method overestimates the values for the refractive index. The intensity of the electric field of the TM_{00} mode inside D^w is shown for $r = 2.0$ micron in figure 4.

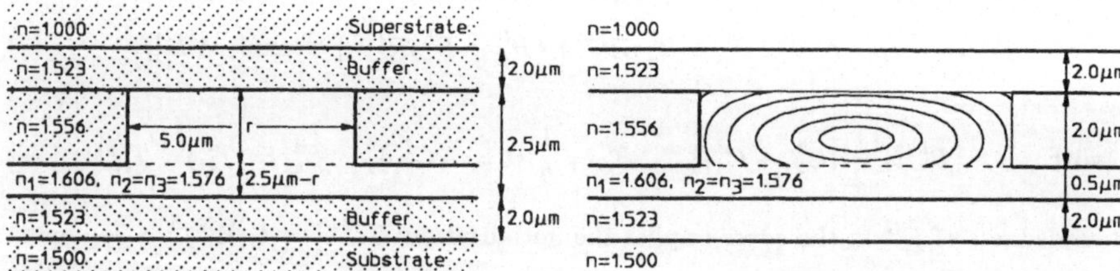

Figure 3: Polymeric rib-waveguide.   Figure 4: Intensity profile in rib.

References

[1] H. de Ruiter (1980),"Integral equation approach to the computation of modes in an optical waveguide", $J.Opt.Soc.Am.$,vol.70,no.12,pp.1519-1524

[2] N.H.G. Baken, M.B.J. Diemeer, J.M. van Splunter and H. Blok (1990),"Computational modeling of diffused channel waveguides using a domain-integral equation",$Journ.Lightw.Tech.$,vol.8, no.4,pp. 576-586

[3] H.J.M. Bastiaansen, N.H.G. Baken and H. Blok (1992), "Domain-integral analysis of channel waveguides in anisotropic multi-layered media", accepted for publication in IEEE Transactions on Microwave Theory and Techniques

DEFORMATION ANALYSIS OF THE EARTH
BY A CONSTRUCTIVE INTERPOLATION MODEL

by

J. Brudowsky, W. Freeden, M. Tücks

Laboratory of Technomathematics, P.O. Box 3049, University of Kaiserslautern,

6750 Kaiserslautern, F.R.G.

ABSTRACT. An interpolation method is given for modelling crustal movements of the earth from discrete data.

1 Introduction

The response of the earth to surface mass loads is a subject of current geophysical interest. Because of the unique opportunity to study loading effects on the earth's crust, the construction of several artificial lakes [la Grande-2, Canada; Blasjø, Norway] was accompanied by a great number of high precision geodetic measurements. Discrete displacement and stress (vector-) data from tilt observations as well as gravity measurements in the reservoir area have clearly detected both vertical and horizontal up- and downlifts (up to decimeter range) and given experimental evidence for the fact of a reservoir induced seismicity (RIS) due to the water load. Approximation methods for the earth's elastic field, based on discrete data given in advance, are of basic importance in understanding these phenomena. Within the framework of a linearized elasticity, the theory of vector spherical harmonics and spherical spline techniques are the basic tools of our investigations undertaken in this note. We focus on the numerical aspect, while the theoretical background and explicit proofs can be found in [2]. An analogous approach in potential theory is given in [1].

Precisely stated, let us consider the following problem: Σ denotes a closed compact surface dividing $\mathbb{R}^3$ into a bounded simply connected inner region Σ_i and an outer region $\Sigma_e = \mathbb{R}^3 \backslash \{\Sigma_i \cup \Sigma\}$, such that $\alpha < \beta = \inf_{x \in \Sigma} |x| \leq \sup_{x \in \Sigma} |x| = \gamma < \omega$ for positive constants α, ω. Let Γ_i, Ω_i denote the interior of the spheres Γ, Ω centered at the origin of radii γ, ω. For a field $f : \Sigma \mapsto \mathbb{R}^3$, given discrete data $X_N = \{(x_k, f(x_k)) \in \Sigma \times \mathbb{R}^3\}$, the solution of the bvp $\tilde{\Delta} u = \mu \Delta u + (\lambda + \mu) \nabla \nabla u = 0$ in Σ_i, $\mu > 0, 3\lambda + 2\mu > 0$, $u|_\Sigma = f$, is to be approximated by a field satisfying the bvp inside Ω_i and which is determined only by use of the discrete set X_N.

2 Vector Spherical Harmonics

The *scalar spherical harmonics* $S_n, n = 0, 1, \ldots,$ are the eigenfunctions of the Beltrami operator Δ^* of the unit sphere for the eigenvalue $n(n+1)$, $n = 0, 1, \ldots$. The linear space of all n'th order scalar spherical harmonics is of dimension $2n+1$, the set $\{S_{n,j}\}, j = 1, \ldots, 2n+1$, is assumed orthonormal with respect to $\int_{|\zeta|=1} dS$, dS: surface element. The (vector-) fields $s_n^{(i)}(\xi), |\xi| = 1, n = 0, 1, \ldots, i = 1, 2, 3$, given by

$$s_n^{(1)}(\xi) = \xi S_n(\xi), \quad s_n^{(2)}(\xi) = (n(n+1))^{-\frac{1}{2}} \nabla_\xi^* S_n(\xi), \quad s_n^{(3)}(\xi) = \xi \wedge s_n^{(2)}(\xi),$$

are called *vector spherical harmonics* (of order n and i'th kind).

The system $s_{n,j}^{(k)}, n = 0, 1, \ldots, j = 1, \ldots, 2n+1, k = 1, 2, 3$, is orthonormal in the sense that $\int_\Omega s_{n,j}^{(i)}(\frac{x}{|x|}) s_{m,l}^{(k)}(\frac{x}{|x|}) \, dS = \delta_{n,m}\delta_{j,l}\delta_{i,k}$, and $s_n^{(i)} \cdot s_n^{(k)} = 0, i \neq k$. The *vectorial addition theorem* [2] reads as follows:

$$P_n^{(k)}(\xi, \eta) = \sum_{j=1}^{2n+1} s_{n,j}^{(k)}(\xi) \otimes s_{n,j}^{(k)}(\eta) \, ,$$

where

$$
\begin{aligned}
P_n^{(1)}(\xi, \eta) &= n(n+1)c(n)P_n(\xi\eta)\xi \otimes \eta \\
P_n^{(2)}(\xi, \eta) &= c(n)[P_n''(\xi\eta)(\eta - (\xi\eta)\xi) \otimes (\xi - (\xi\eta)\eta) + P_n'(\xi\eta)(I - \xi \otimes \xi - (\eta - (\xi\eta)\xi) \otimes \eta] \\
P_n^{(3)}(\xi, \eta) &= c(n)[P_n''(\xi\eta)\xi \wedge \eta \otimes \eta \wedge \xi + P_n'(\xi\eta)((\xi\eta)I - \eta \otimes \xi)],
\end{aligned}
$$

I unit matrix in $\mathbb{R}^3$, and $c(n) = \frac{2n+1}{4\pi\omega^2 n(n+1)}$. $P(\overline{\Omega}_i)$ denotes the space of all vector fields satisfying

$$v \in C^{(2)}(\Omega_i) \cap C(\overline{\Omega}_i), \quad \tilde{\Delta}v = 0 \text{ in } \Omega_i.$$

Furthermore $v_{n,j}^{(k)}, j = 1, \ldots, 2n+1, k = 1, 2, 3$ denote the fields

$$
\begin{aligned}
v_{n,j}^{(1)}(x) &= \sigma_n^{(1)}(\frac{|x|}{\omega})s_{n,j}^{(1)}(\xi) + \tau_n^{(1)}(\frac{|x|}{\omega})s_{n,j}^{(2)}(\xi) && n = 0, 1, \ldots \\[2mm]
v_{n,j}^{(2)}(x) &= \sigma_n^{(2)}(\frac{|x|}{\omega})s_{n,j}^{(1)}(\xi) + \tau_n^{(2)}(\frac{|x|}{\omega})s_{n,j}^{(2)}(\xi) && n = 1, 2, \ldots \\[2mm]
v_{n,j}^{(3)}(x) &= \sigma_n^{(3)}(\frac{|x|}{\omega})s_{n,j}^{(3)}(\xi) && n = 1, 2, \ldots
\end{aligned}
$$

where $x = |x|\xi, |x| \leq \omega$ and $\sigma_n^{(i)}, \tau_n^{(i)}$ are given by

$$
\begin{aligned}
\sigma_n^{(1)}(\frac{|x|}{\omega}) &= (\frac{|x|}{\omega})^{n-1}((\frac{|x|}{\omega})^2 + n\alpha_n((\frac{|x|}{\omega})^2 - 1)) \\[2mm]
\tau_n^{(1)}(\frac{|x|}{\omega}) &= (\frac{|x|}{\omega})^{n-1}(n(n+1))^{\frac{1}{2}}\alpha_n((\frac{|x|}{\omega})^2 - 1) \\[2mm]
\sigma_n^{(2)}(\frac{|x|}{\omega}) &= (\frac{|x|}{\omega})^{n-1}n(1 + n\alpha_n)(n(n+1))^{-\frac{1}{2}}(1 - (\frac{|x|}{\omega})^2) \\[2mm]
\tau_n^{(2)}(\frac{|x|}{\omega}) &= (\frac{|x|}{\omega})^{n-1}(1 - n\alpha_n((\frac{|x|}{\omega})^2 - 1)) \\[2mm]
\sigma_n^{(3)}(\frac{|x|}{\omega}) &= (\frac{|x|}{\omega})^n
\end{aligned}
$$

with $\alpha_n = -\frac{n\tau + 2 + 3\tau}{2(n(\tau+2)+1)}$ and $\tau = 1 + \frac{\lambda}{\mu}$. Then $v_{n,j}^{(k)}$ is the unique solution of the bvp $v_{n,j}^{(k)} \in P(\overline{\Omega}_i)$ wrt the boundary condition $v_{n,j}^{(k)}(x) = s_{n,j}^{(k)}(\xi), x = |x|\xi \in \Omega$.

3 Minimum Norm Interpolation

The linear space H defined by

$$H = \{v \mid v = \sum_{k=1}^{3} \sum_{n=0}^{\infty} \sum_{j=1}^{2n+1} \lambda_n(g, v_{n,j}^{(k)})_{L^2(\Omega)} v_{n,j}^{(k)}|_{\overline{\Gamma}_i}, g \in L^2(\Omega)\} \subset P(\overline{\Gamma}_i)$$

where $\Lambda = \{\lambda_n\}$, $n = 0, 1, \ldots$, $n \mapsto \lambda_n^2$ a nonvanishing rational function, is a separable Hilbert space if the inner product is chosen corresponding to the norm $\|v\|_H = \|g\|_{L^2(\Omega)}$. H has the reproducing kernel

$$K(x,y) = \sum_{k=1}^{3} \sum_{n=0}^{\infty} \sum_{j=1}^{2n+1} \lambda_n^2 v_{n,j}^{(k)}(x) \otimes v_{n,j}^{(k)}(y), \quad x, y \in \overline{\Gamma}_i.$$

For an unknown $v \in H$ and given data $(x_k, v(x_k)) \in \Sigma \times \mathbb{R}^3$, determine the field in the set $I_N = \{w \in H \mid w(x_k) = v(x_k), k = 1, \ldots, N\}$ having minimal H-norm (*smallest H-interpolation problem*). If $S_N = S_N(H; x_1, \ldots, x_N)$ denotes the space of all fields $s \in H$ of the form

$$s(x) = \sum_{i=1}^{3} K(x, x_i) a_i, \quad x \in \overline{\Gamma}_i, a_i \in \mathbb{R}^3$$

the following facts are valid [2]:

1. There exists a unique field $s \in I_N \cap S_N$, briefly denoted by s_N.

2. *Approximation scheme* for determining s_N. Compute the coefficient vector $a_i \in \mathbb{R}^3$ by solving the linear system

$$\sum_{i=1}^{N} K(x_j, x_i) a_i = f(x_j), \quad j = 1, \ldots, N.$$

   This system can be treated by Cholesky decomposition.

   Compute the minimum norm interpolant

$$s_N(x) = \sum_{i=1}^{N} K(x, x_i) a_i, \quad x \in \overline{\Gamma}_i.$$

3. If $f \in H|_\Sigma$, then the solution of the boundary value problem $v \in H_{\overline{\Sigma}_i}, v|_\Sigma = f$ can be approximated by $s_N|_{\overline{\Sigma}_i}$, more explicitly, we are able to show the *stability estimate* [2]

$$\sup_{x \in \overline{\Sigma}_i} |v(x) - s_N(x)| \leq D \, \Theta_N \, \|v\|_H,$$

   for a suitable constant D (depending on the geometry) and the nodal width Θ_N of the system X_N.

For numerical purposes, it is of basic importance that the kernel $K(\cdot,\cdot)$ admits a representation in terms of elementary functions. For this reason we restrict our attention to those Λ–sequences where the denominator of λ_n^2 has only strictly positive (real) roots. We can reduce the evaluation of the kernel to power series of the form

$$f_{\beta,k}(r,t) = \sum_{n=0}^{\infty} \frac{1}{(n-\beta)^k} P_n(t) r^n$$

$(0 \le r < 1; -1 \le t \le 1; r = (|x|/\omega)(|y|/\omega); t = (x/|x|)(y/|y|)$ for the parameters $\beta > 0$ and $k = 0,1,\ldots$. In particular, when $\beta = 1,2,\ldots$ and $k = 0,1$, the series $f_{\beta,k}(r,t)$ represent elementary functions which can be calculated from

$$f(r,t) = \sum_{n=0}^{\infty} P_n(t) r^n = (1 + r^2 - 2rt)^{-1/2}.$$

We define those series $f_{\beta,k}(r,t)$ appearing nontrivially in K_Λ as the *generating functions* of K_Λ. We designate the set of all generating functions for K_Λ by $\mathcal{E}(K_\Lambda)$.

Table 1 exhibits several examples. We omit the straightforward but tedious calculations. Spatial considerations permit us to list only the set of generating functions, $\mathcal{E}(K_\Lambda)$, for the corresponding reproducing kernel, K_Λ.

Table 1

λ_n , $n = 0,1,\ldots$	$\mathcal{E}(K_\Lambda)$
$[(\tau+2)n+1]\sqrt{n+1}$	f
$2[(\tau+2)n+1]\sqrt{\dfrac{n+1}{2n+1}}$	f
$(\tau+2)n+1$	$f,\ f_{-1,1},\ f_{-2,1}$
$\dfrac{(\tau+2)n+1}{\sqrt{n+2}}$	$f,\ f_{-1,1},\ f_{-2,1},\ f_{-3,1}$
$\dfrac{(\tau+2)n+1}{\sqrt{(n+2)(n+3)}}$	$f,\ f_{-1,1},\ f_{-2,1},\ f_{-3,1},\ f_{-4,1}$

References

[1] FREEDEN, W. (1987) *A Spline Interpolation Method for Solving Boundary Value Problems of Potential Theory from Discretely Given Data*, Num. Meth. Part. Diff. Eqns., **3**, 375–398.

[2] FREEDEN, W., GERVENS, T., MASON, J.C. (1990) *A Minimum Norm Interpolation Method for Determining the Displacement Field of a Homogeneous Isotropic Elastic Body from Discrete Data*, IMA Journal of Applied Mathematics, **44**, 55–76.

A BARRIER PENETRATION FORMULA FOR OPTICAL TUNNELLING MODELS

J. BURZLAFF
School of Mathematical Sciences
Dublin City University
Dublin 9
Ireland.

ABSTRACT. Some one-dimensional models for optical tunnelling are discussed. The exponentially small imaginary part of the eigenvalue, which determines the radiation loss, is expressed in terms of an integral of the refractive index.

Radiation loss in a bent optical fibre has been called optical tunnelling because of the similarities to tunnelling in quantum mechanics. We will show that these similarities extend to the form of the barrier penetration formula. This provides us with a useful formula to estimate the amount of the energy loss.

To estimate the energy loss and to find ways of reducing it is of practical importance. A 1-8 splitter, for example, is a very compact waveguide device in which fibres are bent with radii of curvature small enough to make energy loss important. (However, they are large compared to the core radius, which simplifies the calculation.) Another interesting application, which is still in an early product development phase, is the use of energy loss to send out information to a receiver by simply bending the fibre. (In this application, energy leakage is actually put to positive use.) It is hoped that this simple idea can make fibre technology cost effective for local telephone nets with large numbers of receivers and short distances between them.

To calculate the energy loss, we start with the work of Kath and Kriegsmann (1988). For a weakly guiding slightly bent fibre they have derived and studied the following eigenvalue problem:

$$-\left(\frac{\partial^2}{\partial \xi^2} + \frac{\partial^2}{\partial \eta^2}\right) y + V(\xi, \eta, \varepsilon) y = \lambda y \tag{1}$$

together with appropriate boundary conditions. Here, the ξ-axis and the η-axis are orthogonal to the fibre in a co-moving coordinate frame. ε is the core radius over the radius of curvature and very small. V measures the index of refraction and its graph has the shape of a potential well, where the outside of the well is a plane slightly tilted because of the bending. The imaginary part of λ measures the energy loss.

To simplify the calculation and to pin-point the mathematical subtleties, we have studied optical tunnelling from one-dimensional square-well potentials (Burzlaff and Wood (1991)):

$$-y''(x)+V(x,\varepsilon)y(x) = \lambda y(x) \quad \text{for} \quad x \in \mathbf{R} \quad (2)$$

with appropriate boundary conditions, where

$$V(x) = \begin{cases} -\varepsilon x & (|x| > \tfrac{1}{2}d) \\ -2h/d - \varepsilon x & (|x| \le \tfrac{1}{2}d) \end{cases} . \quad (3)$$

Since the refractive index for most splitters is of square-well shape and radiation is anyhow mainly in the plane of the bend, these one-dimensional models should still be relevant.

The main mathematical problem in these calculations is the need for an exponentially improved approximation for the Airy function $\text{Bi}(z)$. The $O(\exp(2z^{3/2}/3))$ terms in its asymptotic expansion lead to terms of the form $\text{Im}\lambda \cdot \exp(2(-\kappa_0)^{3/2}/(3\varepsilon))$, where $\text{Re}\lambda = \kappa = \kappa_0 + \varepsilon\kappa_1 + \cdots$. For consistency we have to include, and therefore know, the leading $O(\exp(-2z^{3/2}/3))$ term. Only recently, this term has been given by Berry (1989) with a rigorous derivation by Olver (1991).

Having overcome the main mathematical difficulty a straightforward lengthy calculation yields

$$\text{Im}\lambda \sim -K(h,d)\exp\left[-\frac{4}{3\varepsilon}(-\kappa_0)^{3/2}\right]. \quad (4)$$

The function K is given in our paper and is plotted as a function of h and d in fig. 1. In fig. 2, we have plotted κ_0, which corresponds to the frequency of the propagating light pulse. A comparison of the two figures shows that for pulses with the same frequency we can have quite different amounts of energy loss. The function K is therefore important. In the following, however, we will concentrate on the argument of the exponential function.

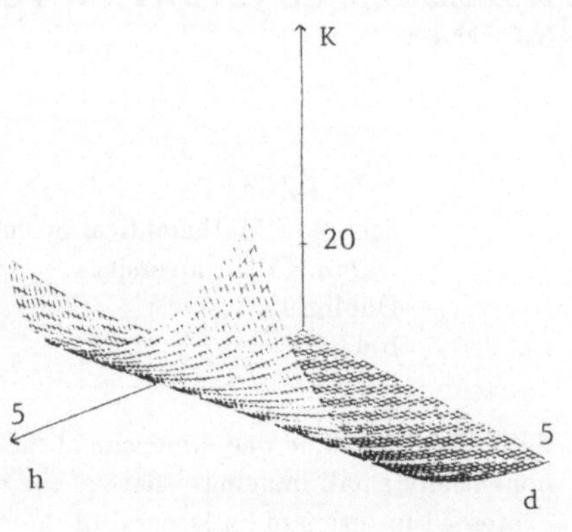

Figure 1. $K(h,d)$ in formula (4).

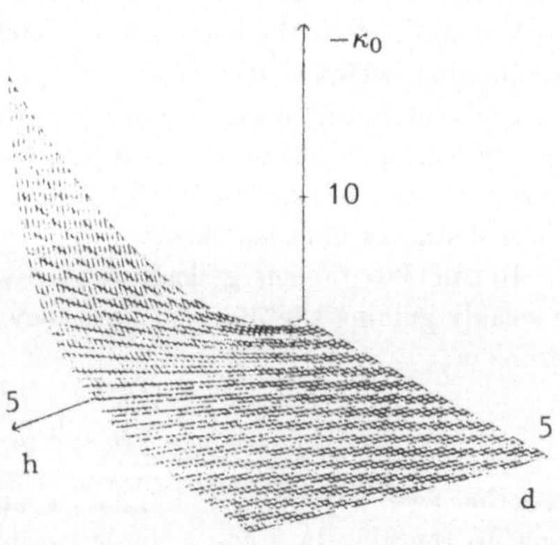

Figure 2. $-\kappa_0(h,d)$, where κ_0 is the eigenvalue for $\varepsilon = 0$.

The mathematical problem with exponentially improved approximations is still present in further simplified models with potentials of the form $V(x) = -2h\delta(x) - \varepsilon|x|^n, x \in \mathbf{R}, n \in \mathbf{N}$. This makes the study of these models (Paris and Wood (1989), Brazel et al. (1992), Liu and Wood (1991)) particularly interesting. In these models, the energy loss is given by the formula

$$\mathrm{Im}\lambda \sim -K(h)\exp[(-2/\varepsilon^{1/n})(-\kappa_0)^{1/2+1/n}\int_0^1 \sqrt{1-v^n}dv]. \tag{5}$$

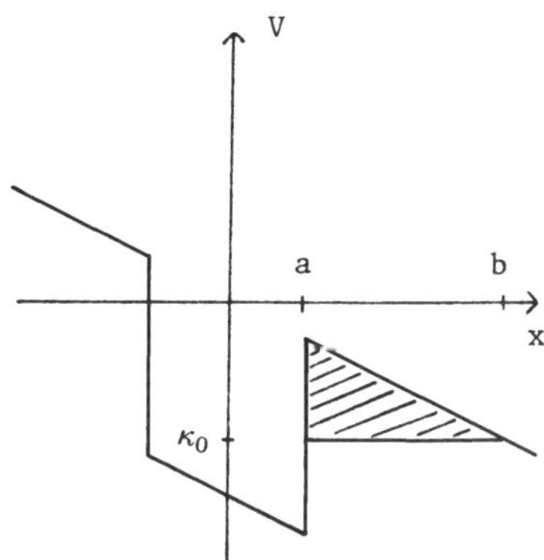

Figure 3. A potential $V(x)$. The shaded area is related to the integral in (6).

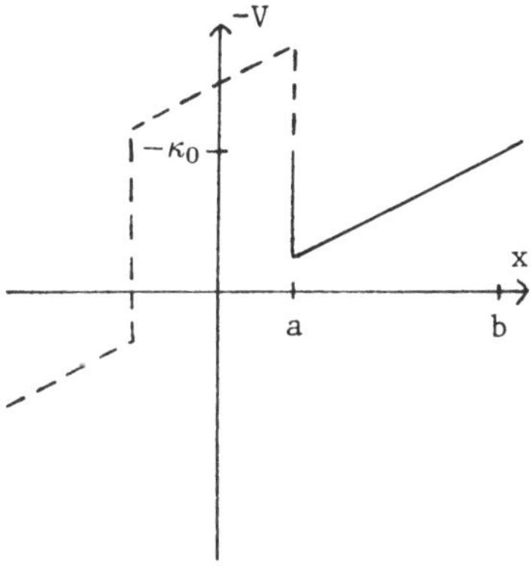

Figure 4. The potential $-V(\bar{x}(t))$ of the bounce (solid part of the graph) for $-T/2 \le t \le T/2$.

The formulae (4) and (5) can be written in the form

$$\mathrm{Im}\lambda \sim -K \exp(-2\int_a^b \sqrt{V(x) - \kappa_0}dx), \tag{6}$$

where the integral is obviously related to the shaded area in fig. 3 for the potential in question. The argument of the exponential function is related to the action

$$S_0 = \int_{-T/2}^{T/2} \left[\frac{1}{4}\dot{\bar{x}}^2(t) + V(\bar{x}(t))\right] dt \tag{7}$$

of the so-called bounce $\bar{x}(t)$. $\bar{x}(t)$ is the finite action solution of the equation of motion which describes the movement of a particle, starting at time $t = -T/2$ at the point $x = a$ with velocity $\dot{x} = 0$ at height $-\kappa$, rolling down and up the slope of the potential $-V$ to reach the point b at $t = 0$ (see fig. 4), and returning to the starting point at $t = T/2$. The relation of (7) to (6) can be established using energy conservation, $\dot{\bar{x}}^2/4 - V(\bar{x}) = -\kappa$, and

$$S_0 = \int_{-T/2}^{T/2} \frac{1}{2}\dot{\bar{x}}^2 dt + \kappa T = 2\int_{d/2}^{\bar{x}(0)} \sqrt{-\varepsilon\bar{x}^n - \kappa}d\bar{x} + \kappa T. \tag{8}$$

Very interesting would be a derivation of formula (6) for a more general class of potentials V, in particular for those in more than one dimension. For similar quantum mechanical models there are non-rigorous arguments which lead to (6), starting from the Feynman-Kac formula, which involves the action S_0 in (7), and using saddle point analysis of function space integrals (see e.g. Coleman (1979)). A similar derivation would be very useful, since formula (6) gives a very simple way of estimating the energy loss. For those potentials for which the formula is valid, all that is required is to find, for an eigenvalue in the potential well V, the escape path over the lowest mountain pass (cf. fig. 3 for the one-dimensional case). The action of the corresponding bounce in the potential $-V$ is then S_0 in formula (7).

We have extended previously published work (Burzlaff and Wood (1991)) into two directions. First, we have studied numerically the function $K(h, d)$ in formula (4), and shown that the ratio of the fiber core to the index difference plays an important role for the energy loss of bent step-index fibers. More importantly, we have given a formula, namely formula (6), which provides us with a simple way of estimating the energy loss in terms of the area under the escape path.

References

1. Berry, M.V. (1989) 'Uniform asymptotic smoothing of Stokes discontinuities', Proc. Roy. Soc. London A422, 7-21.

2. Brazel, N., Lawless, F.R. and Wood, A.D. (1992) 'Exponential asymptotics for the eigenvalues of a problem involving parabolic cylinder functions', Proc. Am. Math. Soc., to appear.

3. Burzlaff, J. and Wood, A.D. (1991) 'Optical tunnelling from one-dimensional square-well potentials', IMA J. Applied Math. 47, 207-215.

4. Coleman, S. (1979) 'The uses of instantons', in Zichichi, A.(ed.), The Whys of Sub-nuclear Physics, Plenum Publ. Co., New York, pp. 805-916.

5. Kath, W.L. and Kriegsmann, G.A. (1988) 'Optical tunnelling: Radiation losses in bent fibre-optic waveguides', IMA J. Appl. Math. 43, 85-103.

6. Liu, J. and Wood, A.D. (1991) 'Matched asymptotics for a generalisation of a model equation for optical tunnelling', Euro. J. Appl. Math. 2, 223-231.

7. Olver, F.W.J. (1991) 'Uniform, exponentially improved, asymptotic expansions for the confluent hypergeometric function and other integral transforms', SIAM J. Math. Anal. 22, 1475-1489.

8. Paris, R.B. and Wood, A.D. (1989) 'A model equation for optical tunnelling', IMA J. Appl. Math. 43, 273-284.

COMMUNICATING CHEMICAL PROCESSES - A TRANSPUTER MODEL OF THE BRITISH COAL TWIN-BED PYROLYSER/COMBUSTOR

A.R. BUTT, M.G. JEPSON, J. MOODIE
British Coal Corporation
Coal Research Establishment
Stoke Orchard, Cheltenham, Glos, GL52 4RZ, UK

ABSTRACT. This paper concerns the development, and transputer implementation, of a mathematical model of a novel coal-fired combustion appliance. The model was devised to help provide an understanding of appliance performance and to provide a facility whereby candidate control strategies can be assessed. A brief introduction is given to the chemical and physical processes occurring in the twin-bed unit, and the design methodology used in model development is described. The model has been used extensively to help specify a control philosophy for the first commercial Pyrolyser/Combustor boiler (8.5 MW_t capacity) installed at the UKAEA Culham laboratory. Comparisons are made between model predictions and experimental data.

1. Introduction

As part of its R&D programme for improved coal-fired equipment, British Coal Corporation (BCC) have developed a novel twin-bed Pyrolyser/Combustor [1]. The firing system comprises two interconnected fluidised beds operated, respectively, as a coal pyrolyser/partial gasifier and a char combustor. The dividing wall separating the char combustor from the pyrolyser spans the width of the combustion chamber, the lower section containing an array of vertical ports to facilitate the interchange of suspended coal/char and sand between adjacent beds. The combustible gas formed in the pyrolyser and the oxidant char combustor off-gas are mixed in a separate combustion chamber where the resultant flame produces gas temperatures in excess of 1250°C. Exhaust gases from this chamber are cooled by passage through the boiler convection passes and economiser and are subsequently cleaned by a bag filter system before discharge via the stack.

On the basis of BCC experience in modelling such systems, the notion of carrying out an exercise using a parallel computing environment appeared attractive. Most real-life systems do indeed comprise a series of parallel processes, as described in [6], that 'communicate' with one another in some sense. (Hence, the title of the paper, obviously adapted from [6].) In addition to giving a more faithful representation of reality, a parallel simulation on a transputer network has the advantage that, as a model is refined, additional computational power can be added if necessary to give acceptable run-times. Thus, the objectives of the modelling exercise were:

(i)   to develop a simulation of the twin-bed unit
(ii) to investigate control strategies

(iii) to evaluate low-cost parallel computing (transputers)

2. Modelling the Process

The complete set of equations describing the system accounts for: char/coal mixing (represented as diffusion processes), combustion in each bed (first order kinetics), calculation of temperature profiles in the beds (Fourier heat conduction), combustion and heat transfer in the second chamber (e.g. modelled as a long furnace [10]), and heat transfer in the convection passes (standard engineering design calculations [10]). A typical formulation of the mathematical equations is given in [2].

In the model described here, only a very simplified version of these equations has been used, the emphasis being on handling the parallelism needed for transputer implementation.

The modelling approach described below comprises a framework that allows the designer to refine the controller as the system becomes more clearly defined. A state-based method has been adopted, the starting point being an heuristic model of the system which is defined as a single 'black box'. Movement from state to state takes place according to the sequence of inputs. Some ideas from the Theory of Finite State Automata (FSA) are useful. A representation of the twin-bed controller using the FSA concept is shown in Figure 1 which illustrates the five principal states (or modes) defined as: 0=Off; 4=Slumped; 7=Pyrolyser slumped; 10=single stage (1S); 13=twin-bed mode (2S); together with the various transitions between them. (See [3] and [4] for further details.)

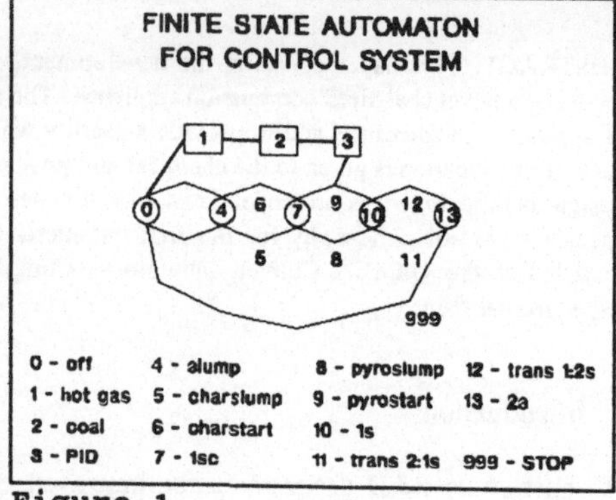

Figure 1

3. The Transputer, Formal Methods, Software Design and Implementation

A transputer is a microcomputer (transistor/computer) with its own local memory and with links to allow one transputer to be connected to another. Thus the transputer can be used in a single processor system or in a network providing the capability of concurrent processing. Various transputer operating systems are available, e.g. INMOS Transputer Development System [7] and Helios [9] which can be supported by a number of hardware options such as IBM/AT or SUN workstation. The work described in this paper was carried out using a TDS/AT/Occam combination.

A methodology has been adopted to aid the design and implementation of a mathematical model on a transputer system. The methodology consists of the following seven steps:

A - initial modelling, using JSD [8] and CSP [6]: Key factors to be included in the simulation are organised (using JSD) into appropriate modules. In the present case, the result is four principle modules: pyrolyser, char combustor, secondary combustion chamber (which also incorporates the boiler convection passes and economiser) and controller.

B - checking the parallelism using CSP: The objective here is to examine algebraically the concurrency aspects of the simplified model from step A and to create a 'harness' that facilitates communication between modules. Initially, the harness was set up with individual modules equipped with only simple routines to send and receive along appropriate channels.

C - specifying the modules using Z [12] & JSP: Each module in the model is defined more precisely using a combination of conventional mathematics, Z and JSP. The modules are essentially self-contained due to the decision to support only communication by channel.

D - refinement [5,12]: Here the modules may be refined by the inclusion of additional detail. In this particular model, many stages of refinement have been carried out as appropriate data have become available.

E - structured text [8]: Jackson's notation is used to write structured text, or pseudocode, for each module.

F - Occam code [11]: At this stage, the procedures are translated into Occam code.

G - transputer implementation [7]: Early work was carried out on a single transputer. However, increasing refinement of the model resulted in degradation of performance. Consequently, a modest network was constructed comprising three transputers, one for each of the two reactor modules and the third shared by the controller and external modules.

4. Results and Discussion

Typical model predictions for two load changes are shown in Figure 2. The first change (at time=190 minutes) is a step increase in load from 30% Maximum Output (MCR) to 70% MCR: in this case, the controller moves from State 10 (1S) to State 13 (2S). The model gives a reasonably accurate prediction of the pyrolyser bed temperature profiles during the transition; the predicted temperature drop in the combustor, however, is greater than that observed. For a second step change, from 70% MCR to 30% MCR, carried out at time=250 minutes, the model slightly overpredicts both bed temperature rises. Some refinement is still required. For example, it is thought that pressure differences between the beds results in a net flow of air from the combustor to the pyrolyser. In practice, this tends to equalise somewhat the temperature difference between the two beds.

In general, the model has provided a good framework within which to assess the performance of larger scale Pyrolyser/Combustor systems, although more experimental data and mathematical analysis are clearly required. The facility to simulate and assess control strategies has been of considerable value, where the model was used in conjunction with boiler tests to optimise the control settings in the Culham unit. In future transputer modelling exercises, it is likely that less emphasis will be placed on formal aspects such as Z and CSP, although the overall framework presented in Section 3 will be retained.

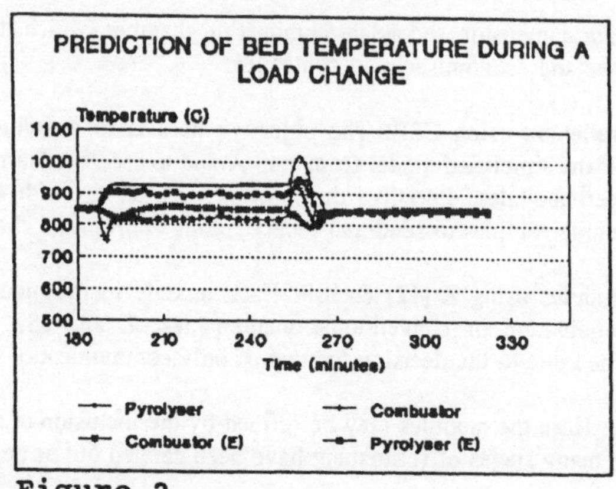

Figure 2

5.Acknowledgements

The authors wish to thank British Coal Corporation for permission to publish this paper. The views expressed are those of the authors and not necessarily those of British Coal.

6. References

1. Butt, A.R., Payne, R.C. and Wareham, G. (1987) 'The application of the twin-bed combustor pyrolyser system to water tube boiler - design and operation', Proc.A.S.M.E. 9th Int.Conf. on FBC, Boston, USA.
2. Butt, A.R., Herring, M.J. and Moodie, J. (1990) 'Some Theoretical Aspects of Mixing in the British Coal Twin-bed Combustor/Pyrolyser', Int.J. of Energy Res., Vol. 14, 449-463.
3. Butt, A.R., Jepson, M., Milner, C.A., and Moodie, J. (1991) 'Scale-up and Transputer Modelling of the British Coal Twin-bed Pyrolyser/Combustor', Proc. A.S.M.E. 11th Int.Conf. on FBC, Montreal, Canada.
4. Butt, A.R., Jepson, M., Milner, C.A., and Moodie, J. (1991) 'Scale-up of the British Coal Twin-bed Pyrolyser/Combustor', 2nd European Conf. on Industrial Furnaces & Boilers, Vilamoura, Portugal.
5. Cohen,B., Harwood, W.T., and Jackson, M.I. (1986) The Specification of Complex Systems, Addison-Wesley, UK.
6. Hoare, C.A.R., (1985) Communicating Sequential Processes, Prentice-Hall International UK Ltd.
7. Inmos Ltd, (1988) Transputer Development System, Prentice-Hall, UK.
8. Jackson, M.J., (1983) System Development, Prentice-Hall, UK.
9. Perehelion Software Ltd (1989) The Helios Operating System, Prentice-Hall, UK.
10. Perry, R.H. and Green, D. (1984) Perry's Chemical Engineers' Handbook, McGraw-Hill, Kosaida Printing Co., Japan.
11. Pountain, D. and May, D. (1987) A Tutorial Introduction to Occam 2, BSP Professional Books, London

Travelling Combustion Waves in Porous Media

H. Byrne
Mathematical Institute
24-29 St. Giles
Oxford OX1 3LB

Abstract. Steady travelling wave solutions of a one-dimensional, time-dependent model of porous medium combustion are sought. There are four dependent variables, the solid and gas temperatures and concentrations, and three dimensionless parameters. By assuming that the reaction is confined to a finite region, it is possible to characterise solutions by the manner in which the combustion reaction is activated and extinguished. Further, the model can be reduced to an equivalent free boundary value problem, defined over the reaction zone. For solutions characterised by short reaction zones and slow flow of the gaseous fuel through the solid, asymptotic expressions describing the regions of parameter space in which the different solutions occur are derived. The dependence of the wave speed on the key parameters is also examined and situations under which it may change sign are discussed.

Model Equations. The model of porous medium combustion which we study is presented below in travelling wave coordinates (derivations of this and the original model are given in [4] and [8]). The model has applications for catalytic reactors used in industrial plants [1], filtration combustion [2] and underground coal gasification [3]. Denoting by z and t the spatial and temporal variables, we introduce the travelling wave coordinate $x = z - ct$, where c is the assumed constant, unknown wavespeed. The dependent variables are $U(x), W(x), Q(x) = c\sigma(x)$ and $G(x)$ and correspond, respectively, to solid and gas temperatures, the product of the solid concentration (σ) and the wave speed (c), and a quantity proportional to the product of gas concentration and temperature. The model equations derive from energy and mass conservation applied to the two phases and are given by:

$$
\begin{aligned}
0 &= U'' + QU' + W - U + r, & (1)\\
W' &= \frac{1}{\mu}(U - W), & (2)\\
Q' &= \lambda r, & (3)\\
G' &= -\frac{a}{\mu}r, & (4)\\
r &= H(U - 1)H(G)H(Q - c\tau)f(W, G). & (5)
\end{aligned}
$$

Here r denotes the reaction rate, $H(.)$ is the Heaviside step-function, and the dimensionless parameters μ and a represent the gas inlet velocity and the rate of gaseous consumption to solid consumption respectively. The parameter λ measures the rate of solid fuel consumption and is proportional to the specific heat capacity of the solid (see [8] for further details). The following boundary conditions are imposed on the system:

$$U(-\infty) = u_a = W(-\infty), G(-\infty) = 1, U(\infty) = u_b, Q(\infty) = c. \tag{6}$$

By assuming that the reaction is confined to a finite region, $(0, L)$ say, of unknown length, and exploiting both the form of the reaction rate and the governing differential equations, it is possible to classify each nontrivial solution according to one of the following groupings:

$$(\text{U,U})\text{-solutions} \Leftrightarrow U(0) = 1 \text{ and } U(L) = 1; \tag{7}$$
$$(\text{U,G})\text{-solutions} \Leftrightarrow U(0) = 1 \text{ and } G(L) = 0; \tag{8}$$
$$(\text{Q,U})\text{-solutions} \Leftrightarrow Q(0) = \tau Q(L) \text{ and } U(L) = 1; \tag{9}$$
$$(\text{Q,G})\text{-solutions} \Leftrightarrow Q(0) = \tau Q(L) \text{ and } G(L) = 0. \tag{10}$$

These equations provide two algebraic relations from which the eigenvalues c and L may be determined, thereby closing the system.

Equations (1)–(6) can be solved outside $(0, L)$, where $r = 0$, giving, for example,

$$U = \begin{cases} u_a + A_1 e^{\alpha_1 x} & x \leq 0, \\ u_b + B_1 e^{\beta_1(x-L)} + B_2 e^{\beta_2(x-L)} & x \geq L. \end{cases} \tag{11}$$

where A_1, B_1 and B_2 are constants of integration and α_1, β_1 and β_2 are roots of the quadratic equations

$$0 = \alpha^2 + (Q(0) + \mu^{-1})\alpha + (Q(0)\mu^{-1}), \tag{12}$$
$$0 = \beta^2 + (Q(L) + \mu^{-1})\beta + (Q(L)\mu^{-1}). \tag{13}$$

Since, in travelling wave coordinates, $Q(L) = c$, we remark that the coefficients (α) and (β) are influenced by the wavespeed. Further, for bounded solutions as $|x| \to \infty$, we observe that $A_1 = 0$ if $\alpha_1 < 0$ (the negative root α_2 has already been discarded) and, similarly, $B_i = 0$ if $\beta_i > 0 (i = 1, 2)$.

To fully determine the solution it remains to solve for (U, U', W, Q, G) when $x \in (0, L)$, with any constants of integration determined by imposing continuity of (U, U', W, Q, G) at $x = 0, L$. Before continuing with the solution, we present (without proof) several interesting results, deriving from conservation of energy for the solid matrix.

Energy Conservation Results. Integrating (1) with respect to $x \in (-\infty, \infty)$ yields the following expression describing conservation of energy in the solid matrix:

$$\int_0^L r[1 - \lambda(U - u_a)]dx = (\mu - c)(u_b - u_a). \tag{14}$$

Using (14) the following results can be derived.

Result 1 *If RHS(14)$\geq$ 0 then λ is bounded above and below: $0 \leq \lambda < (1 - u_a)^{-1}$. Conversely, if RHS(14)$<$ 0 then λ is bounded below: $(U_{max} - u_a)^{-1} < \lambda$, where $U_{max} = \max_{-\infty < x < \infty} U(x)$.*

Result 2 *If $Q(0) \geq \mu$ then only (Q, U) and (Q, G) solutions may occur.*

Result 3 *If $\lambda = 0$ then only (U, U) and (U, G) solutions may occur. Further, $u_b > u_a$ and the wavespeeds are bounded above: $c_{UG} < c_{UU} < \mu$.*

Asymptotic Analysis: ($\lambda = 0$). We now resume our solution for (U, W, Q, G) when $x \in (0, L)$, examining initially solutions characterised by short reaction zones and slow flow of the gas fuel through the porous solid catalyst. Appealing to result 3, we, therefore, consider the limit

$$\lambda = 0, 0 < L = \epsilon \ll 1, 0 < \mu = \epsilon\bar{\mu} \ll 1, 0 < c = Q = \epsilon\bar{c} < \mu.$$

In addition we fix $f(W, G) = f(W)$ in (5), to simplify the analysis, although similar results can be obtained when $f(W, G) = f(W)G^n (n < 1)$. In this limit it is possible to show only (U,G) solutions are realised, with the relevant power series expansion given by:

$$U \sim 1 + \epsilon^2 f(1)y(1 - \frac{y}{2}), \tag{15}$$

$$W \sim 1 + \epsilon^2 f(1)\left\{(1 + \mu_0)(y - \mu_0) - \frac{y^2}{2} + \mu_0^2 \exp(-y/\mu_0)\right\}, \tag{16}$$

$$G \sim 1 - \frac{a_0 f(1)}{\mu_0}y, \tag{17}$$

where $(\bar{\mu}, a, \bar{c}) \sim (\mu_0, a_0, c_0) + \epsilon^2(\mu_1, a_1, c_1)$, $u_b \sim 1$, $f(1) = \mu_0/a_0$ and $c_0 = \mu_0 - f(1)/(1 - u_a)$. The leading order behaviour of c_0 indicates that for sufficiently small gas fluxes $(\mu_0 < f(1)/(1 - u_a))$ $c_0 < 0$. Thus the direction of wave propagation changes from forward (left-to-right) to reverse (right-to-left) combustion as μ_0 decreases through $f(1)/(1 - u_a)$. Practical examples of this phenomenon can be found in [3]. (The stability of these different modes will be analysed in a forthcoming paper.)

By continuing to $O(\epsilon^4)$ a more detailed description of the region of (a, μ)-space in which (U,G) solutions occur is obtained.

$$\frac{\mu_1}{\mu_0} - \frac{a_1}{a_0} = \frac{f'(1)}{f(1)}\left(\frac{(1 + \mu_0)(1 - 2\mu_0)}{2} - \frac{1}{6} + \mu_0^3(1 - \exp^{-1/\mu_0})\right),$$

$$(\mu_1 - c_1)(1 - u_a) = [f(1)]^2\left(\frac{\mu_1}{\mu_0} - \frac{a_1}{a_0} - \frac{1}{2}\right).$$

Asymptotic Analysis: ($\lambda > 0$). Referring to (11)–(13), when we generalise our analysis to $\lambda > 0$ the different cases which arise, and must be examined in turn, are given by:

$$(A) \quad Q(0) < Q(L) < \mu \Leftrightarrow \alpha_1, \beta_1 > 0 > \beta_2 \ (\Rightarrow B_1 = 0);$$
$$(B) \quad \mu < Q(0) < Q(L) < \mu \Leftrightarrow \alpha_1, \beta_1, \beta_2 < 0 \ (\Rightarrow A_1 = 0);$$
$$(C) \quad Q(0) < \mu < Q(L) \Leftrightarrow \alpha_1 > 0 > \beta_1, \beta_2.$$

For each of cases (A)–(C) it is possible to construct power series expansions similar to (15)–(17) in the analogous limit

$$L = \epsilon, \mu = \epsilon\bar{\mu}, \lambda \sim O(1), a \sim O(1), Q = \epsilon\bar{Q}, f(W, G) = f(W), (0 < \epsilon \ll 1).$$

We conclude with a summary of results obtained from such an analysis.
Case (A). Both (Q,G) and (U,G) solutions are realised. As when $\lambda = 0$ forward and reverse (U,G) travelling waves can occur ($c_0 = \mu_0 - f(1)((1 - u_a)^{-1} - \lambda_0)$), whereas for (Q,G) solutions the wavespeed is always nonnegative ($c_0 = \lambda_0\mu_0/a_0(1 - \tau)$).

102

Case (B). In this case only (Q,G) solutions having positive wavespeeds occur ($c_0 = \lambda_0\mu_0/a_0(1-\tau)$).

Case (C). It is possible to show that all four types of solution occur in this case. Restricting attention to the case $u_b = u_a$ and $\tau = 0$, for comparison with [6], yields for each type of solution a forward travelling wave. Further, for all but (Q,G) solutions the leading order behaviour of λ, a parameter proportional to the specific heat capacity of the solid, is identical: $\lambda_0 = (1 - u_a)^{-1}$. For (Q,G) solutions $a_0 < \lambda_0 < (1 - u_a)^{-1}$. Thus we deduce that, in this limit, (Q,U), (U,U) and (U,G) solutions occur only for a particular solid heat capacity whilst (Q,G) solutions are realised under a wider range of conditions. This extends the work of [6], in which $a = 0$ so that only (U,U) and (Q,U) solutions could occur.

The above analysis extends work contained in [6], [7], concerning combustion in a plentiful gas-fuel supply ($a = 0$ in (4)). There the existence of stable (Q,U) and unstable (U,U) solutions was demonstrated for solutions characterised by asymptotically short reaction zones. Preparatory to performing a stability analysis of the type in [7], and guided by the existence of (Q,G) solutions for a range of values of λ, we conjecture that, for the limit being considered here, (Q,G) solutions will prove stable. Thus it should be possible to maintain complete conversion of both gaseous and solid fuel under suitable conditions.

References

[1] Bond, G.C. (1987) *Heterogeneous Combustion*, Clarendon Press, Oxford.

[2] Booty, M.R. and Matkowsky, B.J. (1991) *Modes of Burning Filtration Combustion*, E.J.A.M..

[3] Britten, J.A. (1984) *Activation Energy Asymptotic Modelling of Reverse Combustion Instabilities in Porous Media: Applications to In-Situ Coal Gasification*, Ph.D. Thesis, Univ. of Colorado, Boulder, U.S.A..

[4] Byrne, H. and Norbury, J. (1991) *Modelling Combustion Zones in Porous Media*, submitted to Journal of Mathematical Engineering in Industry.

[5] Krantz, W.B. and Gunn, R.D. (1983) *Underground Coal Gasification: The State of the Art*, A.I.Ch.E. Symp. Ser., 226, Vol. 79, 1-3.

[6] Norbury, J. and Stuart, A.M. (1988) *Travelling Combustion Waves in a Porous Medium. Part I - Existence*, SIAM J. Appl. Math., 48, 155-169.

[7] Norbury, J. and Stuart, A.M., *Travelling combustion waves in a porous medium. Part II - Stability*, SIAM J. Appl. Math., 48 (1988), pp. 374-392.

[8] Norbury, J. and Stuart, A.M. (1989) *A Model for Porous Medium Combustion*, Quart. J. Mech. & Appl. Math., 42, 159-178.

[9] Ohlemiller, T.J. and Lucca, D.A. (1983) *An Experimental Comparison of Forward and Reverse Smolder Propagation in Permeable Fuel Beds*, Combustion and Flame, 54, 131-147.

ESTIMATING NONSTATIONARY TIME SERIES WITH MISSING DATA

D. ČEPAR, Z. RADALJ, B. VOVK
"Jožef Stefan" Institute, Jamova 39
61111 Ljubljana, Slovenia

ABSTRACT. A new method for estimating ARIMA model parameters of nonstationary time series with missing observations is considered. Two approaches are proposed: (i) a nonstationary time series is transformed into stationary one, a stationary time series missing data iterative estimation algorithm is applied and at its end the inverse transformation is performed to obtain the estimates of original time series missing observations, (ii) the inverse transformation and the calculation of original time series missing observations is performed in each loop of the iterative procedure and not only at the end. Use of the methods is illustrated by a case study.

1. Introduction

We introduce two approaches in simultaneous estimation of missing data and ARIMA model parameters of nonstationary time series with missing observations. The first approach described in Sec. 2 is to transform a nonstationary time series into a stationary time series and then apply a stationary time series missing data iterative estimation algorithm to estimate missing observations and ARMA model parameters and at the end to perform the inverse transformation to obtain the estimates of original time series missing observations. The second approach described in Sec. 3 is to perform the inverse transformation and the calculation of original time series missing observations in each loop of the iterative procedure and not only at the end. In Section 4 we illustrate the estimation of ARIMA model parameters by some special cases.

2. The estimator on the stationary time series

We are given a nonstationary time series with missing observations. Using the Box-Jenkins approach we have to identify and estimate a parsimonious ARIMA model. First we transform the nonstationary time series z_i into stationary time series w_i of length N. Some values, $w_{t+1}, w_{t+2}, \ldots, w_{t+m-1}$ ($t + m \leq N$), of the time series w_i ; $i = 1, \ldots, N$, are missing. We estimate the ARMA model parameters and missing observations of the stationary time series and transform the time series back to the nonstationary time series. For the sake of simplicity we discuss the case of a stationary model with the constant equal to zero and the case with only one missing interval (a sequence of missing observations) with $m - 1$ missing observations:

$$\Phi(B)w_i = \Theta(B)a_i , \tag{1}$$

where B is the backward shift operator defined by $Bw_i = w_{i-1}$, and the operators $\Phi(B)$ and $\Theta(B)$ are polynomials of B with the unknown parameters: $\Phi(B) = 1 - \phi_1 B - \phi_2 B^2 -$

$\ldots - \phi_p B^p$ and $\Theta(B) = 1 - \theta_1 B - \theta_2 B^2 - \ldots - \theta_q B^q$; the a_i's are a sequence of independent random variables distributed according to $N(0, \sigma_a^2)$, and w_i; $i = 1, \ldots, N$, are stationary time series values. We assume that the zeros of the polynomials $\Phi(B)$ and $\Theta(B)$ lie outside the unit circle.

We identify and estimate the ARMA model for stationary time series with missing data in four steps, the first being performed only once, and the last three making a loop of the iterative procedure: (i) initialization of the missing values, (ii) identification and estimation of the parsimonious ARMA model, (iii) estimation of the missing observations using estimates of ARMA model parameters obtained in the preceding step and (iv) testing the convergence. Here we list only the basic formulas used for estimation of the missing observations. All details and mathematical background are described in [2].

We estimate the missing value w_{t+l} by

$$\tilde{w}_{t+l} = x(l)\hat{w}_t(l) + y(l)\breve{w}_{t+m}(k) ; \qquad l = 1, 2, \ldots, m - 1 \tag{2}$$

where $l + k = m$ and $\hat{w}_t(l)$ is the "forward" forecast ($\hat{w}_t(l) = \psi_l a_t + \psi_{l+1} a_{t-1} + \psi_{l+2} a_{t-2} + \ldots$, $l = 1, 2, \ldots, m - 1$) and $\breve{w}_{t+m}(k)$ is the "backward" forecast ($\breve{w}_{t+m}(k) = \psi_k e_{t+m} + \psi_{k+1} e_{t+m+1} + \psi_{k+2} e_{t+m+2} + \ldots$, $k = 1, 2, \ldots, m - 1$) forecasts. The values of ψ_j can be obtained from the equation $\Phi(B)\Psi(B) = \Theta(B)$, where $\Psi(B) = 1 + \psi_1 B + \psi_2 B^2 + \ldots$ (see [1], p.134). The e_i are a sequence of independent random variables with the same distribution as a_i. The weights $x(l)$ and $y(l)$ are determined, so as to minimize the variance of the error $w_{t+l} - \tilde{w}_{t+l}$:

$$x(l) = \frac{e(l)(d(l) - c(l))}{d^2(l) - e(l)c(l)}, \qquad y(l) = \frac{c(l)(d(l) - e(l))}{d^2(l) - e(l)c(l)}, \tag{3}$$

where

$$c(l) = \sum_{i=l}^{\infty} \psi_i^2, \qquad e(l) = \sum_{i=m-l}^{\infty} \psi_i^2,$$

$$d(l) = \sum_{i=l}^{\infty} \psi_i^2 - \sum_{i=0}^{m-l-1} \psi_i^2 + \frac{1}{\sigma_a^2} Cov\left(\sum_{i=0}^{l-1} \psi_i a_{t+l-i}, \sum_{i=0}^{m-l-1} \psi_i e_{t+l+i} \right),$$

for $l = 1, 2, \ldots, m - 1$. The covariance between forward and backward forecast errors is estimated by

$$Cov\left(\sum_{i=0}^{l-1} \psi_i a_{t+l-i}, \sum_{i=0}^{m-l-1} \psi_i e_{t+l+i} \right) = \sigma_a^2 \sum_{j=0}^{m-l-1} \sum_{i=0}^{l-1} \psi_j \psi_i \psi_{i+j}^*. \tag{4}$$

where $\psi_i = 0$ for $i < 0$ and

$$\psi*_n = \sum_{i=0}^{\infty} \pi_i \psi_{n+i} , \tag{5}$$

($n = -\infty, \ldots, \infty$) is approximated by a finite sum.

By producing stationary time series from the nonstationary time series by differencing and/or seasonal differencing, we increase the number of missing observations and lose some available information. Each differenciation of time series generates an additional missing observation for each missing interval. A double differenciation generates two missing observations for each given missing interval. A seasonal differenciation might even double the number of missing observations. Suppose the 20th and 40th observations are missing in a time series of length 60 with a trend and a seasonal period of 12. After the differencing and seasonal differencing there are eight missing observations (7, 8, 19, 20, 27, 28, 39, 40) in transformed time series which has now only 47 observations.

3. The missing observations estimator on the nonstationary time series

To avoid the loss of information described in the preceding section, we develop an estimator, which we can use for estimating the missing observations directly on the nonstationary time series.

In Section 2 we estimate the missing observations on each step of iterative procedure by estimator (2). This minimum variance estimator is valid only for stationary time series since only in this case the sequence (5) can be approximated by a finite sum. For nonstationary time series the sequence of $\psi*_n$ does not converge, so we cannot use the theoretical expression of covariance function (4). However, we can still compute an estimate of the covariance function and use the weights (3) since only finite sum of ψ_n in (3) is needed. We have the estimates of a_t and e_s on each step of iterative procedure. Using the usual assumption that a_t and e_t are distributed so that

$$Cov(a_t, e_s) = Cov(a_{t+k}, e_{s+k}) ,$$

where $k = 1, \ldots, \min(N - t, N - s)$, and $t, s = 1, \ldots, N$, an estimate of the covariance function is obtained by

$$Cov \left(\sum_{i=0}^{l-1} \psi_i a_{t+l-i}, \sum_{j=0}^{m-l-1} \psi_j e_{t+l+j} \right) = \sigma_a^2 \sum_{j=0}^{m-l-1} \sum_{i=0}^{l-1} \psi_j \, \psi_i \, Cov \left(a_{t+l-i}, e_{t+l+j} \right) =$$

$$= \sigma_a^2 \sum_{j=0}^{m-l-1} \sum_{i=0}^{l-1} \psi_j \psi_i \frac{1}{N - j - i} \sum_{o=1-t-l+i}^{N-t-l-j} (a_{t+l-i+o} - \bar{a})(e_{t+l+j+o} - \bar{e}) , \qquad (6)$$

where

$$\bar{a} = \frac{1}{N - m + 1} \left(\sum_{i=1}^{t} a_i + \sum_{i=t+m-1}^{N} a_i \right) \quad \text{and} \quad \bar{e} = \frac{1}{N - m + 1} \left(\sum_{i=1}^{t} e_i + \sum_{i=t+m-1}^{N} e_i \right) .$$

We use the same iterative procedure as described in Section 2 except the step (iii). On the each step (iii) of the iterative loop we now estimate missing data on nonstationary time series and not on stationary one.

4. Simulation Study

In order to test the consistency of the methods proposed, we analyze two nonstationary time series given in [3] (time series PC - Petrochemical Consumption and IP - Industry Production). First we analyze the complete time series. Then we omit two observations (65th and 69th) that are supposed by the author of [3] to be inadequate in each time series and analyze these time series according to procedures described in Sections 2 and 3. The model parameter estimates are given in Table 1.

Time series	Transformations	Parameter	Estimates on complete time series	Estimates on series with missing data	
				Sec. 2.	Sec. 3.
PC	$\nabla \nabla_4 \log$	Θ_1	0.91	0.94	0.90
IP	$\nabla \log$	θ_0	0.006	0.005	0.006
		ϕ_1	0.206	0.123	0.241
		Φ_1	-0.194	-0.01	0.013

Table 1: The parameter estimates (Θ_1, θ_0, ϕ_1 and Φ_1) on complete time series and time series with 2 missing observations (65, 69) obtained with different methods proposed in Section 2 and 3.

We compare the differences between ARIMA model parameters estimates obtained on the complete time series and estimates obtained on the time series with missing data. These differences are small, so the proposed methods are consistent. Slightly better results we obtain by the method described in Section 3, especialy for the time series PC. We logged and differenced both time series, seasonal time series PC we seasonaly differenced with a period of 4 to obtain the stationary time series. The total number of missing observations in transformed time series PC is 6, representing 8.5% of the total number of observations of differenced time series PC (76-5=71) whereas the total number of missing observations in transformed time series IP is 4, representing only 5% of the total number of available observations (76-1=75).

5. Conclusion

By the first approach we estimate on each step of iterative procedure more missing observations than necessary. This problem is removed in the second approach by estimating missing observations on the nonstationary time series, since the transformation from non-stationary into stationary time series induces additional missing observations. Athough we use only numerical estimates for the covariance function in the second approach instead of theoretical one as in the first approach, we obtained better results by the second approach in our simulation study. However, these results obtained on the small sample are not sufficient to statisticaly prove the advantage of this approach over the first approach, but nevertheless, we suggest the use of the second approach especially for estimation of the nonstationary time series with more missing observations intervals.

6. Literature

[1] Box G.E.P. and Jenkins G.M., *Time Series Analysis: Forecasting and Control.* Holden Day, San Francisco 1970.

[2] Čepar D., Radalj Z., *ARMA Models of Time Series with Missing Data.* Proceedings of the Fifth European Conference on Mathematics in Industry, ECMI 7, p. 189 - 192, B. G. Teubner Stuttgart and Kluwer Academic Press 1991.

[3] McLeod G., *Box Jenkins in Practice.* Gwilym Jenkins & Partners Ltd. 1982.

MATRIX "STABLE-UNSTABLE" BLOC TRIANGULAR AND DIAGONAL FORMS

A. CHEREMENSKY,   *N. DAKEV*
Institute of Mechanics & Biomechanics
4 "Akad.Bonchev" str., BG-1113 Sofia
Bulgaria

ABSTRACT. The numerically stable Larin's algorithm is used for constructing bloc triangular or diagonal form of a matrix separating its spectrum into the "stable" and "unstable" parts. Various control problems (distinguishing the stable or unstable parts of a linear stationary control system, rational separation, algebraic Riccati equations and polynomial factoring) are solved by using this form. The corresponding software is written in language Turbo C, and the known model example is given.

1. MAIN RESULTS

It is well-known that dichotomy of spectrum, defining the stabilizing solution of algebraic Riccati equation (ARE), orthogonal projection, rational separation and polynomial factoring can be produced with the help of an orthogonal transformation to Schur form of matrix. This form has more detailed structure than is necessary. Here it is sufficient to construct the bloc triangular form, in which its spectrum is partitioned between these blocs into the "stable" and "unstable" parts. For this aim the Larin's frequency approach (Aliev *et al.* 1986, 1987) can be applied because it is more effective than the Schur method (Arnold *et al.* 1985) used for constructing the stabilizing solution of ARE. However the nature of Larin's algorithms proves to have nothing in common with the frequency design.

Let eigenvalues of any $n \times n$ matrix A be in the open left (LHP) and right (RHP) halfplanes, and none of its eigenvalues λ have $Re\ \lambda = 0$; L^+ be the invariant subspace of the matrix A formed by the part of the Schur basis which is associated with all eigenvalues situated in RHP, P^+ be the corresponding orthogonal projection. Let $L^\perp$ be the orthogonal complement of L^+, E be the unit matrix. Then the matrix $P^\perp = E - P^+$ is the orthogonal projection onto $L^\perp$. It is easy to prove that the linearly independent columns of these projections form matrices P_+ and $P_\perp$ such that matrix $S = [\ P_+,\ P_\perp\ ]$ has its inversion $S^{-1} = diag(P_+^T P_+, P_\perp^T P_\perp)^{-1} S^T$.

Theorem 1. Let the eigenvalues of any matrix A be in the LHP and RHP, and none of its eigenvalues λ have $Re\ \lambda = 0$. Then this matrix is represented in the form

$$S^{-1} A S = \begin{bmatrix} A_+ & A_o \\ O & A_- \end{bmatrix} \tag{2}$$

where

$$A_o = (P_+^T P_+)^{-1} P_+^T AP_\perp;\quad A_+ = (P_+^T P_+)^{-1} P_+^T AP_+;\quad A_- = (P_\perp^T P_\perp)^{-1} P_\perp^T AP_\perp \tag{3}$$

and all eigenvalues of matrices A_+ and A_- are in the LHP and RHP, respectively.

Proof. Using the orthogonal projection property $P^+ = (P^+)^T$ and the one of invariance $P^+AP^+ = AP^+$ we have $P^\perp P^+ = O$, $(P^\perp)^T P^+ = O$. Hence $P_\perp^T P_+ = O$, $P_\perp^T AP_+ = O$. That is why we have formulae (2) and (3). ∎

Theorem 2. Let the eigenvalues of any matrix A be in LHP and RHP, and none of its eigenvalues λ have $Re\ \lambda = 0$. Then triangular form (2) can be transformed into the two-bloc "stable-unstable" diagonal form $diag(A_-,$

$A_+)$ with the help of the matrix $\begin{bmatrix} E & X \\ O & E \end{bmatrix}$ where the matrix X is the

unique solution of the Sylvester equation $A_- X - X A_+ + A_o = O$.

The last equation (similarly to the Lyapunov one) can be considered as a particular case of ARE. However here the special structure of Hamilton matrix permits us to simplify the corresponding algorithm (Ikramov 1984).

Corollary 1. Let the eigenvalues of any matrix G be in and out of the unit disk and none of its eigenvalues λ have $|\lambda| = 1$. Then with the help of the matrices $A = (E - G)(G + E)^{-1}$ and S defined above we have

$$S^{-1} G S = \begin{bmatrix} G_+ & G_o \\ O & G_- \end{bmatrix} \tag{4}$$

where $G_o = (P_+^T P_+)^{-1} P_+^T GP_\perp$, $G_+ = (P_+^T P_+)^{-1} P_+^T GP_+$, and $G_- = (P_\perp^T P_\perp)^{-1} P_\perp^T GP_\perp$; all the eigenvalues of matrices G_+ and G_- are in and out of the unit disk, respectively.

Proof. The transform $\lambda = (\rho - 1)/(\rho + 1)$ is the map of the unit disk to the LHP. The same transform defines the relation between the matrices A and $G = (A + E)(A - E)^{-1}$. It follows from the latter formula and formula (4) that the right side of formula (4) is triangular and its blocks have the properties given in the conclusion of the corollary just proved.∎

Corollary 2. Let the eigenvalues of any matrix A be in LHP and RHP, and none of its eigenvalues λ have $Re\ \lambda = 0$. Then this matrix is represented in the bloc triangular form with the help of the orthogonal matrix

$$S_1 = S\ diag(R_+^{-1},\ R_\perp^{-1})$$

Lemma 1. Let the eigenvalues of any matrix A be in LHP and RHP, and none of its eigenvalues λ have $Re\ \lambda = 0$. Then the projection P^+ can be defined as the solution of ARE

$$r(S) = \begin{bmatrix} P, & -E \end{bmatrix} H \begin{bmatrix} E \\ P \end{bmatrix} = 0\ ,\ H = \begin{bmatrix} A^T & -A^T - A \\ 0 & -A \end{bmatrix} \tag{5}$$

with Hurwitz matrix $H_- = \begin{bmatrix} E, & 0 \end{bmatrix} H \begin{bmatrix} E \\ P^+ \end{bmatrix}$.

Proof. Using the orthogonal projection property $P^+ = (P^+)^T$ and the one of invariance $P^+ A\ P^+ = A\ P^+$ we have $P^+ A^T P^+ = A^T P^+$ and ARE (5). Let the matrix S be defined. Then we may produce the transformations $A \Leftrightarrow \begin{bmatrix} A_+ & A_0 \\ 0 & A \end{bmatrix}$, $P_+ \Leftrightarrow diag(\ E_+,\ 0\ )$ where E_+ is a unit matrix with known dimensions. Thus

$$H_- \Leftrightarrow S^{-1} A^T S - S^{-1} (A^T + A)\ S\ S^{-1} P^+ S,\ S^{-1} A^T S = (S^T S)^{-1} S^T A^T (S^T)^{-1} (S^T S),\ \text{and}$$

$$H_- \Leftrightarrow diag\{-A_+,\ (P_\perp^T P_\perp)^{-1} A_-^T (P_\perp^T P_\perp)\}.$$

2. SOFTWARE

The corresponding software has been worked out for computers compatible with IBM/PC/XT/AT/PS2 as a new version of the one developed in the Ukrainian Academy of Science (Aliev *et al.* 1986). For this purpose the program language Turbo C with long double precision operator has been used. Our experience of the computer realization of Larin's algorithms with the balancing procedure has confirmed their higher precision than that of traditional methods, included in Eispak or Linpak (Arnold *et al.* 1985). As an example we may point out table 1 containing the results of computing the known model ARE used in Aliev *et al.* 1986, Arnold *et al.* 1985 for characterizing the algorithms of solving ARE.

TABLE 1. Numerical results of example 1 (Arnold *et al.* 1985)

N	Laub's results		our results	
	δ	k	δ	k
0	1 e-18 [1]	17 [1]	1 e-18	18
2	1 e-14 [1]	14 [1]	1 e-18	18
4	1 e-18	17	1 e-19	17
6	1 e-20	17	1 e-19	17
8	1 e-34	17	1 e-18	18
10	1 e-18 [1]	17 [1]	1 e-40	18
	1 e-18	17		
12	0	17	1 e-49	18
14	degenerated		1 e-51	18
16	unsolved		0	18
20	unsolved		0	18
40	unsolved		0	18
60	unsolved		0	18

$- \varepsilon = 10^{-N}$, $\delta = \|r(P)\|/\|P\|$, where P is an ARE solution,
k is the number of the true digits in a result.
[1] without using the balancing procedure.

ACKNOWLEDGEMENT

This work was held in collaboration with the Ukrainian Academy of Science during one of the phases of this research. The authors gratefully acknowledge many stimulating discussions with Prof. V.B.Larin.

REFERENCES

Aliev, F.A., Bordyug, B.A. and Larin, V.B., 1986, Methods for solving matrix algebraic Riccati equations (in Russian), *Prepr., Inst. Fiz. Akad. Nauk Azerb. SSR*, n. 189.

Aliev, F.A., Bordyug, B.A. and Larin, V.B., 1987, The spectral method of solving matrix algebraic Riccati equations, *Sov. Math. Dokl.*, v. 35, n. 1, 121-125.

Arnold, U.F., and Laub, A.J., 1985, Generalized eigenproblem algorithms and software for algebraic Riccati equations, *Computer-Aided Control Systems Engineering* (Ed. by Jamshidi, M.,and Herget,C.J.), B.V., Elsevier Science Publishers, 279-300.

Ikramov,H.D., 1984, *Numerical Solving Algebraic Matrix Equations* (in Russian), Moscow: Nauka.

OPTIMAL FRICTION LAWS FOR DAMPING THE FORCED VIBRATIONS
OF LINEAR MECHANICAL SYSTEMS

A. CHEREMENSKY   N. DAKEV
Bulgarian Academy of Sciences
Institute of Mechanics & Biomechanics
4 "Akad. Bonchev" str., BG-1113 Sofia
Bulgaria

ABSTRACT. Forced vibrations of linear mechanical systems with dissipation are considered. Some problems for determination of optimal friction laws with regard to new energy criterion are solved. The performance index is connected with the maximizing of the average power dissipated for one period of the forced vibrations. An analytical-numerical algorithm for obtaining optimal periodical damp laws is proposed. It is shown how such an approach can be generalized for the mechanical systems with non-linearities by using the harmonic balance method. For this purpose a model of forced vibrations of one-link articulated manipulator is studied. The results obtained can be used for optimal adjusting the dampers installed at the hinges of articulated manipulators.

1. MAIN RESULTS

It is known [1] that forced vibrations of linear mechanical systems are described by the vector equation

$$A\ddot{x} + B\dot{x} + Cx = g(t) \tag{1}$$

Here x is an n-dimensional vector of generalized coordinates, A, B and C are $n \times n$ constant quadratic matrices, $g(t)$ is 2π-periodic vector function of the excitation and dot denotes differentiation with respect to the time t. Suppose that A, C, and B (corresponding to quadratic forms of kinetic and potential energy and Rayleigh's dissipative function) are symmetric and positive definite matrices. The value of the average power N dissipated for one period of the forced vibrations is estimated with the help of Rayleigh's function Φ as follows:

$$N = \frac{1}{2\pi} \int_0^{2\pi} \Phi(t)\, dt \tag{2}$$

For the system (1) the dissipative function Φ has a form

$$\Phi = \frac{1}{2} \dot{x}^T B x^T \tag{3}$$

Let us state the following optimal *Problem 1*: Define a 2π-periodic positive definite matrix B that provides a maximum value of the functional (2) for the established solution of the system (1) if the corresponding amplitudes are limited by the constant D

$$\int_0^{2\pi} \left[\sum_{\nu=1}^{n} x_\nu^2(\tau) \right] d\tau \leq D \tag{4}$$

Suppose there exists a unique optimal 2π-periodic solution of this problem. The solution of the equation (1) can be expanded in Fourier series

$$x = \sum_{i=1}^{\infty} (u_i \cos(it) + v_i \sin(it)) \tag{5}$$

where u_i and v_i are some n-dimensional vectors. Let us consider the following damp friction laws

$$B = \sum_{j=1}^{m} (B_{j1} \cos(jt) + B_{j2} \sin(jt)) \tag{6}$$

where m is an integer parameter defining the corresponding class of friction laws. Substituting the expressions (5)-(6) into the equation (1) and calculating scalar products with respect to the functions $\cos(jt)$ and $\sin(jt)$ we obtain the following non-linear system

$$F\alpha = \beta \tag{7}$$

where $\alpha = (u_1, u_2, \ldots, u_m, v_1, v_2, \ldots, v_m)^T$, F is a matrix that is linear dependent on the matrices A, B and C, and β is a vector that is linear dependent on the coefficients of the function $g(t)$ in the expansion in Fourier series. Resolving the equation (7) with respect to the vector α we can represent the vectors u_i and v_i as functions of the components of the matrix B

$$u_i = u_i(B_{01}, B_{02}, \ldots, B_{m1}, B_{m2}), \quad v_i = v_i(B_{01}, B_{02}, \ldots, B_{m1}, B_{m2}) \tag{8}$$

According to (8) the problem 1 is reduced to the following standard problem of non-linear programming: Determine matrices B_{01}, B_{02}, ..., B_{m1}, B_{m2} providing a maximum value of the functional (2) on the established solution of the system (1) if the inequality (4) holds. This problem has a unique solution if the monodromy matrix of the homogeneous system (1) has non-zero eigenvalues.

2. OPTIMAL DAMPING THE FORCED VIBRATIONS OF ONE-LINK ARTICULATED MANIPULATOR

Consider the model of one-link articulated manipulator performing forced vibrations under the influence of the kinematic excitation. According to this model the manipulator arm is absolutely rigid and its elastic and dissipative properties are concentrated in the hinge. If the damper has speedy characteristics realizing various friction laws then the forced vibrations of the manipulator are described by the scalar equation [2]

$$I\ddot{\varphi} + h|\dot{\varphi}|^n sgn\dot{\varphi} + c\varphi = f_0\Omega cos(\Omega t + \alpha), \quad f_0 = ms\xi \qquad (9)$$

Here the following notations are used:
I - the moment of inertia of the link with respect to the rotation axis; φ -the angle of link deviation from the equilibrium position (the generalized coordinate); m - the mass of the manipulator link; g - the gravity acceleration; s - the distance from the suspension point of link to its mass center G ; ξ - the amplitude of the forced vibrations; Ω - the frequency of the forced vibrations; α - the initial phase of these vibrations; h - the coefficient of the friction (Coulomb or dry for $n = 0$, linear or damp for $n = 1$ and quadratic for $n = 2$).

With regard to [3] the dissipative function has a form

$$\Phi(\dot{\varphi}) = \frac{h}{n + 1} |\dot{\varphi}|^{n+1} \qquad (n = 0, 1, 2, \dots) \qquad (10)$$

State the problem for optimal adjusting of the damper mounted at the manipulator hinge with regard to the optimal criterion (2). In the other words it is necessary to determine optimal coefficients h providing maximum values for the function N on the established solutions of (1) with limited amplitudes. Consider the solution of this problem assuming that dissipative terms can be linearized by the harmonic balance method.

Introducing the following variables

$$\Omega_0^2 = \frac{c}{I}, \quad \tau = \Omega_0 t, \quad \omega = \frac{\Omega}{\Omega_0}, \quad q = \frac{I}{f_0}, \quad h_n = \frac{f_0^{n-1}\Omega_0^{n-1}}{I^n} h \qquad (11)$$

we transform the equation (9) to the form

$$q'' + h_n|q'|^n sgn(q') + q = \omega^2 cos(\omega\tau + \alpha) \qquad (12)$$

According to the harmonic balance method this equation is replaced by the linear one

$$q'' + k_n q' + q = \omega^2 cos(\omega\tau + \alpha) \qquad (13)$$

where U_n is the amplitude of forced vibrations for given friction law ($n = 0, 1, 2, \ldots$). The coefficient k_n is defined according to [4]

$$k_n = \frac{h_n}{\pi \omega U_n} \int_0^{2\pi} \left[\, |-\omega U_n \sin\psi|^n sgn(-\omega U_n \sin\psi) \sin\psi \, \right] d\psi \qquad (14)$$

The following optimal problem can be formulated: For given value $\omega > 0$ determine the optimal coefficient $k_n = k_n^{(0)}$ that provides a maximum for the functional (2) if the amplitude U_n is limited: $U_n(\omega, k_n) \leq A$. For any $\omega \notin [\omega_{1A}, \omega_{2A}]$ the optimal friction coefficient is defined by

$$k_n^{(0)} = k_*(\omega) = |1 - \omega^2| \, / \, \omega \qquad (15)$$

where ω_{iA} are known functions of A [2]. For $\omega \in [\omega_{1A}, \omega_{2A}]$

$$k_n^{(0)} = k_A(\omega, A) = \frac{1}{\omega A} \sqrt{\omega^4 - A^2(1 - \omega^2)^2} \qquad (16)$$

It can be proved [2] that the damper with dry friction provides maximum level of energy dissipation compared to the dampers with linear or quadratic friction.

3. CONCLUSIONS

The proposed approach can be applied for optimal vibroisolation of the objects with rotational axes, in particular, for optimal damping forced vibrations of articulated manipulators. Optimal coefficients (15) and (16) can be directly used for optimal adjusting the passive dampers installed at the hinges of articulated manipulators. They may also be applied for optimal active damping of forced manipulator vibrations.

REFERENCES

1. Gantmacher, F.R. (1967) Matrix theory, Nauka Publishers, Moscow.
2. Dakev, N.V. (1988) 'Optimal damping the forced vibrations of articulated manipulators', in Proc. of the third conference on application of mechanics in robotics. Varna, 22 - 29 Sept. 1988, pp. 125-130.
3. Lurye, A.I. (1961) Analytical mechanics, Fizmatgiz Publ., Moscow.
4. Kolowsky, M.Z. (1966) Nonlinear theory of vibroisolation systems, Nauka Publishers, Moscow.

IMPLEMENTATION OF PUBLIC KEY and SECRET KEY CRYPTOGRAPHY ON A LOCAL AREA NETWORK

T Coffey
University of Limerick
Ireland

1. Abstract

Cryptography is the art of making communications unintelligible to all except the intended recipients. It provides a means of keeping data secure whether it is stored on a computer or transmitted over a network. A secure computer network must provide protection against unauthorised disclosure, undetected modification, undetected loss or repetition of information, as well as assurances of the correct identity of the sender and receiver of information. Cryptography plays a very important role in providing mechanisms for network security.

This paper introduces a security framework called Ulsecnet, which is concerned with the provision of end-to-end security measures in a local area or global wide network. The paper concentrates on the use of the RSA public key algorithm and the Data Encryption Algorithm,which are used in a hybrid manner in ULsecnet, to implement a number of security services. The empirical performance results achieved in implementing the two algorithms are also presented.

2. An Introduction to ULsecnet

ULsecnet is a secure communication framework, which is being developed at the University of Limerick, concerned with the provision end-to-end security services for local area and global wide networks. The work on ULsecnet has focussed on the provision of end-to end security measures, which rely on the security of equipment only at the source and destination of an association. Some of the security services include: **Authentication** (the corroboration that the source of data received, or the destination of data transmitted is as claimed), **Access Control** (the prevention of unauthorised use of a resource, including the prevention of use of a resource in an unauthorised manner), **Data Confidentiality**, **Data Integrity** (the property that data has not been altered or destroyed in an unauthorised manner), and **Non-repudiation** (prevention of a denial by one of the entities involved in a communication of having participated in all or part of the communication).

The security mechanisms include the development of PC (personal computer) cryptographic co-processor cards, implementation of public key and secret key algorithms, and security protocols to enable the management of the security process.

2.1 SECURE LAN ENVIRONMENT

The ULsecnet security mechanisms outlined above were developed on an IEEE802.3 based LAN environment. The TCP/IP protocol suite is incorporated on the LAN to provide reliable host-to-host communication. TCP/IP is a collection of network protocols (TCP, IP, FTP, SMTP & Telnet) which support internetworking by allowing different network types (networks with different network dependent technologies; for example networks with different data link and physical media layers) to communicate with each other.

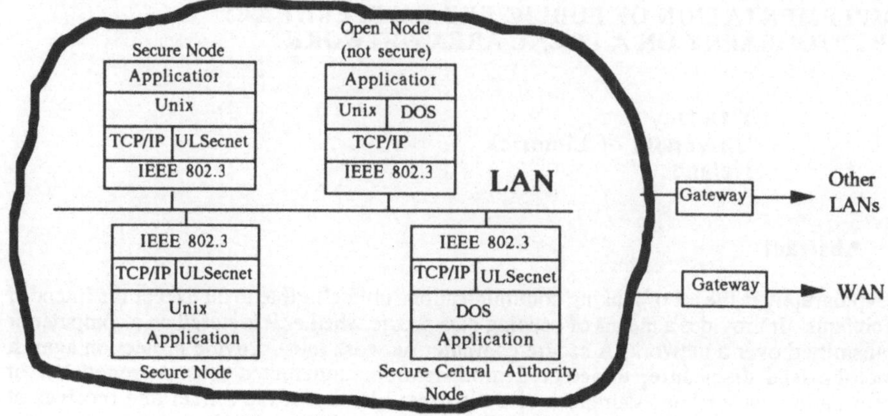

Fig. 1 ULsecnet Secure LAN Environment

The network consists of secure and open (not secure) nodes. Secure nodes are designed to support security services such as authentication, data confidentially, data integrity and non-repudiation. The central authority node is designed to support a range of security management services such as the generation and certification of public key certificates, and public key validation services.

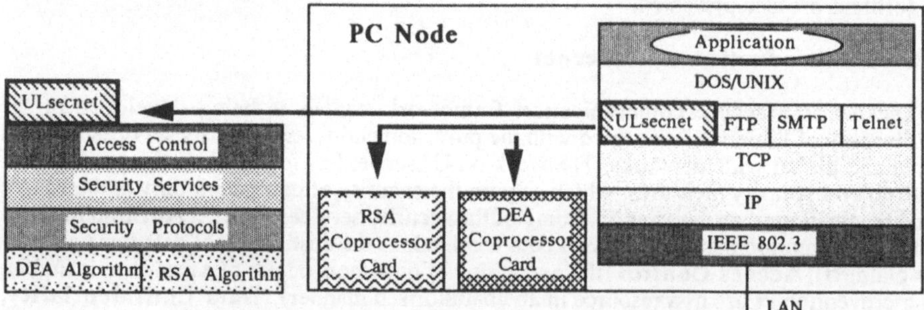

Fig. 2 ULsecnet: Architecture of a Secure Node

The architecture of a ULsecnet node has four layers including: access control, security services, security protocols and the DEA and RSA algorithms. The cryptography based security services and protocols are facilitated by the use of cryptographic co-processor cards to implement the RSA and DEA algorithms. The hardware based cryptographic cards were developed to facilitate the implementation of the algorithms since software solutions are computationally infeasible.

The ULSecnet security element uses a hybrid of both the RSA and DEA algorithms, the RSA algorithm being used to implement services such as authentication, non-repudiation, and key & certificate management services, while the DEA algorithm is used to implement services such as data confidentially and data integrity.

The RSA and DEA algorithms are used in a hybrid manner to realise the advantages of the two algorithms. The RSA Public key algorithm is superior for key distribution and digital signatures, while DEA is used for its superior performance.

3. SECRET and PUBLIC KEY CRYPTOGRAPHY

The two main streams of cryptography are conventional cryptography (some times called symmetric secret key cryptography) and public key cryptography (some times called asymmetric cryptography). Conventional cryptography uses one key for both encryption and decryption, while public key cryptography uses separate keys (a public key and a private/secret key) for encryption and decryption [5]. Two of the most successful current algorithms are The Data Encryption Algorithm(DEA) [1] and RSA (Rivest, Shamir and Adleman) [2] public key algorithm.

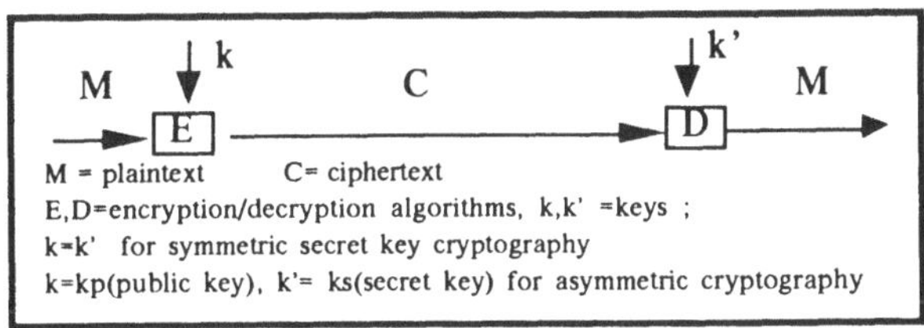

Fig. 3 The Encryption /Decryption Process

3.1 PUBLIC KEY CRYPTOGRAPHY

In public key cryptosystems (PKCS), two keys are used, one key is used for encipherment and the other for deciphering. The requirements for public key systems can be summarised as follows:
- Each user has 2 keys: A public key (PK) and a secret key (SK).
- The public and secret key pair (PK and SK) must be capable of being generated.
- Knowledge of PK must not allow SK to be computed effectively.
- Encryption followed by decryption causes original message (M) to be recovered i.e. $D_{SK}(E_{PK}(M))=M$.
- If a message M is first deciphered and then enciphered, then M is the result. i.e. $E_{PK}(D_{SK}(M)) = M$.
- Secrecy is obtained by encrypting plaintext with PK i.e. Ciphertext (C) = $E_{PK}(M)$.
- Plaintext is recovered by decrypting with SK i.e. Plaintext (M) = $D_{SK}(C)$.

Fig. 4 Use of a PKCS to Exchange Secret Information

3.1.1 THE RSA ALGORITHM

The RSA algorithm, is considered to be the most secure public key algorithm developed to date. The RSA algorithm is being used within the ULsecnet LAN environment, to enable the development of security protocols & services such as authentication, non-repudiation, and key & certificate management

RSA CRYPTOGRAPHIC PARAMETERS

1.   P and Q: Two prime numbers.   (secret)
2.   N: The public modulus (N = P * Q).   (public)
3.   ø (N): Number of primes relative to N. (ø (N)= (P-1) * (Q-1))
4.   PK: Random number less than N.   (public)
5.   SK: The number which satisfies the equation: SK*PK=1 mod (ø(N)).(secret)
6.   M: is the message.   (secret)
7.   C: is the ciphertext.   (public)

RSA ALGORITHM SUMMARY

a.   Two secret prime numbers, P and Q are randomly selected.
b.   The public modulus, N = P*Q is calculated.
c.   The secret Euler Totient Function, ø (N)=(P-1)(Q-1) is calculated.
d.   A secret quantity, K, which is relatively prime to ø(N), is randomly selected. K is defined as the secret key, SK, or the public key, PK.
e.   The multiplicative inverse of K Modulo ø(N) is calculated using Euclid's Algorithm. This quantity is defined to be either the public key, PK, or the secret key, SK, depending on (d).
f.   The plaintext, M, is broken up into blocks $M_1, M_2, .. M_n$, which are less than the modulus number N. The ciphertext (C_i) is then derived using the following encryption function $C_i = M_i^{PK} \text{ Mod } (N)$.
g.   The plaintext, M_i, can be retrieved from the ciphertext (C_i) using the following function: $M_i = C_i^{SK} \text{ Mod } (N)$.

3.2 DATA ENCRYPTION ALGORITHM

The DEA is a block cipher, which operates on 64-bit blocks of plaintext. The algorithm uses 56 bits of the 64-bit key for the encryption process while the remaining 8 bits are used as parity checking bits. The DEA can be used in four different operating modes, Electronic Code Book (ECB), Cipher Block Chaining (CBC), Output feedback (OFB), and Cipher Feedback Mode (CFB).

The main attributes of the DES algorithm are (1) it is cryptographically strong, (2) it can easily be made stronger if required, for example by using multiple encryption techniques, (3) It is relatively inexpensive and fast (see empirical performance results), (4) it has several modes of operation.

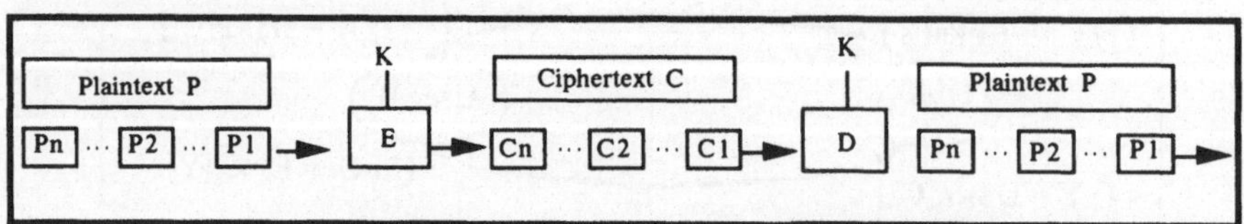

Fig. 5 DEA Electronic Code Book(ECB) Mode

The ECB mode has a fixed block size and each block is encrypted separately. Any data transmission errors occurring in a ciphertext block will not be propagated throughout the message. However, underline plaintext patterns may be exposed, so the ECB mode is normally used to encrypt keys, not message text.

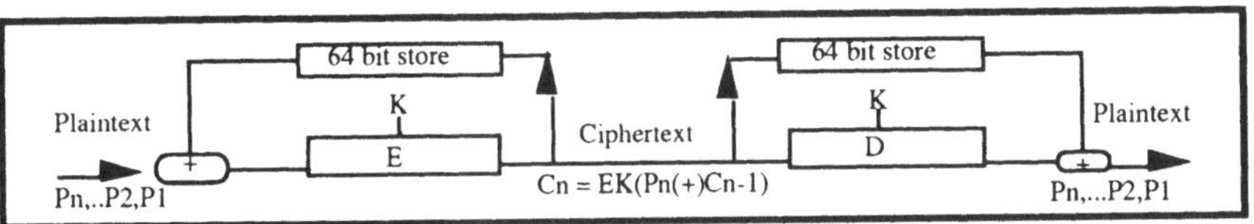

Fig. 6 DEA Cipher Block Chaining(CBC) Mode

In the CBC mode the ciphertext is fedback and added to the next block of plaintext before encryption (E). The CBC has limited error propagation, where an error (e.g in transmission), in the ciphertext will result in two received blocks being in error. The CBC can support authentication/integrity by the generation of a cryptographic checksum or Message Authentication Code (MAC), which is appended to the transmitted message. As can be seen from Fig.6, block n of the ciphertext is defined as $Cn =Ek(Pn(+)Cn-1)$, hence the last block (Cn) can be used as the MAC.

The CFB is used for enciphering streams of characters ranging from 1to 64bits. The character length is typically 1bit or 8bits. The CFB mode has limited error propagation and also supports authentication. However, CFB (excluding CFB-1) does not recover from loss of synchronisation.

In OFB the ciphertext is sum of (Exclusive OR) of the input plaintext and "random" bit stream produced by feedback of the encryption output to its input. Authentication/integrity is not supported in OFB and it is normally only used in environments which have high bit error rates. In these environments OFB, with the aid of suitable synchronisation protocols can minimise the impact of the high error rates.

The data confidentially and data integrity services developed under ULsecnet use the DEA CBC mode of operation.

4. IMPLEMENTATION ISSUES

As previously outlined, the security services provided by ULsecnet are implemented by a series of protocols (such as secure file transfer with integrity, generation & validation of user certificates, generation & distribution of cryptographic keys ,etc.) which use DEA and RSA algorithms functions. These performance of these functions, which include key generation, file encryption, MAC generation, etc., is considerably improved by the use of encrypter co-processor PC cards.

The DEA & RSA encrypters, which are interfaced to the PC, incorporate special purpose DEA & RSA processors with their own instruction sets. The measured performance results of the implementation are provided in the next section.

5. EMPIRICAL PERFORMANCE RESULTS

The measured performance for PC file encryption and decryption using an RSA encrypter co-processor card is given in Figures 7. The performance figures were measured on a 25 Mhz Intel PC, using an 8bit data bus. File encryption rates of the order of 250KBps and decryption rates of the order of 14KBps were realised.

Secure node-to-node message transfer rates are of the order of 10KBps. The addition of message authentication (using a digital signature) does not significantly affect performance.

Modulus Size (bytes)	Encryption (KBps)	Decryption (KBps)
50	214	18
60	247	15.5
70	248	13.2
80	270	11.8

Fig. 7 Performance Measurements of RSA Encrypter Co-Processor Card on a 25Mhz Intel 486 PC

The measured performance for PC file encryption and decryption using an DEA encrypter co-processor card is given in Figures 8. File encryption/decryption rates of 1.8MBps were realised for the ECB, CBC, CFB and OFB modes of operation. A significant result is that triple encryption does not reduce the performance from that obtained with single encryption. This is due to parallelism within the DEA processor. This means that a user can have the extra security of triple encryption without paying any penalty in performance, which would certainly not be the case in a software only implementation.

The generation of a 64 bit MAC or cryptographic checksum, using the CBC or CFB modes, was measured at the rate of 3.6MBps. Thus, message integrity can be added without significantly reducing end-to end performance.

DES OperatingMode	Single Encryption (Mbits/sec)	Triple Encryption (Mbits/sec)
ECB/CBC	1.82	1.82
CFB-64 bit/OFB-64bit	1.82	1.82
CFB-8bit/ OFB-8bit	0.98	0.98
MAC Generation	3.64	3.64

Fig.8 Performance Measurements of DES Encrypter Co-Processor Card on a 25 Mhz Intel 486 PC

6. Conclusions

This paper introduced the architecture of a secure communication environment called ULsecnet. ULsecnet uses a hybrid of symmetric and asymmetric cryptography to realise an effective implementation of its security services.

The hybrid approach uses public key RSA algorithm for tasks such as key and certificate distribution and authentication, and the DEA symmetric algorithm for data confidentiality and integrity.

7. References

1.	Data Encryption Standard, Federal Information Processing Standard (FIPS), Publication 46, NBS, US Dept. of Commerce, Washington ,DC (Jan.1977)
2.	Rivest,R.L., Shamir, A., Adleman,L. " A Method for Obtaining Digital Signatures and Public Key Cryptosystems", Comms of the ACM, 21, No. 2, 120-126 (1978)
3.	Comer, D.E. "Internetworking with TCP/IP: Principles,Protocols and Architecture", Vol 1, Prentice Hall, (1990)
4.	Branstad, D.K., "Considerations for Security in the OSI Architecture", IEEE Network Magazine, Vol. 1 No.2, 34-39, (1987)
5.	Diffie,W. & Hellamn, M.E., "New Directions in Cryptography",IEEE Transactions on information Theory, IT-22, No 6, 644-645 (1976).

INTERFACE PRESSURES IN CONTACT ZONES

W D COLLINS, M S KHALIL, S QUEGAN, D SMITH, D A W TAYLOR

ABSTRACT. A new method of determining inter-face pressures in contact zones between hard and soft bodies by reconstructing the distribution from measurements of strain away from the contact zone is given. Progress in reconstructing theoretical and experimental one-dimensional distributions is described.

Methods of assessing pressures at the interface of two elastic bodies in contact which directly measure pressures in the contact zone are intrusive in that they inevitably disturb the local stress field and so the very pressures being measured. We propose a method of assessment, different in kind from previous methods, for interface pressures under the human foot and the automobile tyre which are similar problems in that one of the bodies in contact, the foot, the tyre, has a modulus of elasticity orders of magnitude less than that of the other, the ground or the sole of a shoe, the road surface. The proposed method aims to make direct measurements of elastic strain in the hard component at locations nearby and remotish from the contact zone and then to calculate the pressure distribution in the zone from these measurements, the reverse of the normal procedure in designing engineering artefacts. Whilst Saint Venant's Principle indicates that strains measured remotely from the load are not at all sensitive to the distribution of load, results so far suggest this method could still be feasible.

We plan to construct a 'force plate' as a nest of closely-spaced parallel beams, each one of which is a carefully contrived single simply supported beam. Figure 1 shows a hypothetical pressure distribution bearing down on the nest of beams. A preliminary investigation however has been carried out solely on one beam considered on its own and separate from the rest of the proposed nest.

A rig has been built for loading this beam both under point loads and also under distributed loadings to known forms of variable distribution, one resembling the sinc function, $\sin x / x$, having been found particularly appropriate because of its spatial frequency resemblance to foot pressure distributions. Figure 2 shows the beam under a combined sinc function and uniformly distributed load.

The reverse calculation of load from strain is based on simple beam theory. For a beam of length l and half depth c the bending moment M at a point, distance x along the beam, is related to the outer surface longitudinal strain e by

$$M = -\frac{Ele}{c}$$

E Young's modulus, I the second moment of area of the beam's cross-section, whilst M is related to the load w per unit length by the equation

$$\frac{d^2M}{dx^2} = w , \quad 0 \leq x \leq l .$$

For a simply supported beam the end conditions are

$$M = 0 \quad at \quad x = 0,l .$$

Integrating gives

$$M(x) = \int_0^l g(x,s) \ w(s)\,ds , \ 0 \leq x \leq l,$$

where the Green's function

$$g(x,s) = \frac{s}{l}(x-l) , \ s < x ,$$

$$= \frac{x}{l}(s-l) , \ x < s .$$

The strain e and hence the bending moment M are known at the stations x_i, $i = 0,1,...N$, $x_0 = 0$, $x_N = l$, along the beam, so that

$$M(x_i) = \int_0^l g(x_i,s) \ w(s)\,ds, \ i = 0,1,..N.$$

$$(1)$$

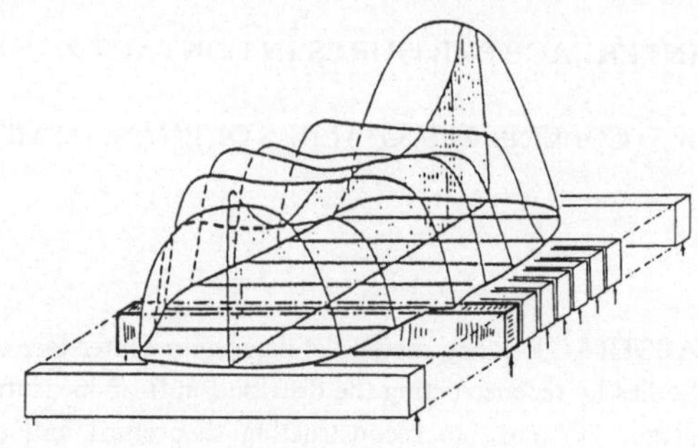

Fig. 1 The Proposed Nest of Beams

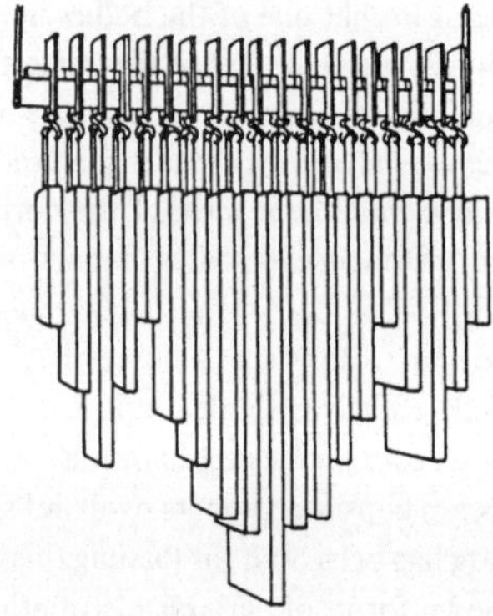

Fig. 2 A Single Beam Under Test Loading

There are an infinite number of functions w which satisfy (1) and a reconstruction of w, to have any hope of success, must rely on actual loads having properties that can be used in the reconstruction. Both foot and tyre pressures and the sinc load are bandwidth limited, being of predominantly low spatial frequency, so that there is some prospect of reconstruction by methods which ignore high frequency components.

Two basic methods have been tried. The first approximates the load as a series of point loads at the stations x_i to obtain the equations

$$M(x_i) = \sum_{j=0}^{N} \alpha_{ij} \left(\frac{l w_j}{N} \right)$$

for the values $w_i = w(x_i)$. Here $\alpha_{ij} = g(x_i, x_j) = \alpha_{ji}$ are well-known influence coefficients. This corresponds to approximating the integral in (1) by the trapezium rule but does not give w at points intermediate to the stations. The method has been extended to overcome this by representing w in terms of cubic splines.

The second method is to formulate the problem in Hilbert space and minimize the norm

$$\| w \|^2 = \int_0^l w^2(s)\, ds$$

subject to the constraint (1). Combining this with singular value decomposition shows that, when the x_i are equally spaced so that $x_i = il/N$, the minimiser $w_0(s)$ is given by

$$w_0(s) = \sum_{m=1}^{N} B_m S_m(s)$$

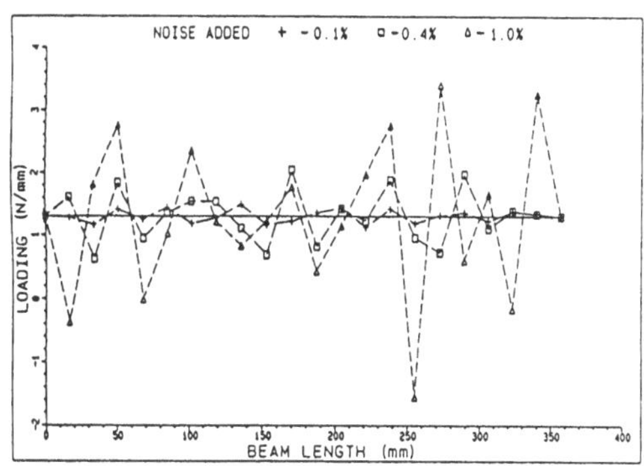

Fig. 3 Sensitivity Analysis to Noise

where

$$S_m(s) = \sin\left[\frac{m\pi s}{l}\right] + \sum_{r=1}^{\infty}\left[\frac{m}{2rN+m}\right]^2 \sin\left[\frac{(2rN+m)\pi s}{l}\right]$$

$$and\quad B_m = -\frac{2}{Nm^2\mu_m}\left(\frac{\pi}{l}\right)^2 \sum_{i=1}^{N} M(x_i)\sin\left[\frac{mi\pi}{N}\right] , \quad \mu_m = \sum_{k=0}^{\infty}[2kN+m]^{-4}.$$

Since real strain gauges cannot be positioned exactly, correction terms have been obtained for the x_i only approximately equally spaced.

Any function w satisfying (1) differs from w_0 by a function in the null space of the operator of equation (1). This null space contains all functions which are the second derivatives of functions vanishing at all the points x_i.

As a first test of these methods theoretical simulations have been carried out in which the values of the bending moments at the points x_i were calculated from known loadings and the loadings then reconstructed from these values by the methods described. This has been done for point loads at various points, a uniformly distributed load and a sinc load. Both methods give good results.

As a second test, different levels of white noise were added to the calculated values of M and the loading then reconstructed. Figure 3 shows the results for a uniformly distributed load represented by the horizontal line. The noise levels of 0.1% and 0.4% of peak M give acceptable reconstructions whilst a 1% noise level does not.

The methods have been experimentally tested by subjecting a beam to test loadings by known weights. The beam is of aluminium alloy, 357 mm long, with a cross section of 10 mm wide by 20 mm deep. Twenty electric resistance gauges were positioned at 20 stations along the underside of the beam at normal spacings of 17 mm. It is possible that the beam was not exactly of parallel sided form and also that the strain gauges were bonded or twisted very slightly out of position. To make some allowance for this and other departures from the ideal, we have first loaded the beam in a closely defined way and used all the strain gauges 'together' in a process of calibration to obtain a value for the modulus of elasticity E of the alloy. The calibrated beam was then loaded with a single 'point' load of 80 N, moved to various positions along the beam, and the corresponding strain readings recorded and processed by the first method. The quite reasonable reconstructed load for one point load position is shown in Figure 4.

We have also imposed a series of point loads as individual weights together on this calibrated beam to step-wise approximate to a distributed loading of a sinc function and a uniformly distributed load. This arrangement gets round the difficulty of devising upward as well as downward loads if a sinc function alone were attempted. Figure 5 shows the loading reconstructed from the strain gauge readings obtained. This test has produced a promising result.

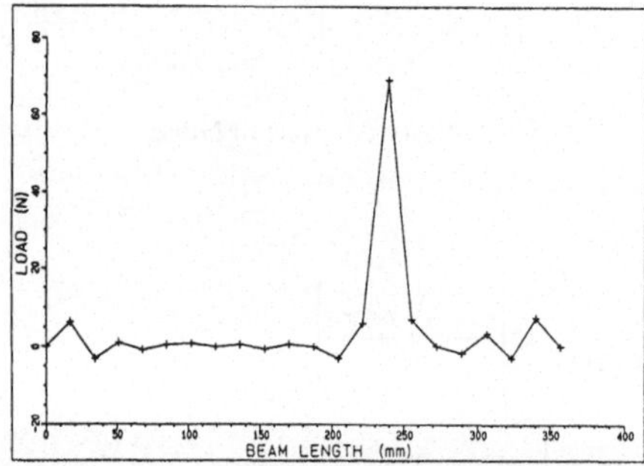

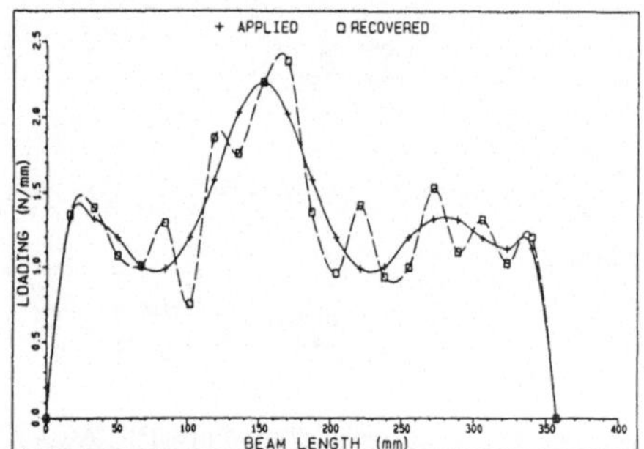

Fig.4 Load Reconstruction from a
Point Load of 80 N

Fig. 5 Load Reconstruction from
a Distributed Loading

The work so far and in particular the experimental results show the need for further investigations into the techniques and methods used. It may be possible to better filter 'noise' in the process, to better allow for individual calibrations of individual strain gauges and to improve the end 'simple' supports to better realize end conditions. There is also much scope for extending the work to consider a single plate element to replace the nest of beams.

One important feature which has still to be tackled is the degree of 'softness' of whatever provides the pressure loading on the nest of beams and how this can be allowed for when the individual beams of the nest deflect under the load by different amounts and so disturb the loading in the 'soft' element.

The University of Sheffield, Sheffield S10 2TN, England, U.K.

MATHEMATICAL MODELS FOR THE TRANSMISSION DYNAMICS OF THE HUMAN IMMUNODEFICIENCY VIRUS (HIV) IN IRELAND.

C.M. COMISKEY
School of Mathematical Sciences
Dublin City University
Dublin 9
Ireland.

ABSTRACT. Despite considerable advances in understanding the basic biology of the human immuno-deficiency virus, the aetiological agent of AIDS, medical,public health and health education planning continues to be plagued by uncertainties. There remain many questions of an epidemiological nature to be answered. The questions to be addressed include, how many people are HIV positive? What groups are most vulnerable? Where should valuable intervention policies be directed? Mathematical models of the dynamics of HIV transmission and its progression to AIDS can clarify what data must be collected in order to predict future prevalence, make predictions about the likely effect of future intervention policies and provide predictions for several decades ahead. This is a good example of how mathematics can be used to assist in the preservation of human life.

1. The Irish Background

The first case of AIDS in Ireland was not identified until 1981. Also in Ireland paediatric AIDS is not uncommon compared with other European countries. This is a direct result of the male:female ratio of HIV cases, which is approximately 5:1. This ratio of HIV infections is well below the estimated ratio of 10:1 for other western European countries. This is explained by the distribution of known cases of the virus. As of October 1991. there were a total of 1138 known cases of the HIV virus in Ireland, 54% of which were identified in the intravenous drug using (IVDU) group, while the homosexual and bisexual group represents only 16% of all cases. This is very different from the rest of western Europe, where, the majority of cases were identified in the homosexual group. In addition, paediatric cases comprise almost 7% of the total identified.

2. Modelling Within The IVDU Group

Research on mathematical models of the virus and its progression to AIDS has to date concentrated on the homosexual community. To-day, researchers are returning to their models and using the insights gained from them to model spread of the disease within the heterosexual population. From the beginning of the AIDS epidemic in Ireland it was observed that we were unlike the rest of western Europe, and would have to proceed immediately to the difficult problem of modelling the spread of the HIV virus within the intravenous drug users. Simple mathematical models for the spread of the HIV virus in the Irish IVDU population are formulated, key model parameters are estimated from a comprehensive HIV survey carried out by us on Irish IVDU's and finally preliminary predictions based on the results of the model are provided.

For practical reasons of parameter estimation and simplicity we restrict our model to routes of transmission within the male and female drug using populations. We assume that the population mixes homogeneously and that all infectious individuals eventually develop AIDS. We make the further assumption that those in the AIDS class do not contribute to the spread of the disease. The assumption that all HIV positive individuals progress to the disease is reasonable. Recent evidence suggests that the proportion of people infected with HIV who eventually develop AIDS is higher than originally thought and may be as high as 80% or even greater as the the duration of the incubation period is seen to increase. We split the total population of drug users at time t into three exclusive compartments of susceptibles, infectious and AIDS, denoted by $X(t)$, $Y(t)$ and $A(t)$ respectively. We denote the different sexes within the population by the subscripts m and f, for male and females . Hence the population of male drug users at time t is denoted by $P_m(t)$ and the population of female drug users at time t is denoted, $P_f(t)$. We formulate a differential equation transmission model to describe the rate of change in the number of susceptible, infectious and AIDS patients in the male and female intravenous drug using populations. We have,

$$\frac{dX_m(t)}{dt} = \Lambda_m - n_m\beta_3 X_m(t)\left[\frac{Y_m(t) + Y_f(t)}{P_m(t) + P_f(t)}\right] - c_m\beta_2 X_m(t)\frac{Y_f(t)}{P_f(t)} - \mu X_m(t), \quad (1)$$

$$\frac{dX_f(t)}{dt} = \Lambda_f - n_f\beta_3 X_f(t)\left[\frac{Y_m(t) + Y_f(t)}{P_m(t) + P_f(t)}\right] - c_f\beta_1 X_f(t)\frac{Y_m(t)}{P_m(t)} - \mu X_f(t), \quad (2)$$

$$\frac{dY_m(t)}{dt} = n_m\beta_3 X_m(t)\left[\frac{Y_m(t) + Y_f(t)}{P_m(t) + P_f(t)}\right] + c_m\beta_2 X_m(t)\frac{Y_f(t)}{P_f(t)} - (\alpha + \mu)Y_m(t), \quad (3)$$

$$\frac{dY_f(t)}{dt} = n_f\beta_3 X_f(t)\left[\frac{Y_m(t) + Y_f(t)}{P_m(t) + P_f(t)}\right] + c_f\beta_1 X_f(t)\frac{Y_m(t)}{P_m(t)} - (\alpha + \mu)Y_f(t), \quad (4)$$

$$\frac{dA_m(t)}{dt} = \alpha Y_m(t) - (\delta + \mu)A_m(t), \quad (5)$$

$$\frac{dA_f(t)}{dt} = \alpha Y_f(t) - (\delta + \mu)A_f(t). \quad (6)$$

Where

$$P_m(t) = X_m(t) + Y_m(t) + A_m(t), P_f(t) = X_f(t) + Y_f(t) + A_f(t),$$

and $X_m(0), X_f(0), Y_m(0)$ and $Y_f(0)$ are known constants and $A_m(0) = A_f(0) = 0$. Λ_m and Λ_f are the recruitment rates to the male and female IVDU populations respectively, $1/\alpha$ is the mean incubation period, $1/\delta$ is the AIDS related death rate, μ is the migration rate, β_1 is the probability of transmission from an infectious male to a susceptible female in a single sexual contact, β_2 is the probability of transmission from an infectious female to a susceptible male in a single sexual contact and finally β_3 is the probability of transmission during a single needle sharing act. This model is based on the one sex model of Anderson et al (1986). We use the model in equations (??) to (??) to describe the spread of HIV in the drug using population given a known number of IV drug users at a certain point in time and we introduce a single infectious male into the population at that time.

3. Transmission Model Parameters and Results

All parameters were estimated from a comprehensive HIV survey. The aim of this survey was to investigate the social, sexual and drug habits of those at risk from HIV. It was our intention to look particularly at the rate of needle use, sharing and needle life in the drug using population and the rate of sexual partner change. In addition to this we aimed to estimate the rate of introduction to each of the populations. To our knowledge very little survey work has been published on this difficult and sensitive information This is the most comprehensive survey of this type, to date, in the Irish population. Research on sexual behaviour has been carried out in America and Britain, Johnson et al (1989), but results from these countries cannot necessarily be applied to the Irish situation. The 187 participants in the survey were drawn from the Genito Urinary Clinic at St. James Hospital Dublin. Results for the parameters of the system described in (??) to (??) are given in Table 1.

TABLE 1. Estimates of Model Parameters.

Group	Parameter	Value
Male IVDU	$\Lambda_m, \mu_m, c_m, n_m$	228, 0.107, 7.440, 4.667
Female IVDU	$\Lambda_f, \mu_f, c_f, n_f$	147, 0.125, 1.007, 6.941

To model the transmission of the virus through the IVDU group we assume a population of 5000 male and 2500 female susceptible IV drug users. These numbers are based on a crude estimation of the size of the IVDU population using the direct approach as discussed by Cox (1988). We introduce one male infectious into this population at time $t = 0$ equivalent to 1980, and model the transmission of the infection through the population. Parameters are as given above, in addition we let $\alpha = 0.125, \delta = 1.000, \beta_1, \beta_3 = 0.020$ and $\beta_2 = 0.010$ in accordance with known estimates. We assume that initially there are no AIDS cases in the population. We then model the spread of the infection over a period of twenty years from 1980 to 2000 given no changes in behaviour or introduction of Government intervention policies, given the introduction of a safe sex campaign in 1991 (equivalent to letting $c_m, c_f = 0$ from 1991) and given the introduction of a needle exchange program in 1991 (equivalent to letting $n_m, n_f = 0$ from 1991).

The total number of HIV infectious drug users from 1991 to the year 2000 expressed as a proportion of the total population size given these three situations are plotted in Figure 1. If we assume that the infection was introduced by a single HIV positive male drug user in 1980 then the model predicts a total of 2663 infectious (but not AIDS) and 327 AIDS cases amongst the IVDU group in 1991. This is over 4 times the number known to the Department of Health at that time. If there is no change in Government policy these decrease to 1584 infectious and 183 AIDS cases by the year 2000. Given the introduction of a safe sex campaign from 1991 this reduces to 1496 infectious drug users by the year 2000. If a large scale needle exchange program were to be introduced instead and drug users stopped sharing needles from that time, the model predicts a total of 1043 infectious by the year 2000. This is a considerable decrease from the 1584 predicted by the model given no intervention policies.

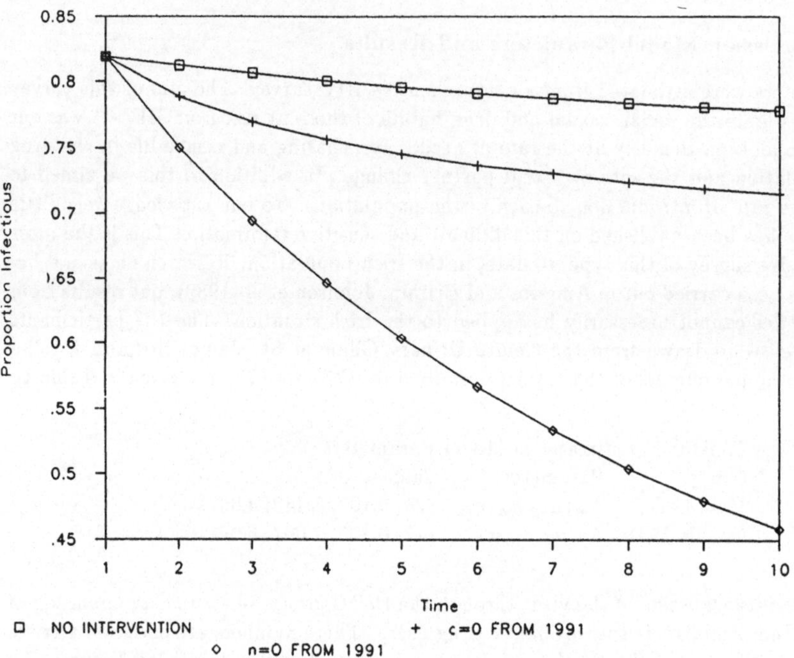

Figure 1. Proportion HIV Infectious.

References

1 Anderson, R.M., Medley, G.F., May, R.M. and Johnson, A.M. (1986). 'A preliminary study of the transmission dynamics of the human immunodeficiency virus (HIV), the causative agent of AIDS.' IMA Journal of Mathematics Applied in Medicine and Biology, 3, 229-263.

2 Cox, Sir D. (1988). 'Short-term prediction of HIV Infection and AIDS in England and Wales.' Report of a Working Group, Her Majesty's Stationary Office, 48-52.

3 Johnson, A.M., Wadsworth, J., Elliott, P., Prior, L., Wallace, P., Blower, S., Webb, N.L., Heald, G.I., Miller, D.L., Adler, M.W., Anderson, R.M. (1989). 'A pilot study of the sexual lifestyle in a random sample of the population of Great Britain.' AIDS, 3:135-141.

NUMERICAL APPROXIMATION IN THE THEORY OF PLATES WITH TRANSVERSE SHEAR DEFORMATION

C. CONSTANDA
Department of Mathematics
University of Strathclyde
Glasgow G1 1XH
Scotland, U.K.

ABSTRACT. A numerical analysis is presented of the bending of a circular steel floor in a wind-driven power generator, which is modelled as a thin homogeneous and isotropic Mindlin-type plate.

1. Introduction

An experimental low-maintenance wind-driven power generator incorporates a circular steel floor whose edge is fixed into the perimeter foundation wall. Since sensitive machinery is installed in a tightly packed underfloor compartment, it becomes important to determine the deflection pattern of the floor when it is subjected to bending caused by various factors, for example, by small settlement movements in the foundation.

The physical situation can be modelled mathematically as a Dirichlet problem for a homogeneous and isotropic circular disk within the framework of the theory of bending of plates with transverse shear deformation [1]. After presenting the main equations of the theory, we apply the method of generalized Fourier series to compute an approximation of the solution, and discuss the accuracy of this approximation in terms of some of the parameters of the scheme.

2. Bending of thin plates

Let $S \subset \mathbf{R}^2$ be a finite domain bounded by a simple closed C^2-curve ∂S, $n = (n_1, n_2)^{\mathrm{T}}$ the unit outward normal to ∂S (the superscript T indicates matrix transposition), and $x = (x_1, x_2)$ a generic point in $\mathbf{R}^2$. In [1] a two-dimensional theory of bending of plates with transverse shear deformation is investigated, which reduces to the equilibrium equations

$$L(\partial_x)v(x) = 0, \quad x \in S, \tag{1}$$

and boundary conditions of the form $v(x) = A(x)$, $x \in \partial S$ (Dirichlet), or $T(\partial_x)v(x) = B(x)$, $x \in \partial S$ (Neumann). Here $v = (v_1, v_2, v_3)^{\mathrm{T}}$ is a vector characterizing the displacements, $L(\partial_x) = L(\partial/\partial x_1, \partial/\partial x_2)$,

$$L(\xi) = L(\xi_1, \xi_2) = \begin{pmatrix} h^2\mu\Delta + h^2(\lambda+\mu)\xi_1^2 - \mu & h^2(\lambda+\mu)\xi_1\xi_2 & -\mu\xi_1 \\ h^2(\lambda+\mu)\xi_1\xi_2 & h^2\mu\Delta + h^2(\lambda+\mu)\xi_2^2 - \mu & -\mu\xi_2 \\ \mu\xi_1 & \mu\xi_2 & \mu\Delta \end{pmatrix},$$

130

$\Delta = \xi_1^2 + \xi_2^2$, $T(\partial_x) = T(\partial/\partial x_1, \partial/\partial x_2)$ is the boundary stress operator defined by

$$T(\xi) = T(\xi_1, \xi_2)$$
$$= \begin{pmatrix} h^2(\lambda + 2\mu)n_1\xi_1 + h^2\mu n_2\xi_2 & h^2\mu n_2\xi_1 + h^2\lambda n_1\xi_2 & 0 \\ h^2\lambda n_2\xi_1 + h^2\mu n_1\xi_2 & h^2\mu n_1\xi_1 + h^2(\lambda + 2\mu)n_2\xi_2 & 0 \\ \mu n_1 & \mu n_2 & \mu(n_1\xi_1 + n_2\xi_2) \end{pmatrix},$$

$\sqrt{12}\,h$ is the constant thickness of the plate, λ and μ are the Lamé coefficients of the (homogeneous and isotropic) material, and $A(x)$ and $B(x)$ vectors prescribed on ∂S. As seems appropriate for the physical situation considered, we assume that the body forces and moments, and the forces and moments acting on the faces, are negligible.

The existence and uniqueness of the solutions of the fundamental boundary value problems associated with the system (1) are discussed in [1]. In particular, in [1] it is shown that the interior Dirichlet problem has a unique solution if $A(x)$ is Hölder continuous.

Let $D(x, y) = [\text{adj}\, L(x)]\, t(x, y)$ be a matrix of fundamental solutions for L, where

$$t(x, y) = a\left[(|x - y|^2 + 4h^2)\ln|x - y| + 4h^2 K_0(h^{-1}|x - y|)\right],$$

$a = \left[8\pi h^2 \mu^2(\lambda + 2\mu)\right]^{-1}$ and K_0 is the modified Bessel function of order zero. If we write $P(x, y) = \left[T(\partial_y)D(y, x)\right]^T$, $(Tu)(x) = \psi(x)$, $x \in \partial S$, and $S^+ = S$, $S^- = \mathbf{R}^2 \setminus \bar{S}^+$, then the Somigliana representation formula [1] in the case of the interior Dirichlet problem yields

$$u(x) = \int_{\partial S} \left[D(x, y)\psi(y) - P(x, y)A(y)\right] ds(y), \quad x \in S^-, \tag{2}$$

$$0 = \int_{\partial S} \left[D(x, y)\psi(y) - P(x, y)A(y)\right] ds(y), \quad x \in S^+. \tag{3}$$

3. Computational algorithm

From (2) and (3) it follows that

$$u(x) = \int_{\partial S} D(x, y)\psi(y)\, ds(y) - F(x), \quad x \in S^+,$$

where

$$F(x) = \int_{\partial S} P(x, y)A(y)\, ds(y), \quad x \in S^+ \cup S^-,$$

is known and ψ is the solution of the integral equation

$$\int_{\partial S} D(x, y)\psi(y)\, ds(y) = F(x), \quad x \in S^-. \tag{4}$$

Let ∂S_* be a smooth closed curve outside ∂S, and let $\{x^{(k)}\}_{k=1}^{\infty}$ be a sequence of points densely distributed on ∂S_*. We set

$$f^{(1)}(x) = (1, 0, -x_1)^T, \quad f^{(2)}(x) = (0, 1, -x_2)^T, \quad f^{(3)}(x) = (0, 0, 1)^T$$

and

$$\theta^{(jk)}(x) = D^{(j)}(x, x^{(k)}), \quad j = 1, 2, 3, \ k = 1, 2, \ldots, \tag{5}$$

where $D^{(j)}(x, y)$ are the columns of the matrix $D(x, y)$.

Theorem 1. *The set* $S = \{f^{(i)}, \theta^{(jk)}\}$, $i, j = 1, 2, 3$, $k = 1, 2, \ldots$, *is linearly independent on* ∂S *and complete in* $L^2(\partial S)$.

From (4) and (5) we see that

$$\int_{\partial S} (\theta^{(jk)})^{\mathrm{T}} \psi \, ds = F_j(x^{(k)}).$$

Orthonormalizing S, we obtain a new complete sequence

$$\{\omega^{(n)}\}_{n=1}^{\infty}, \quad \omega^{(n)} = \sum_{m=1}^{n} k_{nm} \theta^{(m)},$$

where k_{nm} are known numerical coefficients. We now construct the truncations

$$\psi^{(n)} = \sum_{r=1}^{n} p_r \omega^{(r)}, \quad n = 1, 2, \ldots,$$

where

$$p_r = \int_{\partial S} (\omega^{(r)})^{\mathrm{T}} \psi \, ds = \sum_{m=1}^{r} k_{rm} \int_{\partial S} (\theta^{(m)})^{\mathrm{T}} \psi \, ds, \quad r = 1, 2, \ldots,$$

and the corresponding approximations for the solution

$$u^{(n)}(x) = \int_{\partial S} D(x, y) \psi^{(n)}(y) \, ds(y) - F(x), \quad x \in S^+.$$

Theorem 2. $u^{(n)} \to u$ *uniformly on any closed subdomain* $S' \subset S^+$.

4. Computational results

An experiment was conducted, which simulated a boundary movement described by

$$A_1(\theta) = 0.016(1 + \cos 2\theta), \quad A_2(\theta) = -0.016 \sin 2\theta, \quad A_3 = -1.573 \cos \theta - 1.6 \cos 3\theta$$

for a circular steel floor characterized by $\lambda = 1.141 \times 10^6$, $\mu = 8.262 \times 10^5$, $r = 200$, and $h = 5$, with kg and cm as units. For a numerical approximation, we take ∂S_* to be the circle of radius r_* concentric with ∂S and the set $\{x^{(k)}\}$ on ∂S_* to correspond to the polar angles $0, \pi, \frac{1}{2}\pi, \frac{3}{2}\pi, \frac{1}{4}\pi, \frac{3}{4}\pi, \frac{5}{4}\pi, \frac{7}{4}\pi, \ldots$, and use Simpson's rule with 128 strips to evaluate integrals. The table below shows various approximations (corresponding to uniform distributions of the $x^{(k)}$ on ∂S_*) and the measured experimental values of the solution at some points of the disk.

	$(0, 0)$	$(10, 0)$	$(50, \frac{1}{4}\pi)$	$(100, \frac{1}{2}\pi)$	$(150, \frac{3}{4}\pi)$	$(190, \pi)$
	0.0001	0.0002	0.0011	-0.0052	0.0097	0.0338
$u^{(27)}$	0.0000	0.0000	-0.0014	0.0000	0.0100	0.0000
	0.0000	0.0001	0.0057	0.0000	-0.0675	2.7166
	0.0000	0.0001	0.0010	-0.0042	0.0091	0.0352
$u^{(51)}$	0.0000	0.0000	-0.0010	0.0000	0.0092	0.0000
	0.0000	0.0010	0.0057	0.0000	-0.0242	2.6723
	0.0000	0.0001	0.0010	-0.0040	0.0090	0.0362
$u^{(99)}$	0.0000	0.0000	-0.0010	0.0000	0.0090	0.0000
	0.0000	0.0010	0.0049	0.0000	-0.0127	2.6811
	0.0000	0.0001	-0.0010	-0.0040	0.0090	0.0433
$u^{(\exp)}$	0.0000	0.0000	-0.0010	0.0000	0.0090	0.0000
	0.0000	0.0009	0.0048	0.0000	-0.0143	2.7179

It can be seen that $u^{(51)}$ offers a reasonable approximation. The values of $u^{(51)}$ remain stable for r_* between about 250 and 1000, but vary sharply as r_* decreases towards 200, or as it increases beyond 1000. Loss of accuracy also occurs as we approach the boundary ∂S, where the kernels of the integral operators involved in the solution are singular. To maintain the accuracy of the approximation near ∂S, we need to compute higher terms in the sequence $u^{(n)}$.

The figure below illustrates the profile of $u_1^{(\exp)}(x_1, 0)$ and the values of $u^{(51)}$ for $r = 250$.

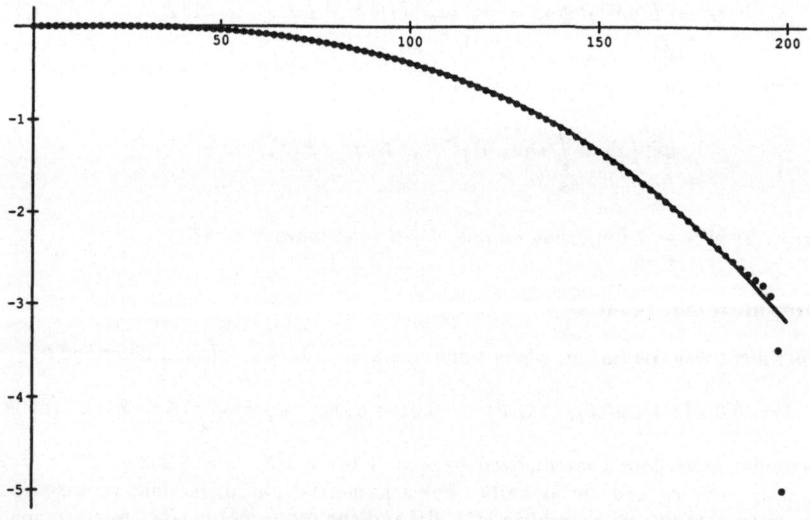

References

[1] Constanda, C. (1990) *A Mathematical Analysis of Bending of Plates with Transverse Shear Deformation*, Longman Scientific & Technical, Harlow, Essex.

THREE DIMENSIONAL CONDUCTIVE HEAT TRANSFER FROM DISCRETE PERIODIC SOURCES TO A CONVECTIVE SINK

M.R.D. DAVIES, J. PUNCH AND P. RODGERS
Department of Mechanical Engineering
University of Limerick
Limerick
Ireland

ABSTRACT. Spatially periodic or quasi-periodic heat sources mounted on a laminate are often found in electronic equipment. Power transistors mounted on a cold plate and thermo-electric cooler cells are prominent examples. The solution presented here models these sources as a periodic temperature wave on one surface of a laminate; the other surface is open to convection and the convective heat transfer is assumed to be constant. An equation describing the temperature field is derived which leads to an exact solution for the heat transfer rate. Implications of the solution are discussed with reference to the design of electronic circuit boards.

NOMENCLATURE.

A,B	Factors in general solution	h	Heat transfer coefficient
a,c	One-quarter periods	K	Thermal conductivity of substrate
b,d	One-half source length	Q	Heat flux
C_{mn},D_{mn}	Factors in general solution	T	Temperature
H	Substrate thickness	x,y,z	Cartesian coordinates

1. INTRODUCTION

Spatially periodic or quasi-periodic heat sources dissipating heat by conduction through a substrate to a convective sink are commonly found in electronic equipment. The elements of a thermo-electric cell and power transistors on a cold plate are examples shown in figure 1.

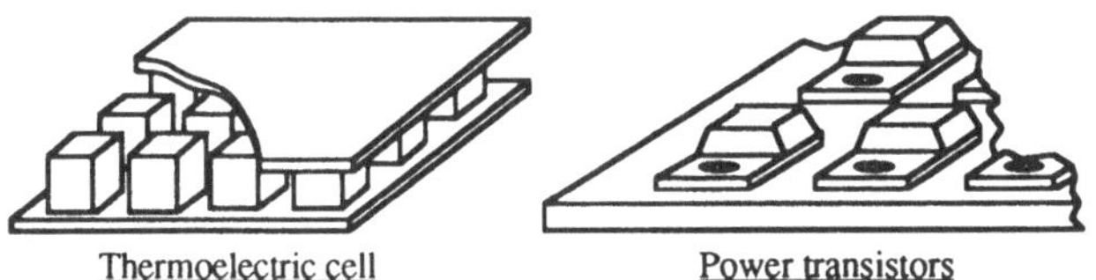

Figure 1.   Examples of discrete periodic sources.

Analytical models of similar substrates have been developed by various authors, reviews of the published works being provided by Pinto and Mikic (1986) and Ellison (1990). All of the models rely upon computer programs for implementation. Parametric studies are possible only by numerical experimentation on the computer and not directly by observation from the analysis.

The solution presented here employs a double Fourier cosine series to provide a continuous expression for the Dirichlet boundary condition imposed by the discrete sources. One layer of a material with isotropic thermal conductivity is modelled. A Robin condition of constant heat transfer coefficient and fluid temperature acts on the non-source side. The influence on the heat flux of the dimensions, the thermal conductivity and the convective sink conditions is illustrated.

2. THEORY

2.1 TEMPERATURE DISTRIBUTION

Surfaces between periods are adiabatic. The simplified domain is shown in figure 2.

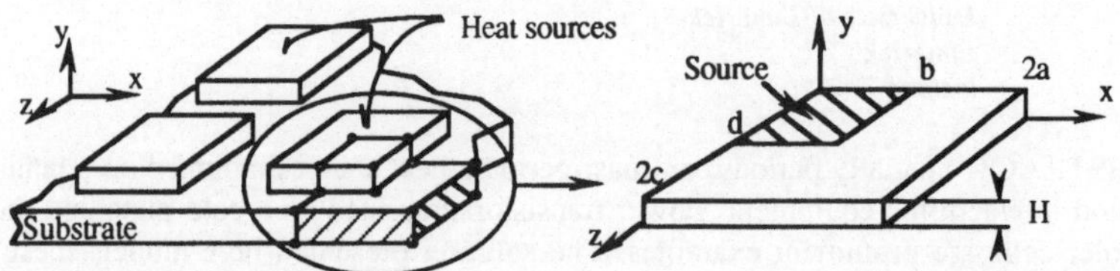

Figure 2.   Application of spatial periodicity to the domain.

The temperature field, with no internal heat generation, is governed by Laplace's equation:

$$\nabla^2 T(x,y,z) = 0 \tag{1}$$

Spatial periodicity implies zero temperature gradients at the edges of the substrate.

$$\left.\frac{\partial T(x,y,z)}{\partial x}\right|_{x\,=\,0,2a} = \left.\frac{\partial T(x,y,z)}{\partial y}\right|_{y\,=\,0,2c} = 0 \tag{2}$$

A Dirichlet boundary condition is imposed on the source side, the temperatures shown in figure 3.

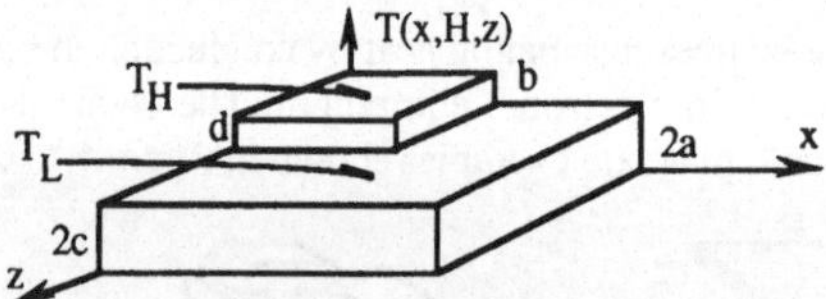

Figure 3.   Source side boundary condition.

Using the equations listed by Carslaw and Jaeger (1959), a continuous representation of the source side temperature distribution that fulfils the edge boundary conditions may be found:

$$T(x,H,z) = T_L + \left(\frac{T_H - T_L}{4}\right)\left[\frac{16}{\pi^2}\sum_{m=0}^{\infty}\sum_{n=0}^{\infty}\left(\left(\frac{1}{m}\right)\text{Sin}\left(\lambda_{m0}b\right)\text{Cos}\left(\lambda_{m0}x\right)\right).\right.$$
$$\left.\left(\left(\frac{1}{n}\right)\text{Sin}\left(\lambda_{0n}d\right)\text{Cos}\left(\lambda_{0n}z\right)\right)\right] \tag{3}$$

The wave numbers of the double Fourier series are given as:

$$\lambda_{mn}^2 = \left(\frac{m\pi}{2a}\right)^2 + \left(\frac{n\pi}{2c}\right)^2 \tag{4}$$

Noting that:

$$\text{Lim}_{m \to 0}\left(\frac{\text{Sin }(\lambda_{m0}b)}{m}\right)\left(\frac{\text{Sin }(\lambda_{0n}d)}{n}\right) = \frac{\pi b}{2a}\left(\frac{\text{Sin }(\lambda_{0n}d)}{n}\right) \tag{5}$$

A Robin boundary condition describes the non-source side. Equating the heat conducted across the surface, given by Fourier's law, to the heat convected away by Newtonian cooling:

$$\int_0^{2c}\int_0^{2a} -K \left.\frac{\partial T(x,y,z)}{\partial y}\right|_{y=0} dx\, dz = \int_0^{2c}\int_0^{2a} -h\left[\, T(x,0,z) - T_\infty\right] dx\, dz \tag{6}$$

Separation of variables yields a general solution which satisfies the Laplace equation and incorporates the double Fourier cosine representation of the source side boundary condition. A linear decay term, $(A.y + B)$, is included to permit a more general solution:

$$T(x,y,z) = (A.y + B) + \sum_{m=0}^{\infty}\sum_{n=0}^{\infty} C_{mn}\text{Cos }(\lambda_{m0}x)\,\text{Cos }(\lambda_{0n}z)\left[e^{\lambda_{mn}y} + D_{mn}e^{-\lambda_{mn}y}\right] \tag{7}$$

The temperature distribution in the substrate is:

$$T(x,y,z) - T_\infty = \left(\frac{K + hy}{K + hH}\right)\left[\left\{T_L + \left(\frac{T_H - T_L}{4}\right)\left(\frac{bd}{ac}\right)\right\} - T_\infty\right]$$
$$+ \left(\frac{T_H - T_L}{4}\right)\frac{16}{\pi^2}\sum_{m=0}^{\infty}\sum_{n=0}^{\infty}\left(\left(\frac{1}{m}\right)\text{Sin }(\lambda_{m0}b)\,\text{Cos }(\lambda_{m0}x)\right).$$
$$\left(\left(\frac{1}{n}\right)\text{Sin }(\lambda_{0n}d)\,\text{Cos }(\lambda_{0n}z)\right).$$
$$\left[\frac{K\lambda_{mn}\text{Cosh }(\lambda_{mn}y) + h\,\text{Sinh }(\lambda_{mn}y)}{K\lambda_{mn}\text{Cosh }(\lambda_{mn}H) + h\,\text{Sinh }(\lambda_{mn}H)}\right] \tag{8}$$

The Fourier expression for $m = n = 0$ is not considered.

2.2 HEAT FLUX

The heat flux passing through the substrate is the conduction through *any* plane of constant y.

$$Q \cdot \frac{1}{4ac}\left(\frac{H}{K} + \frac{1}{h}\right) = \left[\left\{T_L + \left(\frac{T_H - T_L}{4}\right)\left(\frac{bd}{ac}\right)\right\} - T_\infty\right] \tag{9}$$

The heat flux is proportional to a temperature difference, the constant of proportionality being the series combination of the one-dimensional conductive and convective thermal resistances shown in figure 4. The driving temperature is the *area* average temperature of the source side.

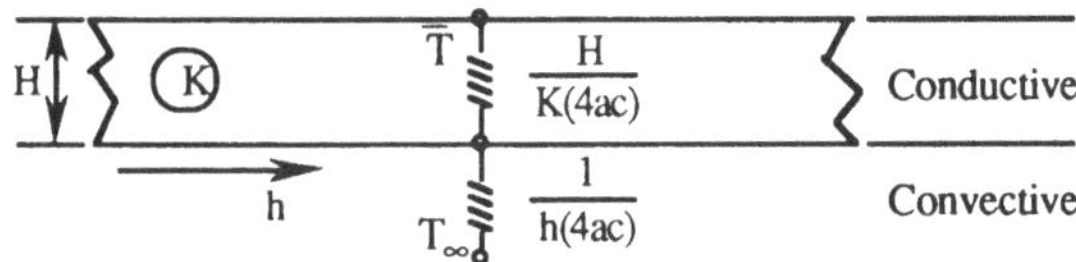

Figure 4. One-dimensional thermal resistances.

3. DISCUSSION

A simple analytical expression has been developed which describes the heat conducted from discrete periodic sources through a substrate to a convective sink. In the chief practical application of this, electronic components mounted on a circuit board, the designer may use the critical junction temperature, the junction-to-casing thermal resistance and the power dissipation of the components to determine the casing temperatures of the components and hence the temperature distribution acting on the source side of the circuit board. Heat flux through the board is proportional to the area average temperature of the source side, a temperature easily obtained from this distribution or measured on an actual circuit board by experimental methods such as infra-red thermometry. The designer may check if a particular combination of board and component dimensions, material properties and convective conditions is sufficient to dissipate the power without the junction temperatures of the components exceeding the critical values. However, it must be noted that the form of solution presented here is more suitable for the *measurement* of heat flux through actual circuit boards than *prediction* during development.

Heat flux is linearly related to the temperature difference between source and sink by the series combination of the one-dimensional conductive and convective thermal resistances shown in figure 4. Any reduction of this series resistance improves the heat transfer through the substrate for a *particular* source side temperature distribution, or, for a particular heat flux, reduces the area average source side temperature. The series resistance may be reduced by increasing the area for heat transfer, reducing the substrate thickness, using a material of higher thermal conductivity or by improving the heat transfer coefficient. The introduction of a heat spreading layer of thermally-conducting material beneath the sources increases the area of the source side at the component temperature and so, for a constant heat flux, reduces the area average temperature. It must be noted that whilst reducing the thickness of a substrate reduces the area average temperature of the source side, *peak* temperatures may increase due to poor heat spreading within the thin substrate. In any case, substrate and source dimensions are usually fixed for a particular application and so any improvements must be obtained in the thermal conductivity or heat transfer coefficient. For most practical applications the internal conductive resistance is usually very much less than the external convective resistance and therefore the largest gains may be achieved by improving the convective heat transfer.

ACKNOWLEDGEMENTS

The Authors gratefully acknowledge the support of Eolas.

REFERENCES

Carslaw, H. S. and Jaeger, J.C. (1959) Conduction of Heat in Solids, Clarendon, Oxford, pp. 180-182.

Ellison, G.N. (1990) 'Methodologies for Thermal Analysis of Electronic Components and Systems', to be published in A. Bar-Cohen and A.D. Kraus (eds.), Advances in Thermal Modeling of Electronic Components and Systems, Vol. III, Hemisphere, New York.

Pinto, E.J. and Mikic, B.B. (1986) 'Temperature Prediction on Substrates and Integrated Circuit Chips', AIAA/ASME 4th Thermophysics and Heat Transfer Conference, HTD, Vol. 57, pp. 199-207.

MATHEMATICAL MODELS FOR ESPRESSO COFFEE PREPARATION

A. FASANO, M. PRIMICERIO, G. BALDINI
Department of Mathematics
University of Firenze
Viale Morgagni 67/A, 50134 Firenze
Italy

ABSTRACT.   Mathematical models are proposed for some basic processes
occurring in the preparation of "espresso" coffee. A good agreement is found with
the experimental data obtained at the laboratories of Illy-Caffé, Trieste, Italy.

1. Introduction

This is a report on a work in progress, carried out in the framework of a
research contract between ECMI and the company illycaffè, based in Trieste, Italy.

The company has designed a compact espresso-coffee machine electrically
operated and using suitably prepared "cakes" of ground coffee. They consist in 6.94
grams of ground coffee, packed and pressed between two sheets of paper sealed in
an almost cylindrical form of 2.2 cm radius and 0.6 cm height.

In order to achieve the best quality of the drink it is of great importance to
combine such elements as the article size distribution in the ground coffee and the
operating values of water pressure and temperature in an appropriate way [1]. This
is the ultimate goal of the present research, but the first step consists in modelling
the basic aspects of the filtration process and in identifying some relevant
parameters.

There are two main processes occurring in the preparation of coffee: (i)
filtration of liquid through the bed, and (ii) extraction of aromatic substances from
the powder. Clearly the two processes are intimately entangled: extraction modifies
the properties of the porous medium (porosity, permeability) and of the liquid
(density, viscosity), whereas it is influenced by filtration since the extraction rate
depends on the flow.

2. The basic processes

In order to investigate the filtration process reducing its coupling with
extraction, we performed filtration experiments at low temperature (4°C). The
discharge Q (i.e. the mass of liquid percolating per unit time and per unit surface)
at constant pressure gradient has been measured. The experimental values
correspond to a law of the type

$$Q = c_1 + c_2 \, e^{-c_3 t} .$$

(1)

We found that this behaviour can be ascribed to the deformation of the porous medium during the filtration which can be taken into account assuming, instead of Darcy's law, the following relationship between the pressure gradient ∇p and the local discharge q

$$|\nabla p| \;=\; a\,q + c\,\frac{\partial q}{\partial t}\,. \tag{2}$$

Laws of type (2) have been considered in the literature to study relaxation phenomena [2].

Coffee preparation consists in the extraction of several aromatic substances present in the powder [4]. Just to sketch a first qualitative model, we consider just one of these extraction processes.

Assuming the phenomenon to be one-dimensional, x being the coordinate parallel to the motion of the liquid, we denote by $r(x,t)$ the concentration of the given substance in the powder (in grams per grams of the powder) and by $e(x,t)$ its concentration in the liquid.

We consider the following kinetics of the process

$$\frac{\partial r}{\partial t} \;=\; -\alpha(r - r_0)^+ + f(r,e) \tag{3}$$

where r_0 is a threshold value. The term $f(r,e)$ models the adsorption by the grains of the substance in solution. Examples of adsorption laws in chemical processes are given by the Langmuir or by the Freundlich law [3], but it is expected that adsorption plays no role in coffee preparation. The linear form of the extraction kinetics is generally adopted, but a dependence of α on the local discharge $q(x,t)$ is often postulated to take into account emulsification of the substance, induced by the motion. Rescaling r, we will write

$$\frac{\partial r}{\partial t} \;=\; -\alpha\,r\,, \tag{4}$$

$$\alpha \;=\; \lambda_0 + g(q)\,, \tag{5}$$

with $g(0) = 0$ and g monotonically increasing.

3. A simplified approach

In the first attempt to model the process, we assume that extraction does not influence the physical properties of the liquid and of the porous medium.

Therefore we write the mass balance of the liquid and of the aromatic substance in the following form:

$$\frac{\partial}{\partial t}(\epsilon \rho \sigma) + \frac{\partial q}{\partial x} = 0\,, \tag{6}$$

$$\rho\epsilon\frac{\partial e}{\partial t} + q\,\frac{\partial e}{\partial x} + (1-\epsilon)\rho_c\,\frac{\partial r}{\partial t} = 0\,, \tag{7}$$

where σ is the saturation, possibly depending on x and t, while the constants ϵ, ρ, and ρ_c denote the porosity, the density of water and the density of the coffee grains, respectively.

Equations (6) and (7) are to be supplemented by state laws. First of all we assume (2) with a and c constants. Finally, we neglect capillarity assuming that

$$\sigma \in H(p) , \tag{8}$$

where H is the Heaviside graph and $p = 0$ is the saturation pressure coincident with the atmospheric pressure.

Using also (4) we have that in the saturated region $\sigma = 1$ the following equations hold

$$q_x = 0 , \tag{9}$$

$$-p_x = aq + cq_t , \tag{10}$$

$$\rho\epsilon e_t + qe_x + (1-\epsilon)\,\rho_c r_t = 0 , \tag{11}$$

$$r_t = -\alpha(q)\,r . \tag{12}$$

We studied system (9)-(12) in two different cases: (i) assuming that the coffee powder is initially saturated with pure water, and (ii) assuming that the powder is initially dry. In both cases the pressure on the two sides of the cake is given, i.e.

$$p(0,t) = p_0 > 0 , \qquad p(\ell,t) = 0 . \tag{13}$$

In case (i) it is easily found that

$$q = \left(q_0 - \frac{p_0}{\ell a} \right) e^{-(a/c)t} + \frac{p_0}{\ell a} , \qquad t > 0 , \tag{14}$$

where q_0 is the initial "peak" discharge.

Moreover $e(x,t)$ can be found integrating

$$\rho\epsilon e_t + q(t)\,e_x = \alpha(q(t))(1-\epsilon)\rho_c\,r_0 \exp\!\left(-\int_0^t \alpha(q(\tau))\,d\tau \right), \tag{15}$$

with $q(t)$ given by (14) and with the initial and boundary conditions:

$$e(x,0) = 0 , \quad x \in (0,\ell) ; \qquad e(0,t) = 0 , \quad t > 0 . \tag{16}$$

The problem is globally well-posed. In particular the quantity $\gamma(t) = \int_0^t q(\tau)e(\ell,\tau)\,d\tau / \int_0^t q(\tau)\,d\tau$, i.e. the concentration of the substance "in the coffee cup" has been computed numerically and compared with experimental data;

140

the results, as well as the details of the proof of the well-posedness, will appear in a forthcoming paper.

Case (ii) is more complicated because there is a time $(0,\beta)$ in which a free boundary $x = s(t)$ separates the dry region $x > s(t)$ from the saturated part and a singular second order O.D.E. has to be studied.

4. Averaging on space

A simpler version of the model neglects the space dependence of the relevant quantities.

Let $R(t)$ and $E(t)$ denote the concentration of the substance in the coffee powder and in the liquid at time t averaged over the cake.

Mass concentration is expressed by

$$\rho \epsilon \, V \, \dot{E} + \rho_c(1-\epsilon) \, V \, \dot{R} = -Q\Sigma E , \quad t > 0 , \tag{17}$$

where V and Σ are the volume and the cross section area of the cake respectively and $Q(t)$ is the discharge.

Assuming for the extraction a linear law like in (4), (5),

$$\dot{R} = -\lambda(Q) \, R , \tag{18}$$

and assuming for the filtration the form (1), we obtain a linear first order ODE for $E(t)$ which can be analized and computed with standard methods. The quantity $\int Q(\tau)E(\tau) \, d\tau / \int Q(\tau)d\tau$ has been calculated and compared with the measured values of the concentration of the coffee as a function of percolation time. The agreement is surprisingly good and this simplified version of the model has been used for the identification of the parameters.

Acknowledgements. The authors are grateful to Dr. E. Illy and to Dr. M. Petracco of Illy-Caffé S.p.A. for stimulating discussions and for the assistance in the laboratory work.

References

[1] E. ILLY, T. BULLO, *Considération sur le procédé d'extraction*, "Café, Cacao, Thé" 7, 395-399 (1963).

[2] S. IRMAY, *Theoretical models of flows trhough porous media*, RILEM Symp. Transf. Water in Porous Media, Paris (1964).

[3] P. KNABNER, *Mathematische Modelle für den Transport gelöster Stoffe in sorbierenden porösen Medien*, Rep. n. 121 Inst. Math. Univ. Augsburg (1989).

[4] M. PETRACCO, *Physico-chemical and structural characterization of espresso coffee brew*, 13 Colloquium ASIC, Paipa 1989.   ASIC = Association Scientifique Internationale du Cafe, 42, rue Scheffer, Paris.

APPLICATION OF DENAVIT AND HARTENBURG CO-ORDINATE FRAME TO ROBOT KINEMATICS IN A FLEXIBLE MANUFACTURING CELL

M.FINN, R.HOWARD-HILDIGE and M.HILLERY
University of Limerick

ABSTRACT To obtain control path operation the Denavit and Hartenburg co-ordinate frame method was applied to reduce the number of variables associated with each joint of a four axis cylindrical robot. This requires careful development of joint transformations. It was found that alternatively an additional co-ordinate frame may be assigned to overcome imprecise development of forward kinematic transformations. Inverse kinematics may then be developed and related to motion to achieve robot programming in the tool co-ordinate frame.

Introduction
For the Flexible Manufacturing (F.M) system at the University of Limerick, materials handling is to be achieved by a four axis cylindrical robot. Three types of strategy are currently used to control a robot. These are point to point, controlled path and continuous path. Of these, controlled path operation is the only option which permits programming in the tool co-ordinate frame, this requires a kinematic analysis of the robot.

 To move the robot end effector (E.E.) along any axis of the tool co-ordinate frame requires linear interpolation, achieved by storing the recorded points specifying the E.E.path as cartesian co-ordinates and wrist orientation angles i.e. (joint variables). Joint variables are related to the E.E. path using forward kinematics which describes the relative origin position and orientation of successive joint co-ordinate frame which have been previously assigned to each joint of the robot. To reduce the effort in relating a vector in the tool co-ordinate frame, the Denavit and Hartenburg technique [1] can be used.

Denavit and Hartenburg Co-ordinate Frame
Each joint is assumed to comprise of a rotation and a translation on the "z" and "x" axes only. To permit this assumption, the co-ordinate frames must be assigned to the robot joints in a prescribed manner.Fig 1 describes the application of these rules graphically for joint "i" and indicates the variables associated with the joint. The transformation relating these frames is

$$t_{i-1}^{i} = \text{Rot}[x_i].\text{Trans}[x_i].\text{Rot}[z_{i-1}].\text{Trans}[z_{i-1}] \qquad \text{Eq 1}$$

In 4x4 matrice form [2],the first three columns give the orientation of the ith co-ordinate system in relation to the i-1 co-ordinate frame depicted in Fig 1.Translation between the origins of each matrix is described by the fourth column.

Forward Kinematic Equations
Fig 2(a) shows the assignment of the co-ordinate frames to the robot using the Denavit and Hartenburg method. Subsequent development of the joint transformations indicated that assignment of the co-ordinate frames would not give the correct robot transformation. This stemmed from the $_3T^4$ transformation. Element 2,4 (row 2,column 4) describes the position of the origin of the fourth co-ordinate frame as being a distance d_3 from the origin of the third co-ordinate frame along the y_3 axis. This is incorrect as can be seen from fig 2(a), the fourth co-ordinate frame is located a distance d_3 from the third co-ordinate frame along the z_3 axis. To correct this, an additional co-ordinate frame was assigned to the robot structure as shown in Fig 2(b) as the C co-ordinate frame. Table 1 describes the parameters associated with each frame. Adaption of planar robot kinematics [3] avoids the use of the correction frame, the method of assignment of frames to the robot is shown in Fig 4, and table 2 summarises the parameters of each joint .Noting that the result is the same as the result derived by using the correction co-ordinate frame method. In developing the forward kinematics for this robot, attention was confined to the three major axes and the yaw axis of the wrist. The roll axis was not considered because only two angles of roll are permitted which

prevents the interpolation process being applied to this axis. The origin of the fifth co-ordinate frame is positioned at the tool centre point (T.C.P.), to obtain an expression for the T.C.P. in the robot co-ordinate axis, the vector in the tool frame D_5 must be considered to be zero, i.e.

$$[D_0] = [_0T^5].[D_5] \qquad Eq\ 2$$

Where D_0 is the D_5 vector referenced to the robot co-ordinate frame.
Using the translational part of the matrix the forward kinematics can be specified .

Inverse Kinematics
Inverse kinematics positions the end effector at a specific location in the robot co-ordinate frame The process of solving for joint variables involves finding the inverse of the robot transformation matrices, a process made simple as the rotational part of the transformation matrix is orthonormal.The inverse of this matrix is given by transposing the rotational part of the matrix and forming the dot products for the translation column [2]. The inverse kinematics are derived using the following

$$[T^r] = [_0T^1].[_1T^2].[_2T^3].[_3T^4].[_4T^5] = [_0T^5] \qquad Eq\ 3$$

$$[_1T^0].[T^r] = [_1T^2].[_2T^3].[_3T^4].[_4T^5] \qquad Eq\ 4$$

Equating elements [1,4], [2,4] and [3,4] of the resulting matrices produces three equations,

$$D_x.Cos\theta_1 + D_y.Sin\theta_1 = a_5.Cos\theta_4 + a_3 \qquad Eq\ 5$$
$$D_y.Cos\theta_1 - D_x.Sin\theta_1 = a_5.Sin\theta_4 + d_3 \qquad Eq\ 6$$
$$D_z = d_2 \qquad Eq\ 7$$

Parameters d_3 and d_2 can be obtained from Eqs 6 and 7. It can be deduced that four variables must be defined to completely describe the orientation of the robot arm i.e. D_x, D_y, D_z and θ_4.

Kinematics Related to Motion
It is usual for the EE of a robot to follow a straight line path at constant velocity. To achieve this, the relationship between the straight line path velocity and the joint variable velocities and accelerations must be established. This is a non linear relationship, but it must be ascertained to determine if the joint velocities and accelerations exceed their finite values. The velocity of θ_4 will be given as part of the trajectory specification. Given a position "L", related to joint variables " θ" by a function "f", where J is the Jacobian matrix of the robot

$$[\dot{L}] = [J] \times [\dot{\theta}] \qquad Eq\ 8$$

The relationship between cartesian position in the robot co-ordinate frame and joint variables is established as given in Eq 9. Noting that Eq 9 only deals with the translational velocities, rotational velocities of the tool coordinate frame relative to the robot co-ordinate system being ignored.

$$\begin{bmatrix} D_x \\ D_y \\ D_z \\ \theta_4 \end{bmatrix} = \begin{bmatrix} a_5Cos(\theta_1+\theta_4)+a_3Cos\theta_1-d_3Sin\theta_1 \\ a_5Sin(\theta_1+\theta_4)+a_3Sin\theta_1+d_3Cos\theta_1 \\ d_2 \\ \theta_4 \end{bmatrix} \qquad Eq\ 9$$

The joint variables are pre multiplied into this matrix on the right hand side because it is not possible to form a column matrix of the joint variables on the right hand side of a forward kinematic matrix which multiplied together give the cartesian position matrix. The joint velocities can be related to the straight line velocity by manipulating Eq 8,but this is possible only in the case where "J" is non singular.Having established velocity equations, acceleration can be determined by differentiating the velocity equations with respect to time [4].

Robot Programming in the Tool Co-ordinate Frame
If motion is required along the "y" axis of the tool co-ordinate frame, then a number of points along this axis must be transformed into robot cartesian co-ordinates using Eq 2. Obtaining the inverse kinematic solutions of the robot co-ordinates describes the required trajectory in terms of joint variables,thus making it possible to achieve the desired path. Using the inverse kinematic equations requires θ_4 to be specified, to overcome this, it is noted that motion along the "x" or "y" axis of the tool co-ordinate frame demands that

$$\theta_1 + \theta_4 = \gamma \qquad Eq 10$$

if the tool cartesian system is to maintain its orientation with the robot cartesian frame. θ_1 is determined using the following [4].

$$\theta_1 = Tan^{-1}\left[\frac{D_y - a_5 Sin\gamma}{D_x - a_5 Cos\gamma}\right] + Tan^{-1}\left[\frac{\pm\sqrt{(D_x - a_5 Cos\gamma)^2 + (D_y - a_5 Sin\gamma)^2 - a_3^2}}{a_3}\right] Eq 11$$

θ_4 is then determined from Eq 10. The constant γ specifies the orientation of the tool co-ordinate frame to the robot co-ordinate system which is determined from the joint variables on the wrist and robot axes just prior to performing the move.

Conclusions
The Denavit and Hartenburg co-ordinate frame method has been applied to a four axis cylindrical type robot and it has been found that assignment of the co-ordinate frames requires careful analysis of the individual joint transformations to ensure that the correct relationship exists between subsequent co-ordinate frames. Alternatively, an additional co-ordinate frame must be assigned to the robot to provide the correct relationship between the robot and tool co-ordinate frames.
The Denavit and Hartenburg method has been used to develop the robot transformation which relates the robot co-ordinate frame to the tool co-ordinate frame. This permits programming in the tool co-ordinate frame which is achieved by trajectory planning in this co-ordinate frame and converting this path to robot co-ordinates which are subsequently converted to joint variables using inverse kinematics.

References
[1] Denavit, J. and Hartenburg, R.S. "A Kinematic Notation for Lower Pair Mechanisms Based on Matrices" Journal of Applied Mechanics,pp 215-221,(1955)

[2] Ranky,P.G. and Ho,C.Y. "Robot Modeling Control and Application Software" IFS (Publications) Ltd UK (1985)

[3] Wolovich,W.A. "Robotics: Basic Analysis and Design" CBS College Publishing N.Y. (1987)

[4] Finn, M. "M.Tech Thesis"University of Limerick (1991)

144

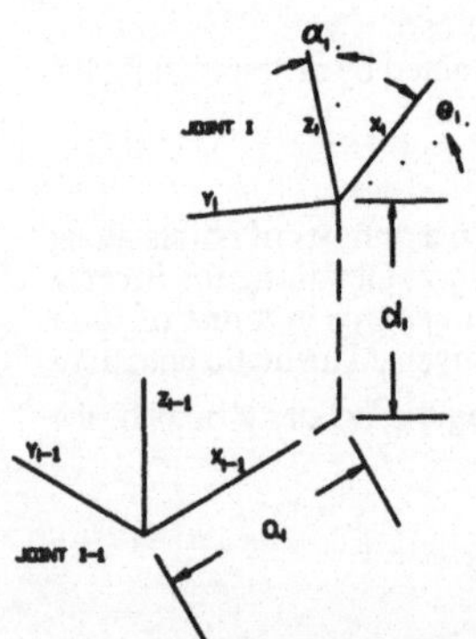

θ_i = Angle of rotation about the +ve z_{i-1} axis which is measured between the x_{i-1} and x_i axis.

α_i = Angle of rotation about the +ve x_i axis, being measured between the z_{i-1} and z_i axes.

d_i = Translation along the z_{i-1} axis, the distance between the x_i and x_{i-1} axes.

a_i = Translation along the x axis, the distance from the i-1 and i co-ordinate frames [3].

Fig 1. Assignment of Denavit and Hartenburg frames.

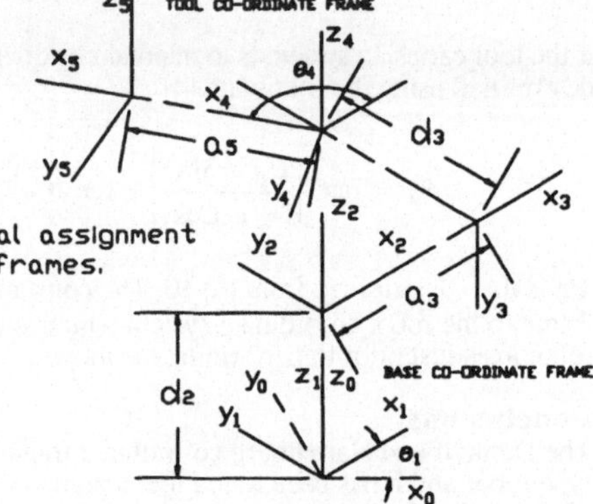

Fig 2(a). Initial assignment of frames.

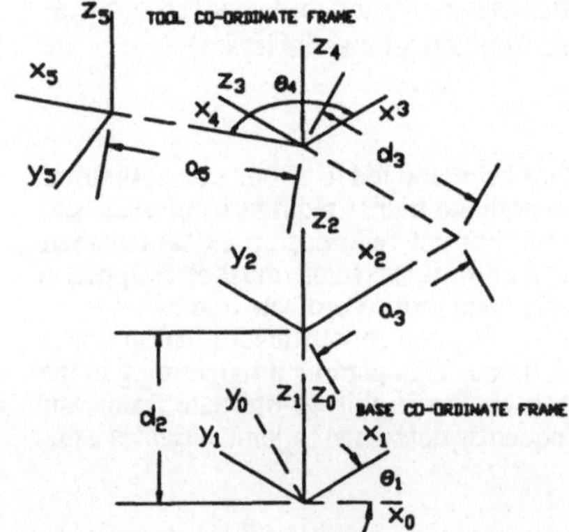

Fig 3. Alternative assignment of frames.

Fig 2(b). Assignment of correction frame.

TABLE 1

Frame i	1	2	3	C	4	5
Variables						
α_i	0	0	-90	0	90	0
θ_i	θ_1	0	0	0	θ_4	0
d_i	0	d2	0	d3	0	0
a_i	0	0	a3	0	0	a5

TABLE 2

Link i	1	2	3	4	5
Variables					
α_i	0	0	-90	90	0
θ_i	θ_1	0	0	θ_4	0
d_i	0	d2	d3	0	0
a_i	0	0	a3	0	a5

HYPERBOLICITY MAPS FOR TWO-PHASE FLOW PROBLEMS

A.D. Fitt

Faculty of Mathematical Studies
University of Southampton
Southampton SO9 5NH, UK

ABSTRACT. The governing equations for gas/particulate two-phase flow with added terms are analysed. Making extensive use of symbolic algebra to perform the lengthy calculations involved, parameter regimes are determined within which the conservation laws are hyperbolic and the problem is therefore well-posed.

1. The Equations of Motion for Gas/Particulate Two-phase Flow

The precise formulation of systems of conservation laws governing multiphase flow is a topic which has long caused controversy. Put plainly, the nub of the problem is as follows : if we proceed using a traditional 'control-volume' approach to formulate the equations for inviscid flow, a method which is known to prove successful for single phase flows, then the matrix A in the resulting system of first order partial differential equations

$$w_t + Aw_x = 0$$

where w is the vector of unknowns, may prove to possess complex eigenvalues. This is contrary to physical intuition and renders the problem ill-posed, so that correct specification of boundary data, computation of stable numerical solutions and accurate wave speed predictions are all impossible. This defect has a simple cause; the dependent flow variables in the equations represent quantities arising from (both ensemble and cross-sectional area) averaging. The averaging procedure is unavoidable if we wish to avoid the hopelessly complicated task of tracking each phase boundary, but frequently is not carried out with sufficient care, leading to the neglect of important physical terms. Drew & Wood (1985) showed how to carry out the required averaging rigourously, and emphasized the need for a careful non-dimensional analysis to decide the dominant terms in the equations of motion for any given two-phase flow regime.

Concentrating in particular on gas/particulate flow where grains of reactive solid are ignited and gasified inside a tube of constant cross-sectional area, the equations of motion may be shown to be (subscripts indicating differentiation, x along the tube axis)

$$(\alpha_1\rho_1)_t + (\alpha_1\rho_1 u_1)_x = \dot{m}$$

$$(\alpha_2\rho_2)_t + (\alpha_2\rho_2 u_2)_x = -\dot{m}$$

$$(\alpha_1\rho_1 u_1)_t + (\alpha_1\rho_1 u_1^2 C_{u1})_x + \alpha_1 p_{1x} = -C_s\rho_1(u_1 - u_2)^2\alpha_{1x}+$$

$$(\alpha_1\phi_T\alpha_2\rho_1(u_1 - u_2)^2)_x + C_{vm}\alpha_2\rho_1[(u_{1t} + u_1 u_{1x}) - (u_{2t} + u_2 u_{2x})] + \dot{m}u_2+$$

$$C_D \alpha_2 \rho_1 (u_1 - u_2)^2 - \alpha_1 \rho_1 g \sin \theta$$

$$(\alpha_2 \rho_2 u_2)_t + (\alpha_2 \rho_2 u_2^2 C_{u2})_x + \alpha_2 p_{1x} = \alpha_2 [C_s \rho_1 (u_1 - u_2)^2]_x +$$

$$(\alpha_2^2 \rho_2 \phi_T (u_1 - u_2)^2)_x - C_{vm} \alpha_2 \rho_1 [(u_{1t} + u_1 u_{1x}) - (u_{2t} + u_2 u_{2x})] -$$

$$(\alpha_2 \rho_1 c_1^2 \alpha_0^2 \left(\frac{1}{\alpha_1} - \frac{1}{\alpha_0} \right))_x - \dot{m} u_2 - C_D \alpha_2 \rho_1 (u_1 - u_2)^2 - \alpha_2 \rho_2 g \sin \theta$$

$$(\alpha_1 \rho_1 (e_1 + u_1^2/2))_t + (\alpha_1 \rho_1 u_1 C_{e1} (e_1 + u_1^2/2))_x + (p_1 u_1 \alpha_1)_x + p_1 \alpha_{1t} =$$

$$C_s \rho_1 (u_1 - u_2)^2 \alpha_{1t} - (d_L \alpha_2 (u_1 - u_2)^3)_x + (\alpha_1 u_1 \phi_T \alpha_2 \rho_1 (u_1 - u_2)^2)_x -$$

$$d_T (\alpha_2 \rho_1 (u_1 - u_2)(e_1 - T_2))_x + u_2 C_{vm} \alpha_2 \rho_1 [(u_{1t} + u_1 u_{1x}) - (u_{2t} + u_2 u_{2x})] +$$

$$C_D \alpha_2 \rho_1 u_2 (u_1 - u_2)^2 - \alpha_1 u_1 \rho_1 g \sin \theta + \dot{m}(e_2 + u_2^2/2)$$

Space permits only the briefest explanation of the terms in these equations; a subscript 1 refers to the gas and 2 to the solid phase, α represents void fraction ($\alpha_1 + \alpha_2 = 1$), whilst u, p, ρ, e and T represent respectively velocity, pressure, density, internal energy and temperature. C_{u1}, C_{u2} and C_{e1} are profile coefficients which arise from the non-commutativity of the operations of averaging and multiplication, $\dot{m}$ characterizes the interfacial mass transfer which arises from the burning, α_0 is the settling porosity at which elastic intergranular waves may propagate at speed c_1, whilst g is the acceleration due to gravity and θ the inclination of the tube to the vertical. C_s, ϕ_T, C_{vm}, C_D, d_L and d_T are coefficients of added terms, representing the effects respectively of interfacial pressure differences, turbulence, virtual mass, interphase drag, laminar interphase heat transfer and turbulent interphase heat transfer. Throughout we assume that since the solid phase is composed of incompressible particles, ρ_2 is constant. The system is closed by selecting the relevant constitutive equation for the gas. In this case the perfect gas law

$$e_1 = \frac{p_1}{\rho_1 (\gamma - 1)}$$

has been used, where γ, assumed to be a known constant, is the ratio of specific heats.

2. Hyperbolicity Maps

Our goal is to examine the effect of different values of the added terms upon the hyperbolicity of the resulting system. In principle, this is easy to undertake; the equations are written as a 5×5 first order system for the unknown quantities p_1, ρ_1, u_1, u_2 and α_1 and the eigenvalues are examined to determine whether the imaginary parts are all zero so that the system is hyperbolic. In practice however the calculations involved are very large and a symbolic algebra system (MAPLE was used in the current study) together with a custom-written package for performing the calculations (See Fitt (1987)) was employed.

When all the added terms are included the eigenvalues are given by the roots of a quintic equation containing 4759 terms, and although some analysis is possible for this case we prefer here to examine only the effect of the interfacial pressure, virtual mass and turbulence coefficients. Accordingly the values $C_{u1} = C_{u2} = C_{e1} = 1$ and $c_1 = d_L = d_T = C_D = 0$ have been used. In this case the eigenvalues are determined by a quintic equation which contains C_{vm}, C_s and ϕ_T as parameters, and depends on $R = \rho_1/\rho_2$, $V^2 = (u_1 - u_2)^2/c^2$, (where $c^2 = \gamma p_1/\rho_1$) and α_1. For fixed R and γ (unless otherwise stated, the physically realistic values of 1/5 and 6/5 respectively were used in all the calculations reported below) hyperbolicity regions for the equations may be examined by plotting α_1 against V^2.

First we consider the case when there are no added terms so that $C_{vm} = C_s = \phi_T = 0$. For these values it may be shown that the equations are hyperbolic only if $\alpha_1 = 0$ or $V = 0$, or if $V^2 > (1 + q^{1/3})^3$ where $q = \alpha_2 \rho_1/(\alpha_1 \rho_2)$. Each of the four figures 1(a) - 1(d) indicates the boundary of this region (denoted L) as a solid line. In this case, just over 32% of the region $D = \{(\alpha_1, V^2) : \alpha_1 \leq 1, 0 \leq V^2 \leq 6\}$ comprises parameter space where the equations are hyperbolic. Moreover, any flows which start from rest must pass through elliptic regions.of parameter space. Figure 1(a) shows the hyperbolicity map for the values $\phi_T = -1/4$ and $C_{vm} = 1/2$ (values previously suggested by diverse authors) and $C_s = -7/16$ (spherical solid particles). The hyperbolicity map is completely changed (hyperbolic regions indicated by H) but still only 38% is hyperbolic.

Some more general conclusions may be drawn by considering the effects of specific parameters; whilst no situations have been observed in which the virtual mass term helps the situation, in general the smaller the value of ϕ_T, the greater the hyperbolic region. Figure 1(b) shows the case $C_{vm} = 0$, $C_s = 0$ and $\phi_T = -5$, when 87% of parameter space is hyperbolic. An even better result may be obtained by the choice $C_{vm} = \phi_T = 0$, $C_s = -1$, when it is possible to show that the equations are hyperbolic everywhere. Figure 1(c), where $C_s = -99/100$ and the other coefficients are zero, shows that this is a somewhat singular limit however; here only 73% of parameter space is hyperbolic.

Finally, the influence of the density ratio R may be examined. Figure 1(d) shows a hyperbolicity map for the case $C_{vm} = 0$, $C_s = -7/16$ and $\phi_T = -1$ so that the particles are spherical, virtual mass effects are ignored and the turbulent effects are fairly strong. The density ratio in this case is $R = 1/100$ (heavy particles) and 56% of parameter space is hyperbolic. For the same values of C_{vm}, C_s and ϕ_T, but a density ratio of 1/5, 68% of parameter space is hyperbolic.

3. Conclusions

Two phase flows are undoubtedly highly complicated phenomena and any attempt to construct mathematical models must reflect this complexity. The diversity of possible flow regimes and qualitative behaviours makes it unrealistic to expect simple models to be successful, and this manifests itself in the non - hyperbolicity of the governing equations. The work outlined above provides a process by which 'incorrect' models may be identified and discarded. However, even if submodels and correlations are used which render the equations totally hyperbolic, the hyperbolicity in itself does not *prove* that the equations are necessarily correct. As in all highly involved modelling problems, careful dimensional analysis, comparison with experiment and shrewd physical intuition are obligatory requirements for successful predictions.

References

Drew, D.A. & Wood, R.T. (1985) 'Overview and Taxonomy of Models and Methods for Workshop on Two-phase Flow Fundamentals' National Bureau of Standards Report, Gaithersburg, Maryland

Fitt, A.D. (1987) 'Symbolic Computation of Hyperbolicity Regions for Systems of Two-Phase Flow Conservation Laws Using Maple', J. Symbolic Computation 8 305-308

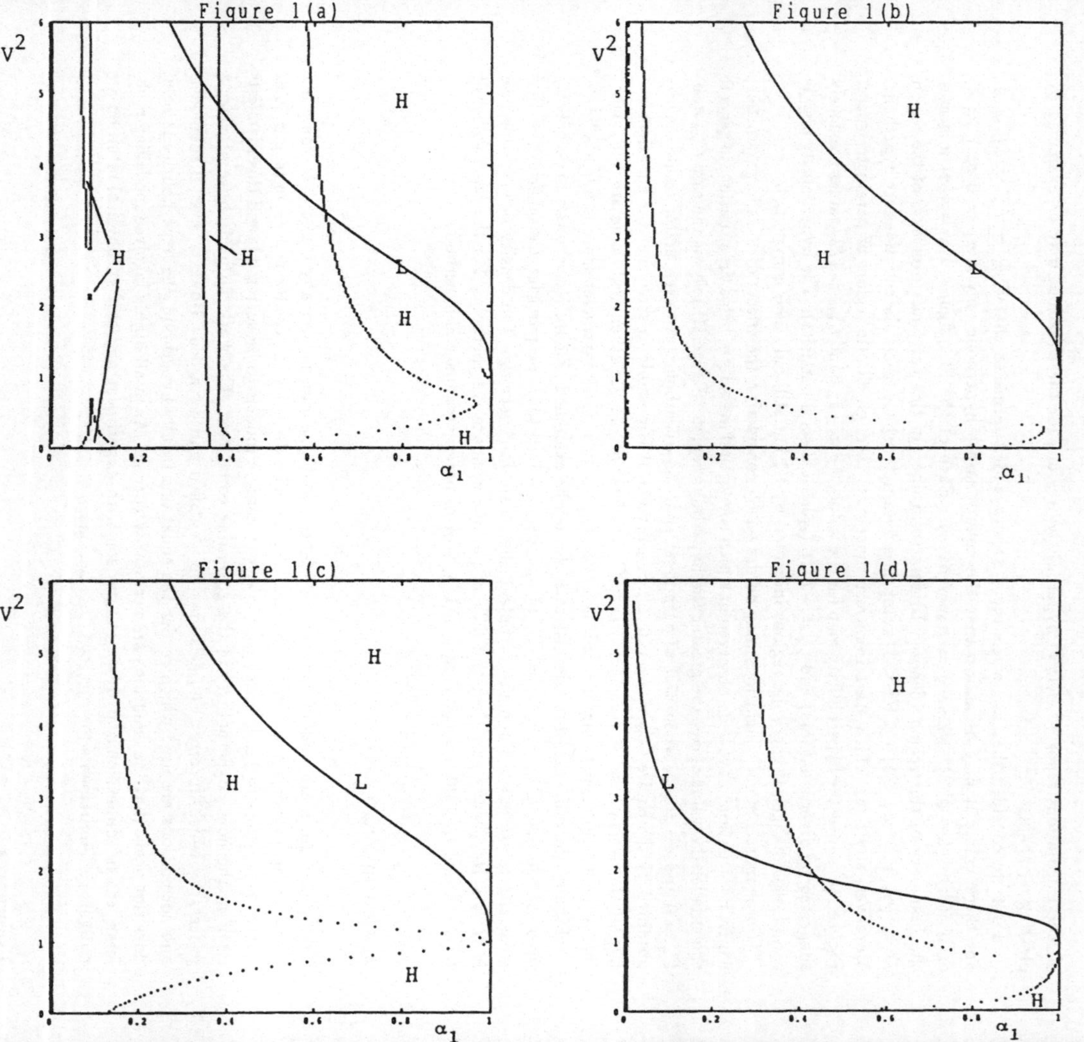
Figure 1(a)
Figure 1(b)
Figure 1(c)
Figure 1(d)

OBSERVATION OF OCEAN WAVES WITH A SYNTHETIC APERTURE RADAR

TONE GAMMELSÆTER
Norsk Hydro a.s
3900 Porsgrunn
Norway

ABSTRACT. The imaging of the ocean surface by a Synthetic Aperture Radar (SAR) is affected by a number of mechanisms; both linear and nonlinear effects. S. and K. Hasselmann has developed a closed integral transform which describes the nonlinear mapping of the surface wave spectrum into a SAR-image spectrum. A simplified derivation of this transform is presented and discussed. The discussion deals with linearizations, series expansion and asymptotic behaviour of the transform. Finally, the transform is used in an example with real data from NORCSEX'88 (Norwegian Continental Shelf Experiment 1988).

1. Introduction

Like a traditional radar (RAR), the Synthetic Aperture Radar (SAR) consists of an antenna, a receiver and a transmitter. The radar transmits electromagnetic pulses which are reflected at the surface and then received by the SAR.

The radar is being moved by an airplane or a satellite. Because of the relative velocity between the radar and the imaged object the radar can 'build up' a large synthetic antenna. By convolution between the RAR-data and the phase-history of the signal caused by this relative velocity, we obtain the final SAR-image.

This article is based on my Master's Thesis (1991), which I finished at The Norwegian Institute of Technology in March 1991. My advisor is H. E. Krogstad.

2. Derivation of the SAR-spectrum

The modelling of the ocean surface is based on the conservation laws for mass and momentum. The ocean surface is modelled as a homogeneous, stochastic field of pointscatterers, each with surface elevation

$$\eta(x,t) = \int_{k,\omega} e^{i(kx - \omega t)} dZ(k,t) \tag{1}$$

k is the wavenumber, ω the angular frequency, t the time and Z is a stochastic field with properties given in Krogstad (1985).

The imaging of the ocean surface is affected by a number of mechanisms, described by Krogstad (1991), Hasselmann et.al. (1985) among others, and is thoroughly discussed in my

Master's Thesis.

Since SAR is a scanning instrument, there will be some scanning distortion due to the motion of the SAR during the scanning. The most important mechanism is the surface tilt; the variation in the reflection at the surface. Combined with the hydrodynamic modulation caused by the short waves that are riding on the longer waves, the tilt modulation is the mechanism used by a traditional radar (RAR) (Krogstad (1991) and Hasselmann et.al. (1985)). When the imaged object has a nonzero velocity, there will be an apparent shift in the location of the object. By combination of the RAR-modulation and the azimuth shift we obtain the velocity bunching transformation which again leads to the SAR-image.

The RAR-image is denoted $I^R(x)$ and may be written

$$I^R(x) = I_0[1+\delta^R(x)] \tag{2}$$

where $I_0 = E[I^R(x)]$. The azimuth shift, Δx, is given by $\Delta x = Ru_r/V$, assuming a point scatterer at a position x_0 is travelling with the velocity u (Fig. 1). The stochastic field of such displacements is denoted $\xi(x) = (R/V)u_r(x)$. A thoroughly derivation of the SAR spectrum is shown in my Thesis, using a conservation law for point scatterers, the Fourier transform and the characterisic function. The derivation results in the following expression for the SAR spectrum (Eqn. 3).

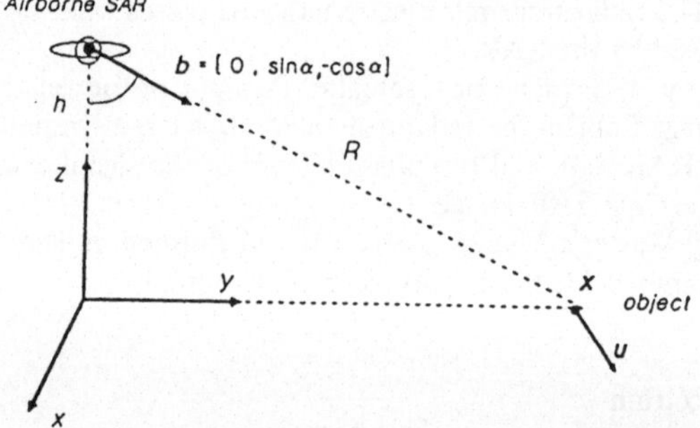

Figure 1. A pointscatterer observed by an air-borne SAR with velocity V.

$$S^V(k) = \int_x e^{ikx}G(x,k_x)d^2x - I_0^2\delta(k) \tag{3}$$

where $G(x,k_x)$ is given by Eqn. 4.

$$G(x,k_x) = I_0^2 \exp[-k_x^2(\rho_{\xi\xi}(0)-\rho_{\xi\xi}(x)]$$
$$*[1+\rho_{II}(x)+ik_x(\rho_{I\xi}(x)-\rho_{I\xi}(-x)) + k_x^2(\rho_{I\xi}(0)-\rho_{I\xi}(x))(\rho_{I\xi}(0)-\rho_{I\xi}(-x))] \tag{4}$$

This is the same expression as obtained by Hasselmann (1990), Hasselmann and Hasselmann (1991) and Krogstad (1991). $\rho_{I\xi}$ is the cross covariance function for I^R and ξ, and ρ_{ii} is the

151

covariance function for i.

We observe that the expression for the spectrum is rather complicated. The integral is a Fourier Stiltjes integral, and the Fourier transform includes the unknown covariance functions.

3. Discussion of the spectrum.

For small $|k_x|$, that is $|k_x|<1$, the SAR-spectrum is similar to the RAR-spectrum;

$$S^V(k) \approx S^R(k) \tag{5}$$

that is; the SAR image is similar to the image made by a traditional radar.

For large $|k_x|$, the spectrum may be approximated by

$$S^V(k) \approx \frac{1}{k_x^2} |A^{-1}|^{1/2} e^{\frac{-(A^{-1})_{11}}{4}} \tag{6}$$

where $\rho_{\xi\xi}$ is assumed to be smooth around $x=0$ and A is a positive definite matrix. This asymptotic behaviour of the SAR spectrum; $S^V(k)\sim 1/k_x^2$, is illustrated in the Thesis (1991).

Linearization of the second part of the exponential function in S^V ($\exp\{k_x^2\rho_{\xi\xi}\} \approx 1 + k_x^2\rho_{\xi\xi}(x)$) and rearranging, yields

$$S_{lin}^V(k) = I_0^2 \left[|T(k_1,\omega_1)|^2 \left|\frac{\partial k_1}{\partial k}\right| \frac{\Psi(k_1)}{2} + |T(k_2,-\omega_2)|^2 \left|\frac{\partial k_2}{\partial k}\right| \frac{\Psi(-k)}{2} \right] \tag{7}$$

where T is the resulting transfer function which combines the tilt-, hydrodynamic- and velocity bunching transfer functions and Ψ is the wavenumber spectrum.

The G-function may be expressed as a series expansion of $\exp(k_x^2\rho_{\xi\xi}(x))$ (the Thesis (1991)), and this expansion may be used in further analytic and numerical analysis of the spectrum. By use of this full series expansion there is found a strong connection between the azimuth wavenumber, k_x, and which terms of the series expansion that have to be calculated for the given k_x. Thus, it is not necessary to to compute the whole sum, but we can choose the terms that gives the largest contribution to the spectrum for a specific wavenumber. Further discussion is made in the Thesis (1991).

4. Exampel

Krogstad and Schyberg (1990) have developed a computer program which computes the SAR spectrum using the series expansion. In Fig. 2, we show the measured and the computed SAR spectrum based on the same directional spectrum. The data is from the Norwegian Continental Shelf Experiment 1988.

We observe that the computed SAR-spectrum is quite similar to the measured spectrum of the ocean waves.

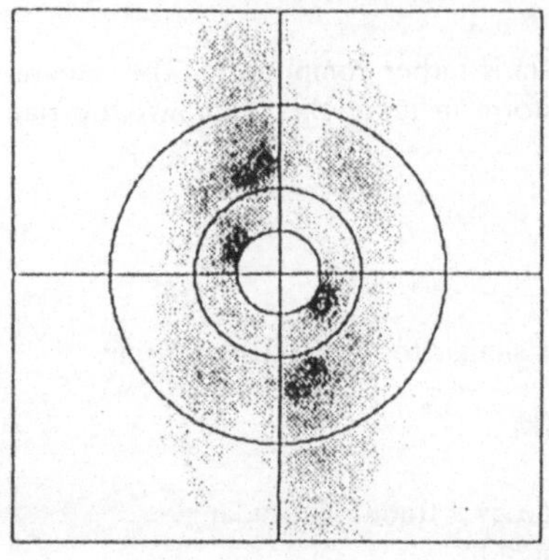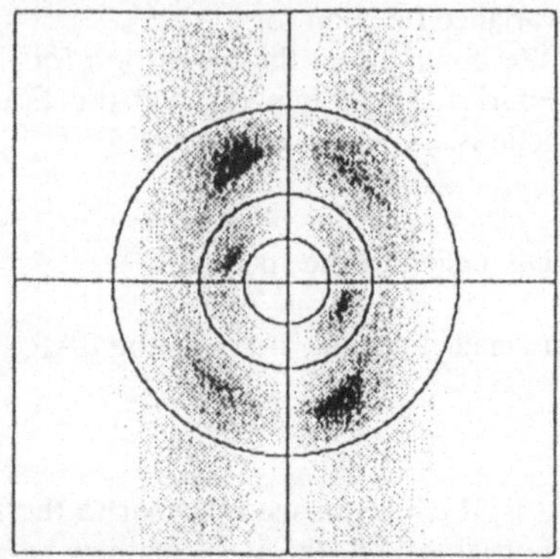

Figure 2. A measured (left) and a computed (right) SAR spectrum based on the same directional spectrum.

5. Conclusions

We have obtained a non-linear transform for the imaging of ocean waves with a Synthetic Aperture Radar, and this transform can help us to better understand the SAR-imaging. This transform includes the non-linear velocity bunching effect on the spectrum, which is rather complicated and therefore difficult to discuss. Through various approximations and simplifications, the spectrum may be discussed analytically, and by applying the series expansion of the transform on real data, we observe that the computed spectrum is similar to the measured spectrum of the ocean waves. To obtain further knowledge about the SAR spectrum, it is necessary with numerical computations.

6. References

Gammelsæter, T. (1991): An Investigation of Hasselmann's Nonlinear Spectral Transform in SAR-imaging of Ocean Waves, Master's Thesis, The Norwegian Institute of Technology, Trondheim (unpublished).

Hasselmann, K et.al. (1985): Theory of Synthetic Aperture Radar Ocean Imaging: A MARSEN View, J. Geophys. Res. 90, no C3, 4659-4686.

Hasselmann, K. et.al. (1990): Application of a New Closed Ocean Wave-SAR Spectral Transformation Relation to LEWEX Data, to appear in APL Digest, Johs. Hopkins University.

Hasselmann, K. and Hasselmann, S. (1991): On the nonlinear mapping of an ocean wave spectrum into a SAR image spectrum and its inversion, J. Geophys. Res.

Krogstad, H. E. (1985): Stochastic Models, Lectures in the course 75048 Mathematical Modelling, NTH, Trondheim.

Krogstad, H. E. and Schyberg, H. (1991): SAR imaging of ocean waves, Report STF A1002, SINTEF, Trondheim.

Krogstad, H. E. and Schyberg, H. (1990): Computer Simulation of SAR Imaging of Ocean Wave Fields, Report OCN-R90004, 066431.00, Oceanor A/S, Trondheim.

ON THE SOLUTION OF THE HYDRO-THERMAL COORDINATION PROBLEM BY DECOMPOSITION

A. Gangl, J. Gutenberger and Hj. Wacker

Institute for Mathematics

University of Linz

A-4040 LINZ/Auhof

1. Introduction

Short-term scheduling in electric power systems requires making two interdependent decisions. The first and integer decision involves determining commitment states (on/off) for the generating units at each hour of the scheduling horizon (1 day - 1 week). The second decision is known as economic dispatch (ED) problem. It involves the allocation of the demand to the on-line units. The arising mixed-integer problem is commonly referred to as unit commitment (UC) for an all-thermal and as hydro-thermal coordination (HTC) for a hydro-thermal system, respectively.

Existing methods for solving real-life systems are based on decomposition, for instance, HTC is decomposed into a hydro scheduling and UC problem and both subproblems are decomposed further. During the last decade Lagrangian Relaxation turned out to be an effective method for generating nearly optimal solutions for UC (cf. [2], [6]). Its structure allows using parallel computing techniques. Investigations are reported in [7]. For a survey of other methods we refer to [3].

However, the Lagrangian Relaxation approaches for solving HTC still have some shortcomings, as they 1) use very simple hydro models, 2) ignore head-dependent hydro generation and 3) do not consider systems with high share of hydro power. Therefore, they do not fit to the Austrian situation.

In this report we focus our attention on the thermal aspects and on the commitment schedule decision of HTC. For details of hydro modelling and ED we refer to [5].

2. The Model

We assume that the stochastic quantities like load and hydro inflows have been forecasted. The resulting deterministic model leads to a large-scale, nonlinear mixed-integer programming problem with the following different sets of variables

- commitment decision variables $u = (u_i^t)$, indicating if thermal unit i is on ($u_i^t = 1$) or off ($u_i^t = 0$) at period t

- thermal dispatch variables $g = (g_i^t)$, describing at what level a unit is generating

- hydro variables x, including storage contents, water releases, etc.

2.1 Constraints

Constraints can be divided into the following three types:

(C1) COUPLING CONSTRAINTS

The sum of the power generated by the thermal units plus the total hydro generation must satisfy the demand

$$\sum_{i=1}^{I} g_i^t + PH^t(x) \geq D^t \qquad t = 1, \ldots, T \tag{1}$$

For security reasons a certain amount of extra power, known as spinning reserves, must be available in order to meet large variations in forecasted load or to cover the loss of a unit

$$\sum_{i=1}^{I} r_i^t(u_i^t, g_i^t) + RH^t(x) \geq R^t \qquad t = 1, \ldots, T \tag{2}$$

where $RH^t(x)$ is the total hydro reserve at period t. The contribution of a thermal unit to reserves is modelled as

$$r_i^t(u_i^t, g_i^t) = u_i^t \min(g_i^{max} - g_i^t, r_i^{max}) \tag{3}$$

(C2) THERMAL UNIT CONSTRAINTS for each individual thermal plant i

The commitment variables $u_i^t \in \{0, 1\}$ are restricted by minimum-up and minimum-down times. For keeping track of a unit's evolution we introduce state variables s_i^t, which record how long a unit has been on or off.

The thermal dispatch variables are subject to generation and ramping limits

$$u_i^t g_i^{min} \leq g_i^t \leq u_i^t g_i^{max} \tag{4}$$

$$-\Delta g_i \leq g_i^t - g_i^{t-1} \leq \Delta g_i + (1 - u_i^{t-1})g_i^{min} \tag{5}$$

The initial status g_i^0, s_i^0 is given.

(C3) HYDRO CONSTRAINTS include limits on storage and flow (discharge, spillage, etc.) variables and water balance equations. For details see [5].

2.2 Cost function

The objective in a hydro-thermal power system is typically that of using the available water resources so as to minimize thermal production costs

$$J(u, g, x) = \sum_{i=1}^{I} \sum_{t=1}^{T} C_i(g_i^t) + S_i(s_i^{t-1}, u_i^t) \tag{6}$$

The cost function consists of two components, namely the fuel costs C_i which are a function of output power and the start-up costs S_i, which depend on the number of hours a unit has been down. The longer this duration the higher the associated costs for heating up. We are now able to state the primal problem

$$\min_{u,x,g} J(u, g, x) \text{ s.t. (C1), (C2) and (C3)} \tag{7}$$

3. Lagrangian Relaxation

Describing Lagrangian Relaxation means to answer 1) how to solve the 'relaxed' problem, 2) how to compute 'good' multipliers and 3) how to use the information provided by the relaxed problem to construct a good feasible solution for the original primal problem.

3.1 Relaxed problem

By adjoining the coupling constraints (C1) onto the objective using multipliers $\lambda^t \geq 0$, $\mu^t \geq 0$ the dual functional decomposes in view of the separable structure into

$$\phi(\lambda, \mu) = \phi_{Th}(\lambda, \mu) + \phi_{Hy}(\lambda, \mu) + \sum_{t=1}^{T} (\lambda^t D^t + \mu^t R^t) \tag{8}$$

where the relaxed subproblems are

$$\phi_{Th}(\lambda, \mu) = \sum_{i=1}^{I} \left(\min_{u_i, g_i} \sum_{t=1}^{T} C_i(g_i^t) + S_i(s_i^{t-1}, u_i^t) - \lambda^t g_i^t - \mu^t r_i^t \right) \tag{9}$$

$$\phi_{Hy}(\lambda, \mu) = \min_{x} \sum_{t=1}^{T} -\lambda^t PH^t(x) - \mu^t RH^t(x) \tag{10}$$

and minimization is understood w.r.t. constraints (C2) and (C3), respectively. To evaluate the dual one has to solve I subproblems involving just a single thermal unit, which can be done by dynamic programming very efficiently. The hydro subproblem is by far more difficult to solve. It is nonconvex and the single hydro plants may be coupled via the water balance equations hydraulically, too. For its solution we propose a decomposition-convexification technique (cf. [4]).

3.2 Duality

It is a well-known result of duality theory that the dual problem provides a lower bound on the optimal value J^* of the primal problem, i.e.

$$\max_{\lambda \geq 0, \mu \geq 0} \phi(\lambda, \mu) = \phi^* \leq J^* \tag{11}$$

Equality is only guaranteed if the primal is a convex problem, which is not the case here due to the discrete variables u and the nonconvex hydro generation. So a strictly positive duality gap $J^* - \phi^*$ is likely. A reasonable criterion for accepting a feasible primal solution $(\bar{u}, \bar{g}, \bar{x})$ is that

$$\frac{J(\bar{u}, \bar{g}, \bar{x}) - \phi^*}{\phi^*} \tag{12}$$

is sufficiently small.

A common approach to solve the nonsmooth dual problem is to use the subgradient method which is a generalization of the steepest descent method. A subgradient of ϕ w.r.t. λ^t, μ^t is easy to compute,

$$\frac{\partial \phi}{\partial \lambda^t}(\lambda, \mu) = D^t - PH^t(x) - \sum_{i=1}^{I} g_i^t, \quad \frac{\partial \phi}{\partial \mu^t}(\lambda, \mu) = R^t - RH^t(x) - \sum_{i=1}^{I} r_i^t(u_i^t, g_i^t) \tag{13}$$

where u, g and x are the solution of the relaxed subproblems at (λ, μ). The subgradient method to update λ, μ is

$$\lambda_{k+1}^t = \max\left(0, \lambda_k^t + \alpha_k \frac{\partial \phi}{\partial \lambda^t}(\lambda_k, \mu_k)\right), \quad \mu_{k+1}^t = \max\left(0, \mu_k^t + \alpha_k \frac{\partial \phi}{\partial \mu^t}(\lambda_k, \mu_k)\right) \quad (14)$$

where α_k is a postive stepsize. The stepsize rule is of crucial importance. To assure convergence the stepsize must satisfy $\alpha_k \to 0$ as $k \to \infty$ and $\sum_{k=1}^{\infty} \alpha_k \to \infty$. However, even a good choice of the stepsize (e.g. Polyak-rule) cannot overcome the slow convergence behaviour of the subgradient method. That is why we have tested some other methods of nondifferentiable optimization. Our numerical expierence favours the use of Shor's method of space dilation in the direction of the difference of two successive subgradients (cf. [8]).

3.3 Construction of feasible primal solutions

As a first attempt we have tried to generate a feasible primal solution at each step of the dual maximization process. Subtracting hydro contribution we have balanced demand and reserves using a heuristic (similar as in [1]) for changing the load without violating (C2). The heuristic has proven to work well for an all-thermal system (relative duality gaps less than 1%). However, the higher the portion of hydro power the worse are the feasible solutions found. The reason for this is the sensitivity of the hydro subproblem with respect to the multipliers. Small changes can cause completely different hydro schedules leading to strong oscillations which destroy the load pattern when hydro contribution is subtracted (instead of shaving the peaks only, new valleys are generated). To avoid these oscillations we have added a penalty term to the objective, which considers the deviation of two successive hydro solutions. This penalty term is also used for convexification of the hydro subproblem (cf. [5]). With this heuristic we have obtained duality gaps ranging between 1% and 2%.

Another more promising heuristic is based on first solving the dual problem to completion and then using the relative probabilities for $u_i^t = 1$ to fix the commitment states. After checking the feasibility of the commitment we finally solve ED as described in [5].

4. Numerical Results

The system we consider is hypothetical and consists of 10 thermal and 7 partially coupled hydro units which should be scheduled over a 24 hours period. After 25 iterations of the space dilation method we obtained a commitment within 0.5% of optimum. The CPU-time on a DEC-3100 is 1 min per iteration on average. About 90% of the time is spent for solving the hydro subproblems. The figure in the appendix shows the allocation of the load among the hydro and thermal subsystem. Hydro power is used to shave the peaks of the demand.

Acknowledgement

This work has been supported by the "Fonds zur Förderung der wissenschaftlichen Forschung" (FWF) under grant P7780.

We are greatly indebted to our supervisor Prof. Hj. Wacker. His death in May 1991 was a great loss to us and our institute.

References

[1] BARD J.F. (1988) 'Short-term scheduling of thermal- electric generators using Lagrangian Relaxation', Operations Research, Vol. 36, No.5, 756-766.

[2] BERTSEKAS D.P. ET AL. (1983) 'Optimal short-term scheduling of large-scale power systems', IEEE Transactions on Automatic Control, AC-28, 1-11.

[3] COHEN A.I. AND SHERKAT V.R. (1987) 'Optimization-based methods for operation scheduling', Proceedings of the IEEE, Vol.75, No.12, 1574-1591.

[4] GUTENBERGER J., GANGL A. AND WACKER HJ. (1991) 'Decomposition method with superlinear convergence for power plant systems', M. Heiliö (ed.), Proceedings of the Fifth ECMI, B.G. Teubner Stuttgart and Kluwer Academic Publishers, 95-98.

[5] GUTENBERGER J., GANGL A. AND WACKER HJ. 'On the scheduling of hydro power plant systems in the frame of the unit commitment problem', P.F. Hodnett (ed.), submitted to the Proceedings of the Sixth ECMI conference at Limerick, B.G. Teubner Stuttgart and Kluwer Academic Publishers.

[6] MERLIN A. AND SANDRIN P. (1988) 'A new method for unit commitment at Electricite de France', IEEE Transactions on Power Apparatus and Systems, PAS-102, 1218-1225.

[7] OLIVEIRA P., MCKEE S. AND COLES C. 'Optimal scheduling of a hydro-thermal power generation system', P.F. Hodnett (ed.), submitted to the Proceedings of the Sixth ECMI conference at Limerick, B.G. Teubner Stuttgart and Kluwer Academic Publishers.

[8] SHOR N.Z. (1969) Minimization methods for nondifferentiable functions, Springer Series in computational Mathematics 3 , Springer, Berlin.

Appendix

HYDRO-THERMAL ALLOCATION

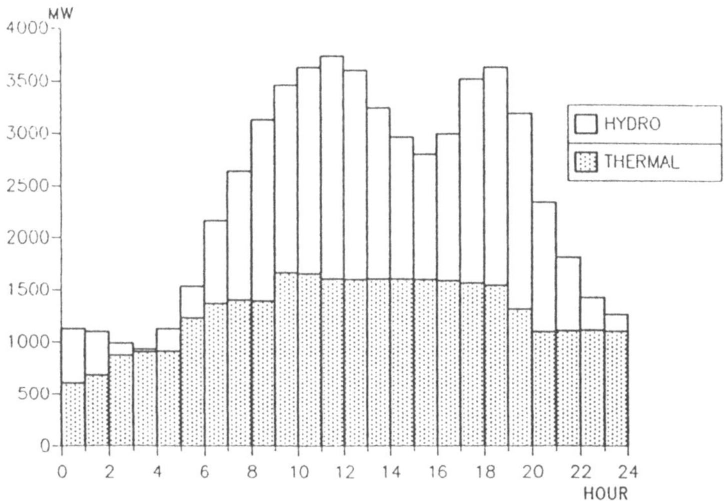

VIBRATIONS OF SPACE STRUCTURES

W.A. GREEN
Department of Mathematical Sciences
Loughborough University of Technology
Loughborough, Leicestershire
LE11 3TU, U.K.

ABSTRACT. The space structure is a circular cylindrical shell, constructed from a diamond mesh lattice skeleton of fibre composite material with a multi-layered outer skin of fibre composite material. This paper is concerned with the effects of the lattice periodicity on the vibrational modes of the structure and attention is directed to the simpler problem of the bending vibrations of a planar reticulated structure. The vibration frequencies of the plane skeleton are evaluated using the Bloch theory for periodic structures and these are compared with the frequencies of a continuum model which replaces the lattice structure by an equivalent plate. For the material constants employed here, it is shown that the continuum model provides an overestimate of the values which can amount to as much as 14%. Results are also presented for the skeleton with cladding, where the discrepancy between the two models is found to be much lower.

1. Introduction

This investigation is concerned with evaluating the effect of structural periodicity on the vibrational behaviour of the JET-X telescope shell. The JET-X telescope is an X-ray monitoring device which is to be launched into space by a consortium of European collaborators. The basic structure of the telescope consists of three circular cylindrical shells, whose lengths vary between 0.9m and 2.35m and whose diameters vary between 0.9m and 1.0m. The sections are assembled end-on to form a tube of overall length 4.3m. Each of the sections is constructed of a diamond mesh lattice skeleton which is clad with a skin of multi-layered fibre composite material. The lattice structure is itself constructed of 20 plies of carbon fibre pre-preg, laid with the fibre direction along the length of the lattice. The outer skin consists of 11 plies for the lower section, 8 plies for the central section and 5 plies for the top section, with the fibre direction in the various layers having specified orientations relative to the cylinder axis. Attached to this basic telescope shell are the X-ray equipment and various ancillary devices, which result in a complex geometrical configuration for the structure as a whole.

It is important that the transient loading experienced under launch conditions does not set up resonant vibrations of the structure and one of the design requirements is to evaluate the natural frequencies. Because of the geometrical complexity, it becomes necessary to adopt a finite element approach to evaluating these natural frequencies and this in turn necessitates modelling the reticulated structure of the shell as a smeared out continuum. The object of this study is to estimate the effect on the frequencies of replacing this periodic lattice by the continuum model.

In order to simplify the problem, attention is here restricted to the bending vibrations of a planar reticulated structure, having the same basic cell as that for the JET-X cylinders. The basic cell may be regarded as a cruciform, consisting of four beams meeting at a central node, as shown in Figure 1. The outer ends of each of the four beams are located at the four adjoining nodes and the overall structure forms a periodic two dimensional configuration. The bending motion of this structure produces a coupling between the bending and the torsional modes of motion of the individual beams which form the basic cruciform.

The influence of the skeleton periodicity is examined in two stages. Section 2 deals with the skeleton on its own and a comparison is given between the predicted frequencies for bending vibrations of flat plate-like structures, using the periodic lattice model and an equivalent continuum plate model. It is shown that, in all cases, the continuum model predicts frequencies that are higher than those predicted by the lattice model and the discrepancy can amount to as much as 14%, in the examples given. The second stage in this examination is dealt with in Section 3, which treats the combination of the skeleton and multi-ply skin. Attention is again focussed on bending vibrations of plane structures and results are obtained using a periodic unit cell model and a continuum plate model. These results relate to the JET-X skeleton structure with a 5-ply skin and to the skeleton structure with an 11-ply skin. It is shown that the effect of the 5-ply skin is to give lower vibration frequencies than those for the skeleton alone. For the 11-ply skin, the frequencies are in some instances lower and in others higher, than those for the unclad skeleton structure.

2. Lattice Skeleton Structure

For a plane lattice structure, the bending modes of vibration and the in-plane modes of vibration are independent and consideration is here restricted to the bending modes only, although the technique is equally applicable to the in-plane vibration modes. The approach which is adopted in this section closely resembles that employed by Anderson [1]. The first step is to derive the general solutions of the equations of bending vibrations and of torsional vibrations of a beam of length L, which constitutes one arm of the basic cruciform cell. These solutions involve six arbitrary constants, which may be expressed in terms of the normal displacement w, the out of plane deflection ψ and the twist θ at each end of the beam. It is then possible to write the shear force Q, the bending moment M and the torsional couple C at each end of the beam, in terms of w, ψ and θ at the two ends. This has the form of the matrix equation,

$$F = BA, \tag{1}$$

where F is the column vector of the stresses and couples, A is the column vector of the displacements and rotations and the elements of the matrix B are functions of the vibration frequency and of the elastic and geometric parameters of the beam. The column vectors F and A are given by

$$F = (Q_a Q_b M_a M_b C_a C_b)^T, \quad A = (w_a w_b \psi_a \psi_b \theta_a \theta_b)^T$$

where T denotes the transpose and suffices a and b relate to the two ends of the beam. The elements of the matrix B are given in reference [2].

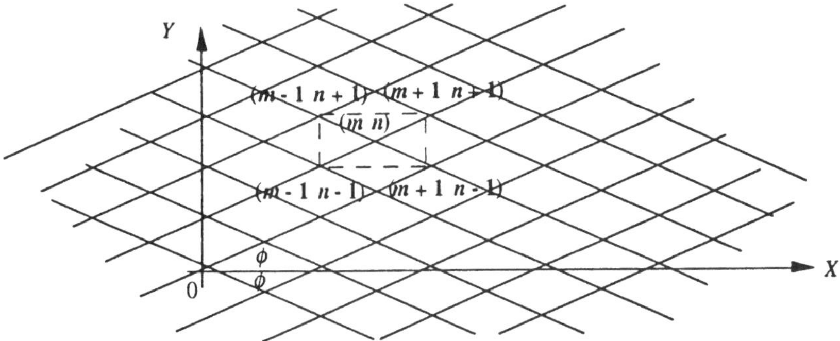

Figure 1. Lattice structure, showing basic cruciform cell and node numbering.

Figure 1 shows, a typical cruciform cell with the node numbering. Using equation (1) for each of the four beams which meet at node (m, n) it is possible to write the shear force, bending moment and torsional couple at the end (m, n) of each beam in terms of the quantities w, ψ and θ at both its ends. At each of the nodes it is useful to define new (global) quantities W, Ψ and Θ, where W is the normal displacement at the node, Ψ is the rotation of the node about the X-axis and Θ is the rotation of the node about the Y-axis. The local displacement and rotations at the end of each beam which meet at a given node are easily expressed in terms of these global nodal values. For example, the quantities w, ψ and θ at the end (m, n) of the beam joining (m, n) to $(m + 1, n + 1)$ are given in terms of W, Ψ and Θ at the node (m, n) by the expressions

$$w = W, \qquad \psi = c\Psi - s\Theta, \qquad \theta = s\Psi + c\Theta$$

where $s = \sin\phi$ and $c = \cos\phi$. Similar expressions hold for each of the other three beams which meet at node (m, n). Thus, it is possible to write the forces and moments at the end (m, n) of each of the beams which meet at this node in terms of the nodal displacements and rotations at each of the five nodes in the cruciform. Writing the conditions for the equilibrium of these forces and moments at the node (m, n) then gives three relations between W, Ψ and Θ at these five nodes. These relations may be written in the matrix form

$$K_0 U_{mn} + K_1 U_{m-1n+1} + K_2 U_{m-1n-1} + K_3 U_{m+1n-1} + K_4 U_{m+1n+1} = 0 \qquad (2)$$

where $U = (W\,\Psi\,\Theta)^T$ and $K_\alpha (\alpha = 0, 1, \ldots, 4)$ are 3×3 matrices. Equations (2) constitute a doubly periodic system of difference equations for the vectors U_{mn}, for which a solution is sought in the form of the doubly periodic vector

$$U_{mn} = L e^{ipm} e^{iqn}$$

where L is some unknown vector, p and q are parameters and $i^2 = -1$. Equation (2) then reduces to the form

$$KL = 0,$$

where the elements of the matrix K are given in reference [2]. The condition for this system to have non-trivial solutions for the vector L is that the determinant of the matrix K should be zero

$$\det K = 0. \tag{3}$$

Equation (3) gives a relation between the parameters p and q and the vibration frequency f, which enters the equation through the elements of matrix B in equation (1). Choosing values of p and q is equivalent to specifying the wavelength of the disturbance in the X and Y directions respectively. When this is done, equation (3) yields the frequency f. This calculation has been carried out for a lattice structure with dimensions equal to those of the JET-X cylinder. Details of the lattice parameters and dimensions together with the results are presented in Table 1.

Reference [2] contains details of the method adopted for replacing the discrete lattice structure by an equivalent continuum plate. This equivalent plate relates specifically to bending behaviour and this model has been employed to calculate vibration frequencies for comparison with those derived by the lattice method. The results are again shown in Table 1.

The two sets of results shown in Table 1 are for $p = q = 0.157$ and the other for $p = 0.0412$ and $q = 0.116$. The first set corresponds to a rectangular plate consisting of ten unit cells in length and ten unit cells in width which, for the dimensions appropriate to the JET-X cell, gives an overall plate length of 1.16m and overall width of 0.58m. The second set of values correspond to a plate of. length 4.42m (38 unit cells) and of width 1.57m (13.5 unit cells). These latter values are chosen so that the length is equal to the overall length of the JET-X composite cylinder and the width is equal to one half of the circumference (diameter 1m). (This choice is consistent with the circumference containing one complete wavelength which is the situation that prevails for bending vibrations of the cylinder). The entries in the table correspond to modes (r, s) where r is the number of half wavelengths in the length of the structure and s the number of half wavelengths in the breadth. For each mode the results for the frequency associated with the lattice structure are derived by solving equation (3) and the table also contains values of the frequency obtained from the expression in reference [2] for the equivalent plate structure.

Both sets of results show that the equivalent continuum plate gives a higher frequency than the lattice in every case for the two examples tabulated. The percentage discrepancy increases markedly with reduction in wavelength along the breadth of the structure (increasing s) but decreases gradually with reduction in wavelength along the length (increasing r).

3. Combined Skin and Lattice Structure.

The considerations in this section relate to the combined structure formed by overlaying the lattice by a multi-ply skin. Here, the approach is first to calculate the bending stiffnesses of the multi-ply skins using the basic data for the prepreg material and adopting the method presented by Jones [3]. The results for the 5-ply, 8-ply and 11-ply skins are presented in reference [2].

Two different methods are adopted in order to model the combined system. In the first of these, the combined system is treated as consisting of a repeating pattern of unit cells, with the basic cell consisting of a single cruciform of the lattice and the associated rectangle of skin. The procedure employed is to calculate the total potential energy of this unit by adding the strain energy of bending and twisting of the beams to the bending energy of the rectangle of skin. Using Rayleigh's principle (see [4]) this total potential energy is equated to the combined kinetic energy of bending and twisting of the beams and the kinetic energy of the skin to yield the vibration frequencies. In order to calculate the bending energy and kinetic energy it is assumed that the rectangular portion of skin undergoes a normal displacement of the form

$$W(X, Y) = W_0 e^{\frac{ipX}{cL}} e^{\frac{iqY}{sL}} e^{2\pi ift}$$

where W_0 is the normal displacement of the central node of the cruciform. This form of displacement is consistent with the periodic displacement assumed for the lattice nodes. This procedure leads to an expression for the frequency which is given in reference [2].

The second method of treating this problem is to replace the lattice by an equivalent continuum plate and to assume that this plate and the skin undergo the same deformation. It is further assumed that the bending moments and shearing forces in the plate and in the skin are generated independently of each other and that they add together to give the resultant motion. This is quite different from simply adding an equivalent thickness to replace the lattice and treating the combined system as a single equivalent plate. The approach used here is based on the assumption that the coupling between the cruciform and the skin serves to constrain the displacements to be equal at the points of contact but otherwise makes little contribution to the deformation energy. It is felt that this most closely represents the actual situation. The resulting expression for the frequency of this overall continuum structure is also presented in reference [2].

These two approaches have been applied to calculate the vibration frequencies for the same two lattice structures as reported in Table 1, carrying out the calculations for a 5-ply skin and for an 11-ply skin, each laid up according to the specification for the JET-X tube. Results for the structure with the 5-ply skin are shown in Table 2 and for the structure with 11-ply skin, the results are presented in Table 3. These show that here also the smoothing approach gives rise to higher frequencies than those derived from the cellular model but the discrepancies never exceed 5% for the first set of results in each table ($p = q = 0.157$) and they are less than 1% for the second set of results ($p = 0.0412, q = 0.116$).

Of more interest is the comparison between the results for the lattice skeleton and for the skeleton with a 5-ply skin and an 11-ply skin. A study of entries 2(a) in each of Tables 1, 2, and 3 shows that the addition of the skin leads to a drop in the frequency of the fundamental mode and of the higher longitudinal modes, showing that the longitudinal bending stiffness of the skin is not sufficient to compensate for the extra mass. For the higher harmonics in the width direction on the other hand, the frequencies for the structure with the 5-ply skin are lower than those for the skeleton. For the structure with the 11-ply skin, the frequencies are higher than those for the skeleton. Thus, the 11-ply skin produces a stiffening effect as regards bending in the width direction whereas the 5-ply skin produces a reduction in the bending stiffness in both directions.

Acknowledgement

This work was carried out in part under a consultancy agreement with Leicester University Space Physics Department and I am grateful to Mr Alan Wells for his support.

164

References

[1] Anderson, M.S. (1982) 'Vibration of Prestressed Periodic Lattice Structures', AIAA Jour. 20, 551-555.
[2] Green. W.A. (1990) Consultancy Report to Leicester University Physics Department.
[3] Jones, R.M. (1975) Mechanics of Composite Material, McGraw-Hill, New York.
[4] Rayleigh, J.W. Strutt, 3rd Baron (1945) The Theory of Sound, Vol 1, Dover Publishing, New York.

Tables

TABLE 1. Vibration frequencies for lattice skeleton using (a) discrete lattice model (b) equivalent plate model. Results 1 for $p = q = 0.157$, results 2 for $p = 0.0412, q = 0.116$.

mode	(1.1)	(1.2)	(1.3)	(1.4)	(2.1)	(3.1)	(4.1)
1(a)	52.07	118.2	215.8	348.0	117.8	214.5	345.5
1(b)	53.25	125.6	235.7	386.6	118.8	216.2	348.8
2(a)	13.45	44.20	95.01	166.0	20.82	30.32	41.64
2(b)	14.62	49.79	108.1	189.6	21.63	30.94	42.18

TABLE 2. Vibration frequencies for lattice with 5-ply skin using (a) periodic cell model (b) equivalent plate model. Results 1 for $p = q = 0.157$, results 2 for $p = 0.0412, q = 0.116$.

mode	(1.1)	(1.2)	(1.3)	(1.4)	(2.1)	(3.1)	(4.1)
1(a)	34.01	79.77	149.2	243.9	74.70	134.6	215.3
1(b)	34.55	82.20	155.3	255.7	77.07	140.5	226.9
2(a)	9.60	32.93	71.62	125.7	14.00	19.88	27.00
2(b)	9.63	33.03	71.86	126.2	14.08	20.06	27.33

TABLE 3. Vibration frequencies for lattice with 11-ply skin using (a) periodic cell model (b) equivalent plate model. Results 1 for $p = q = 0.157$, results 2 for $p = 0.0412, q = 0.116$.

mode	(1.1)	(1.2)	(1.3)	(1.4)	(2.1)	(3.1)	(4.1)
1(a)	35.52	99.51	204.6	351.2	72.43	130.6	210.5
1(b)	35.82	100.7	207.2	356.1	73.85	134.2	217.5
2(a)	12.69	47.48	105.4	186.6	15.74	20.45	26.64
2(b)	12.70	47.52	105.5	186.8	15.79	20.55	26.84

EARTHQUAKE LOADING ON PIPEBRIDGES

MARTIN GREENHOW
Department of Mathematics and Statistics
Brunel University
Uxbridge, UB8 3PH
England

ABSTRACT. We here consider the design loads due to earthquake-induced oscillations of a novel type of tubular bridge/tunnel (called a pipebridge), which contains a road within a tube of diameter 10m, and is submerged below the surface of the water at depths of up to 40m. The highest loads occur for sections near the free surface, which radiate waves caused by the pipebridge motions. We here compare the loads predicted by a very simple added-mass model with those from the numerical solution of the fully nonlinear problem.

1. Added-Mass Model.

When a body accelerates in water, it must also accelerate that water around it, and this can be thought of as an added mass of the body. The presence of other boundaries will also affect the fluid flow, and hence the added mass. In this case we have the free surface relatively close to the cylinder (within a few radii) and this complicates the problem since one should apply nonlinear boundary conditions here. This is done in section 2. However, analytical progress may be made if we restrict the cylinder motions to be small and of high frequency, as will be the case for motions which may arise from earthquakes in the Norwegian fjords (the proposed pipebridge sites). We can then apply a rigid-lid condition on the free surface, i.e. $\Phi = 0$ on $z = 0$. The solution to Laplace's equation for Φ satisfying this condition and $\partial\Phi/\partial n = v_n$ on the cylinder surface has been derived by Greenhow and Li (1987), to give the added mass as

$$\frac{A(z)}{\rho\pi a^2} = \frac{(1-q^2)^2}{q^2}\left[\frac{1}{12} + \frac{1}{3}(1+m_1')\frac{K^2(m_1)}{\pi^2} - \frac{E(m_1)K(m_1)}{\pi^2}\right.$$
$$\left. -\frac{1}{6} - \frac{2}{3}(1+m_2')\frac{K^2(m_2)}{\pi^2} + 2\frac{E(m_2)K(m_2)}{\pi^2}\right] - 1 \tag{1}$$

where z is the distance from the boundary to the cylinder centre, ρ = density, a = radius, E and K are complete elliptic integrals, $m_1+m_1' = 1$, and $q = \exp[-\pi K(m_1')/K(m_1)] = \exp(-\sigma)$ where $z^2-a^2 = a^2\sinh^2(\sigma)$. Also m_1 and m_2 are related by

$$K(m_1')/K(m_1) = K(m_2')/2K(m_2). \tag{2}$$

The added masses are equal for a circular cylinder for both vertical and horizontal modes of motion, and are shown in figure 1. Once these are known, the forces in each of the modes i due to velocity U_i can be calculated as

166

$$F_i = A_{i,i}(z) \frac{\partial U_i}{\partial t}.$$

(3)

These forces are compared with those from the fully nonlinear calculations in figure 6.

2. Numerical Solution of The Nonlinear Boundary-Value Problem.

We consider loading arising from earthquakes, which may exert extreme, or design, loads on the pipebridge. We treat the fully-nonlinear boundary-value problem shown in figure 2, under the assumption of two-dimensional flow around the cylinder in a vertical plane, here called the x-y plane . The method outlined below is due to Vinje and Brevig (1981) and uses the fact that in the fluid region the complex potential $\beta(z,t)=\Phi(x,y,t)+i\Psi(x,y,t)$ is an analytic function of the variable $z = x + iy$. Here Ψ is the stream function. The complex fluid velocity is then $w=d\beta/dz=u-iv$ and so the pressure condition (shown in fig 2) may, after using the transport equation, be written as

$$\frac{D\Phi}{Dt} = \frac{1}{2} ww^* - gy$$

(4)

Here the time derivative is a material derivative because we wish to follow free-surface particles around in space at each time step, i.e a Lagrangian description of the free surface: this automatically then satisfies the kinematic free-surface condition. The required velocities are calculated by numerical differentiation of $d\beta/dz$, where β is determined below.
The problem then may be expressed as an initial-value problem where we specify the free-surface profile and value of Φ, and the body position and motion which specifies Ψ on the body; $\Psi=0$ by definition on the seabed. The situation is shown in figure 3 the contour C being split up into C_Φ and C_Ψ according to the known part of the β. Since β is analytic within this contour, Cauchy's theorem holds, according to which

$$\beta(z_0, t) = \frac{1}{i\pi} \oint_C \frac{\beta(z,t)}{z - z_0} dz$$

(5)

where z_0 lies on any smooth part of the contour C. Numerically the discretised contour is not smooth and we take account of the exact angle α_k between the collocation points, as shown in figure 2. By taking either real or imaginary parts of equation 5 we then obtain equations for the unknown parts of $\beta(z,t)$ on C

$$\alpha_k \Psi(z_k, t) + Re[\int_C \frac{\Phi + i\Psi}{z-z_k} dz] = 0$$

(6)

when z_k is on C_Φ, and

$$\alpha_k \Phi(z_k, t) + Re[i\int_C \frac{\Phi + i\Psi}{z - z_k} dz] = 0 \qquad (7)$$

when z_k is on C_{γ}.
Equations 6 and 7 may be discretised around the contour to give a matrix equation for the unknown part of β and this then specifies the solution at t=0. To step forward in time we need to solve identical equations but with β replaced by dβ/dt which is also analytic within the fluid domain. The required value of dΦ/dt on the free surface is given from the free-surface dynamic condition, while dΨ/dt on the body is determined by the body accelerations (which are obtained by integrating the hydrodynamic pressures over the body and solving the body equation of motion). The above technique has been used to study wave development through overturning until jet touch down (Vinje and Brevig (1980)) and nonlinear motions of surface-piercing bodies (Greenhow (1987)) for which the intersection of free and body surfaces poses special problems. For the submerged body considered here we avoid this problem, but we do need to account for the branch cut shown in figure 3. Full details of its treatment are given in Brevig, Greenhow and Vinje (1981).
We use the above technique to look at the loading due to earthquake-induced oscillations of the pipebridge. The design specification for such oscillations of offshore structures in the North Sea is an amplitude of 3m at a period of 2s. The results, examples of which is shown in figures 4, 5 and 6 for horizontal and vertical motions, demonstrate that free-surface nonlinearity is relatively unimportant for these cases as far as the total forces on the cylinder are concerned. We conclude that the simple added mass model above is adaquate for design load calculations during earthquakes - a fact which may be of some importance for oil rig design.

3. References

1) Brevig P, Greenhow M and Vinje T. (1981) " Extreme wave forces on submerged cylinders" Paper E2 BHRA 2nd Int Symp Wave and Tidal Energy, Cambridge, pp 143-166.
2) Greenhow M (1987) "Wedge entry into initially calm water" Applied Ocean Res. Vol. 9 no. 4 pp 214-223.
3) Greenhow M and Li Y (1987) "Added masses for circular cylinders near or penetrating fluid boundaries - Review, extension and application to water-entry, -exit and slamming" Ocean Engng Vol 14 No 4 pp 325-248.
4) Vinje T and Brevig P (1980) " Breaking waves on finite water depths. A numerical study." SIS report, Div Marine Hydrodynamics, Norwegian Institute of Technology, Trondheim, Norway.
5) Vinje T and Brevig P (1981) "Nonlinear, two-dimensional ship motions" Report R-112.81 Dept Marine Tech, Norwegian Institute of Technology, Trondheim, Norway.

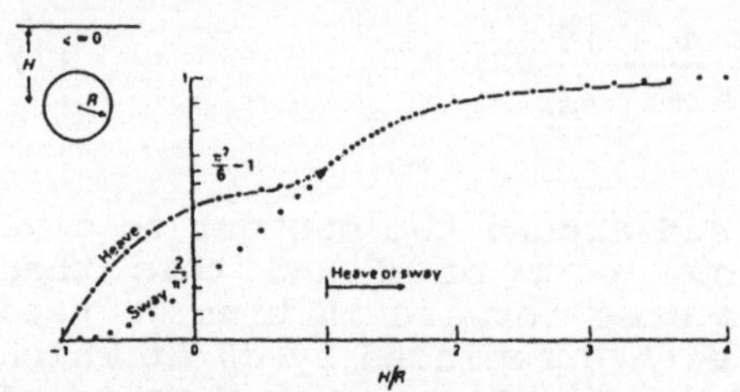

Fig 1. Variation of added mass near a boundary with $\Phi = 0$.

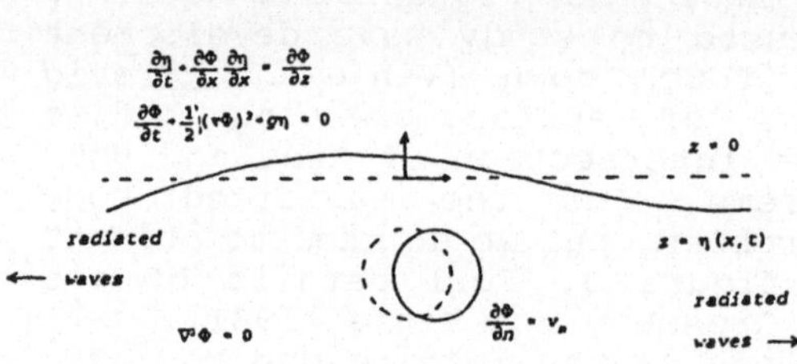

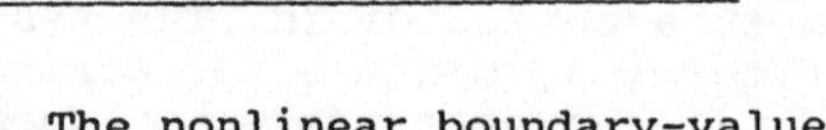

Fig 2. The nonlinear boundary-value problem.

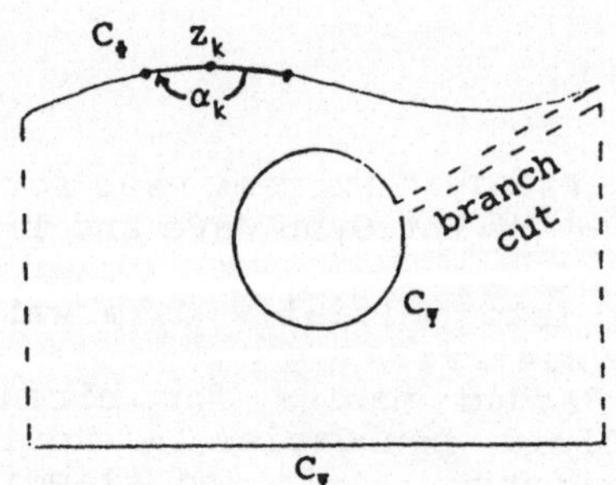

Fig 3. Contour of integration showing branch cut.

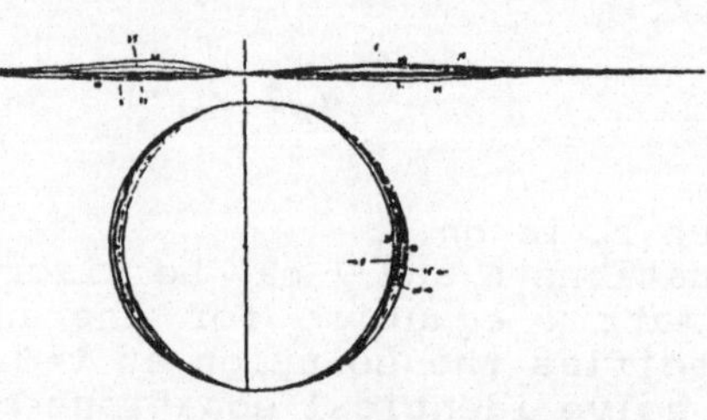

Fig 4. Free-surface deformation - horizontal motion.

Fig 5. Free-surface deformation - vertical motion.

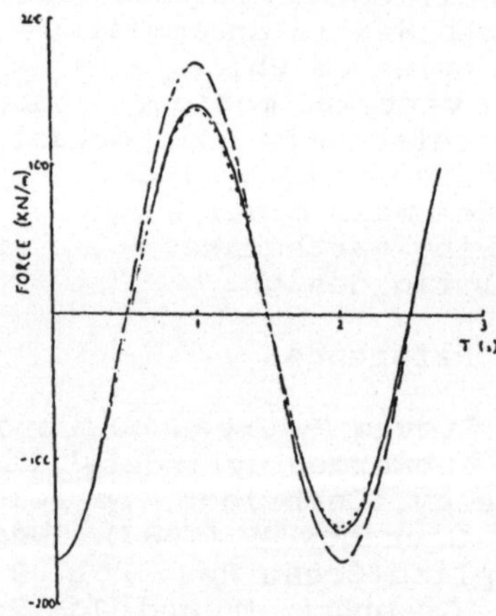

Fig 6. Forces due to horizontal (solid) and vertical (dashed) motions, compared with the added mass model (chain).

NUMERICAL TREATMENT OF A STOCHASTIC PROGRAMMING MODEL FOR OPTIMAL POWER DISPATCH

N. Gröwe and W. Römisch
Humboldt-Universität zu Berlin
Fachbereich Mathematik
D-O-1086 Berlin, Germany

ABSTRACT. The economic dispatch of electric power with uncertain demand is modelled as a stochastic program with simple recourse. The unknown distribution functions of the demand are approximated by smooth nonparametric estimates. We discuss the numerical treatment of the model and report on computational results.

1. The stochastic power dispatch model

We consider the optimal scheduling of power generation to the units of an energy production system, aiming at minimizing the total generation costs while the power demand is met and certain operational constraints of the system are satisfied. This process is often divided into a three-stage procedure: (1) unit commitment for the base-load-plants, (2) power dispatch for the operating cycle and (3) short-term load dispatch for the base-load-plants (optional). The peculiarities of the energy production system and of the power dispatch model we shall consider are the following:

(a) The system consists of thermal power stations *(tps)*, pumped hydro storage plants *(psp)* and an energy contract with connected systems,
(b) *tps* and *psp* serve as base-and peak-load plants, respectively,
(c) the model is designed for a daily operating cycle and assumes that a unit commitment stage has been carried out before,
(d) the cost functions of the *tps* are taken to be strictly convex and quadratic,
(e) the transmission losses are modelled by means of adjusted portions of the demand,
(f) the model takes explicit account of the uncertainty of the electric power demand.

The mathematical formulation of the optimal power dispatch model reads

$$\min\{g(x) : x \in C, Ax = z\}. \tag{1.1}$$

Here, the components of x are the outputs (inputs) of the *tps* and *psp* at each interval of a discretization of the time-horizon and the levels of electric power which correspond to the energy contract at each time interval. The total generation cost function g is convex quadratic, the portion which corresponds to the contract is a linear function. The set C is a nonempty bounded convex polyhedron formed by the operational constraints of the system, e.g. bounds for the power output of the plants, balances between generation and pumping

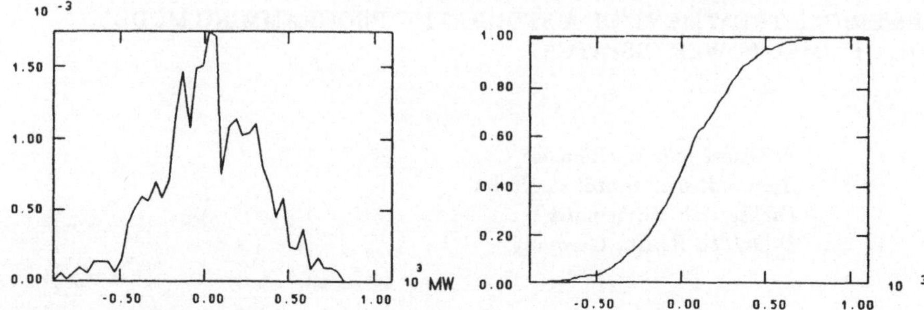

Figures 1 and 2. Estimates for the density and distribution function

in the *psp*, balances for the *psp* over the whole time-horizon, fuel quotas for the *tps* etc. At this instant the equation $Ax = z$ in (1.1) means that the total generated output $[Ax]_r$ meets the electricity demand $z_r, r = 1, \ldots, N$, at each time interval. A detailed description of the model can be found in [2] and [6].

Now, we consider the demand z as a random vector and denote by F_r the probability distribution function of $z_r, r = 1, \ldots, N$. As in [1], [2] we introduce penalty costs for the deviation of the scheduled output from the actual demand for under- and overdispatching, respectively. Adding the expected penalty (or recourse) costs to the deterministic objective function g we obtain the following stochastic power dispatch model:

$$\min\left\{g(x) + \sum_{r=1}^{N} \int_{-\infty}^{+\infty} Q_r(t - \chi_r)dF_r(t) : x \in C, Ax = \chi\right\}, \qquad (1.2)$$

where

$$Q_r(t) := \begin{cases} q_r^+ t, & t \geq 0 \\ -q_r^- t, & t < 0 \end{cases}$$

and q_r^+, q_r^- denoting the recourse costs for under- and overdispatching at the r-th time interval, respectively. For a discussion of the interpretation and choice of the recourse costs we refer to [1]. More information on various aspects of power dispatch can be found in [8]. It is well-known that (1.2) is a particular stochastic program with simple recourse (cf. [3]). Under weak assumptions, (1.2) is a (large scale) convex nonlinear program having linear constraints and C^1- data (if all distribution functions F_r have densities). For our application of the model to the electricity sector of East Germany the dimension of the vector x is 840. For the uncertain demand, a set of empirical data is given. In [1] it is suggested that the distribution functions F_r can be chosen as (trimmed) normal. However, our tests with the available empirical data have not justified this hypothesis. The Figs. 1 and 2 show estimates for the density and distribution function of the centered demand during 1 p.m.-2 p.m. of a day of normal category. The estimates are obtained by a nonparametric kernel estimator according to (2.1).

2. Numerical treatment, results and conclusions

For the numerical solution of the stochastic power dispatch model (1.2) we first replace the distribution functions F_r by the following smooth nonparametric estimates $\hat{F}_r$ $(r =$

$1, \ldots, N$). Let $z_{1r}, \ldots, z_{nr}, \ldots$ be an independent sample from the distribution F_r ($r = 1, \ldots, N$), $k : \mathrm{R} \to \mathrm{R}$ a nonnegative function having the property $\int_{\mathrm{R}} k(t)dt = 1$ ("*kernel*"), and (b_n) a sequence of positive numbers tending to zero ("*smoothing parameters*"). Then we consider *kernel estimates* of F_r ($r = 1, \ldots, N$)

$$\hat{F}_r(t) := \frac{1}{nb_n} \sum_{i=1}^{n} \int_{-\infty}^{t} k\left(\frac{x - z_{ir}}{b_n}\right) dx \quad (t \in \mathrm{R} \,;\, n \in \mathrm{N}). \tag{2.1}$$

For more information and background on kernel-type estimators, especially on possible choices of k and (b_n), and on asymptotic results for $n \to \infty$, we refer to [7],[9]. These asymptotic arguments together with stability results for stochastic programs yield a theoretical foundation of our approach (see [2]). The use of the estimates $\hat{F}_r$ for the distribution functions of the electricity demand leads to the following convex program having a continuously differentiable objective and linear constraints:

$$\min\{g(x) + \hat{Q}(\chi) : x \in C, Ax = \chi\} \tag{2.2}$$

Explicit formulas for $\hat{Q}$ and its gradient can be derived easily (see Sect. 5 in [2]). They show the dependence of $\hat{Q}$ on the recourse costs, the samples and smoothing parameters, and on the real functions $\mathcal{K}_1(t) := \int_{-\infty}^{t} k(u)du$, $\mathcal{K}_2(t) := \int_{-\infty}^{t} uk(u)du$. Since, for most kernels k, $\mathcal{K}_1$ and $\mathcal{K}_2$ can be calculated explicitly, no numerical integration has to be performed when evaluating $\hat{Q}$ and its gradient.

The second step in our treatment of (1.2) is the solution of (2.2) by standard NLP-techniques. A program system STOCHOPT following this approach has been developed by using the NLP-code MINOS (see [5]). The system is written in FORTRAN 77 (algorithm) and TURBO-PASCAL (user-interface). The first version of STOCHOPT was implemented on an IBM PC 386 and first tested for a stochastic aircraft allocation problem (see [2]). Recently, a series of test runs for the stochastic power dispatch model (of East Germany) were performed. During all experiments the running time on a PC 386 did not exceed 5 minutes and had the same order of magnitude as for the corresponding deterministic model. The reason for this surprising effect is the low number of iterations of the nonlinear programming algorithm, which is (probably) due to the strong convexity of the recourse cost function $\hat{Q}$. Another surprising result is illustrated in Fig.3, showing the quotient QUO of the optimal costs of the stochastic model over that for the deterministic one as a function of the prescribed reserve level RL (%) for the demand, and elucidating the fact that the optimal solution of the stochastic model is superior to that of the (corresponding) deterministic one even if a reserve level of only 3% of the demand is adjusted.

QUO	RL
1.05	0%
0.97	3%
0.94	5%

Figure 3.

A more detailed comparison of (stochastic and deterministic) output schedules of all generation units shows that, for the "stochastic solution", high-cost (low-cost) units operate (as long as possible) at the lower (upper) output level. Fig.4 shows the daily output schedule for a high-cost unit (Boxberg2). An additional observation is that the total number of

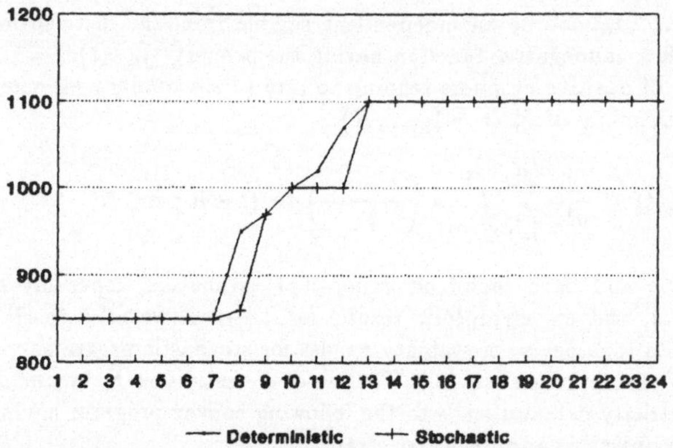

Figure 4. Comparison of a stochastic and deterministic output schedule

regulations of all units (during the whole time period) is smaller in case of the "stochastic solution".

References

[1] D.W. Bunn and S.N. Paschentis, Development of a stochastic model for the economic dispatch of electric power, European Journal of Operational Research 27 (1986), 173-181.

[2] N. Gröwe and W. Römisch, A stochastic programming model for optimal power dispatch: stability and numerical treatment, in: Stochastic Optimization: Numerical Methods and Technical Applications (K. Marti Ed.), Lecture Notes in Economics and Mathematical Systems, Springer-Verlag, Berlin, 1992 (to appear).

[3] P. Kall, Stochastic Linear Programming, Springer-Verlag, Berlin, 1976.

[4] K. Mai and J. Polzehl, Regression-based short term prediction of electrical load for a power system, Humboldt-Universität Berlin, Fachbereich Mathematik, Preprint Nr. 91-14 (1991).

[5] B. Murtagh and M. Saunders, Large-scale linearly constrained optimization, Mathematical Programming 14 (1978), 41-72.

[6] W. Römisch and R. Schultz, Distribution sensitivity for certain classes of chance constrained models - with application to power dispatch, Journal of Optimization Theory and Applications 71 (1991) (to appear).

[7] B.W. Silvermann, Density Estimation for Statistics and Data Analysis, Chapman and Hall, London, 1986.

[8] Hj. Wacker (Ed.), Applied Optimization Techniques in Energy Problems, Teubner, Stuttgart, 1985.

[9] B.B. Winter, Convergence rate of perturbed empirical distribution functions, Journal of Applied Probability 16(1979), 163-173.

ON THE SCHEDULING OF HYDRO POWER PLANT SYSTEMS IN THE FRAME OF THE UNIT COMMITMENT PROBLEM

J. Gutenberger, A. Gangl and Hj. Wacker
Institute for Mathematics
University of Linz
A-4040 LINZ/Auhof

1. Introduction

The hydro-thermal coordination problem consists in finding a balance between production and consumption for every moment. The aim is to control hydro power in a way such that thermal fuel costs are minimized. In this report we preassume the thermal unit commitment to be already established (cf. [2]). The remaining problem is to find an optimal schedule of the load among the generating hydro and thermal units, which is known as the Economic Dispatch problem. Here we concentrate on the hydro part of the system. We introduce the mathematical model of hydraulically coupled systems of storage power plants. As against other authors (cf. [1], [5]) the model contains the full nonlinear dependence of generation from head. For the mathematical solution technique we apply a decomposition - convexification procedure proposed by Gfrerer [3].

2. Mathematical model - System of pump storage power plants

The generated energy h of storage power plant j at time period t is given by

$$h_j^t = c \cdot H_j^t \cdot Q_j^t \cdot \eta_j(H_j^t, Q_j^t)$$

where Q denotes the discharge through the turbines, H the head difference between water surface of the reservoir and downstream water level, η the efficiency and c a physical constant. The head H depends nonlinearly on the storage content. It is modelled as a function of the volume V, called reservoir capacity function: $H = f(V)$. In general the efficiency depends on H and Q. For first calculations it may be assumed constant.

In order to pump the amount P of water back to the reservoir l, the following energy is necessary: ($\tilde{\eta}$ is the efficiency for pumping)

$$h_l^t = -c \cdot H_l^t \cdot P_l^t \cdot \frac{1}{\tilde{\eta}_l(H_l^t, P_l^t)}$$

The available reserves are equal to generating at maximum level minus actual generation.

• *Water flow balance equation:*

For each reservoir changes in the volume are equal to total inflow minus outflow. The inflow is composed of natural influx Z, upstream releases (discharge Q, spillage U) and pumping quantities P. The outflow is composed analogously. The natural inflow is assumed to be known in advance. The water flow balance or state equation describes how far the individual plants are coupled hydraulically.

$$V_k^t = V_k^{t-1} + Z_k^t + I_k^t - O_k^t \qquad k = 1, \ldots, K \qquad t = 1, \ldots, T$$

$$I_k^t = \sum_{j \in \mathcal{Q}_k} Q_j^{t-\tau_j} + \sum_{j \in \mathcal{U}_k} U_j^{t-\tilde{\tau}_j} + \sum_{j \in \mathcal{P}_k} P_j^t$$

$$O_k^t = \sum_{j \in \tilde{\mathcal{Q}}_k} Q_j^t + \sum_{j \in \tilde{\mathcal{U}}_k} U_j^t + \sum_{j \in \tilde{\mathcal{P}}_k} P_j^t$$

$\mathcal{Q}_k, \mathcal{U}_k, \mathcal{P}_k, \tilde{\mathcal{Q}}_k, \tilde{\mathcal{U}}_k, \tilde{\mathcal{P}}_k$ are index sets corresponding to controlled discharge flows, spilled flows, pumping quantities, into and out of node k, respectively. τ and $\tilde{\tau}$ denote water travel delays, K the number of reservoirs and T the scheduling horizon.

- *Storage and flow limitations:*

$$\begin{aligned} V_k^{min} &\leq V_k^t \leq V_k^{max} & k = 1, \ldots, K \quad t = 1, \ldots, T \\ Q_j^{min} &\leq Q_j^t \leq Q_j^{max} & j \in \cup_{k=1}^{K} \mathcal{Q}_k \\ 0 &\leq P_j^t \leq P_j^{max} & j \in \cup_{k=1}^{K} \mathcal{P}_k \\ U_j^{min} &\leq U_j^t \leq U_j^{max} & j \in \cup_{k=1}^{K} \mathcal{U}_k \end{aligned}$$

- *Boundary conditions:*

$$V_k^0 \ldots initial\ \ content$$
$$V_k^T \ldots target\ \ content$$

Note, that the content of the reservoir at the end of the considered time horizon T affects future operation.

- *Economic Dispatch Problem:*

The aim is to use the available water resources in a hydro-thermal system in order to minimize thermal fuel costs $\mathcal{C}$, meeting load and reserve requirements. The thermal commitment states are already established.

$$\begin{aligned} minimize \quad & \mathcal{C}(g) \\ subject\ to \quad & PH^t(x) + PT^t(g) \geq D^t \\ & RH^t(x) + RT^t(g) \geq R^t & t = 1, \ldots, T \\ & x \in \Omega_H \\ & g \in \Omega_T \end{aligned}$$

PH^t, PT^t, RH^t and RT^t denote total hydro generation, thermal generation, hydro reserves and thermal reserves at time period t. D^t is the required demand, R^t the spinning reserve. The sets Ω_H and Ω_T describe the local constraints for the hydro systems and thermal units, respectively. The thermal generation variables are denoted by the vector g, the hydro variables (V,Q,U,P) are summarized in the vector x. For details on the thermal part of the model see [2].

Mathematically we get a high dimensional, nonlinear and nonconvex control problem with many constraints.

3. Solution technique

We apply a dual method, since it exploits the natural decomposition of the problem ideally. In the Lagrangian the coupling demand and reserve constraints are relaxed, whereas the local constraints are treated directly. Unfortunately the problem is not convex with respect to the hydro variables. In order to be able to use classical duality theory, we convexify the problem by adding a quadratic penalty term to the Lagrangian. It consists of a convexification parameter y and a matrix C, which is chosen in a way, such that the problem is convex with respect to x and such that the separable structure is preserved. Now the dual function ϕ is well defined as the minimum of the Lagrangian, subject to local constraints.

$$\phi_y(\lambda,\mu) = \sum_{t=1}^{T}[\lambda^t D^t + \mu^t R^t] + \min_{g\in\Omega_T}\{C(g) + \sum_{t=1}^{T}[-\lambda^t PT^t(g) - \mu^t RT^t(g)]\}$$

$$+ \min_{x\in\Omega_H}\{\sum_{t=1}^{T}[-\lambda^t PH^t(x) - \mu^t RH^t(x)] + \frac{1}{2}(x-y)^T C(x-y)\}$$

We recognize, that the minimization problem for evaluating the dual splits into a thermal and a hydro subproblem. The thermal subproblem is continuous, convex and can be solved efficiently, since it breaks down further into subproblems involving one unit only. The hydro subproblem decomposes further, too, into subproblems for each system of hydraulically coupled plants.

A maximum of the dual provides optimal shadow prices $\lambda(y)$ and $\mu(y)$ and an optimal solution $(g(y), x(y))$ of the penalized primal problem. Gfrerer [3] has shown, that a fixed point of the function $x(.)$ fulfills the first order necessary condition for a local minimum. This suggests the following updating rule for y:

$$y_{k+1} = x(y_k)$$

The algorithm consists of three nested stages:

$$
\begin{array}{ll}
stage\ I: & updating\ of\ y \\
stage\ II: & maximizing\ the\ dual\ function\ \phi_y \\
stage\ III: & solving\ the\ subproblems
\end{array}
$$

The main effort lies in stage III. However, since the problem decomposes into smaller subproblems the effort is bearable. This stage is also well suitable for being parallelized on a computer. For stage II and III we use Newton-like methods. Only the highest level, the iteration in y, converges linear. An acceleration like in [4] is here not useful.

4. Numerical results

We performed a 24 hourly scheduling of an electric power system consisting of 10 thermal units and 7 partially hydraulically coupled hydro plants. The hydro system contains 2 single plants, whereby one allows for pumping, a chain of 2 serially connected plants and a more complex system of 3 plants. The thermal commitment has already been determined using the technique from [2]. Here we calculate the remaining economic dispatch.

The input data are heuristic, however fitted to realistic situations in Austria. The portion of hydro generation amounts more than 50 percent.

The resulting mathematical problem contains 528 variables and about 1800 constraints. The calculations have been performed on a DECsystem 3100 consuming 8 minutes CPU-time.

In the appendix the load demand of a day is sketched, divided into hydro and thermal generation. The hydro plants supply the peaks of the demand occurring at midday and evening, whereas the thermal plants generate almost constant base load. Observing the hydro generation in more detail, we can see, that each hydro system (marked as CHAIN, BRANCH, SINGLE and PUMPED) operates differently. This shows the great influence of power plant characteristics (fall of head, storage shape, plant capacity and influx) on the optimal control of the whole system.

Acknowledgement

This project has been supported by the "Fonds zur Förderung der wissenschaftlichen Forschung" (FWF) under grant P7780.

We are greatly indebted to our supervisor Prof. Hj. Wacker. His death in May 1991 was a great loss to us and our institute.

References

[1] AOKI A. ET. AL. (1987) 'Unit Commitment in a Large Scale Power System including Fuel Constrained Thermal and Pumped-Storage Hydro', IEEE Transactions on Power Systems, Vol. PWRS-2, No.4, pp. 1077 - 1084.

[2] GANGL A., GUTENBERGER J. AND WACKER HJ. 'On the solution of the hydro-thermal coordination problem by decomposition', P.F. Hodnett (ed.), submitted to the Proceedings of the Sixth conference of ECMI at Limerick, B.G. Teubner Stuttgart and Kluwer Academic Publishers, pp. 1-5.

[3] GFRERER H. (1984) 'Globally convergent decomposition methods for nonconvex optimization problems', Computing 32, 199-227.

[4] GUTENBERGER J., GANGL A. AND WACKER HJ. (1991) 'Decomposition method with superlinear convergence for power plant systems', M. Heiliö (ed.), Proceedings of the Fifth ECMI, B.G. Teubner Stuttgart and Kluwer Academic Publishers, 95-98.

[5] SANDELL N.R. ET AL. (1982) 'Optimal scheduling of large scale hydrothermal power systems', Proceedings of IEEE Large Scale Syst. Conf., Virginia Beach, VA.

APPENDIX

HYDRO-THERMAL ALLOCATION

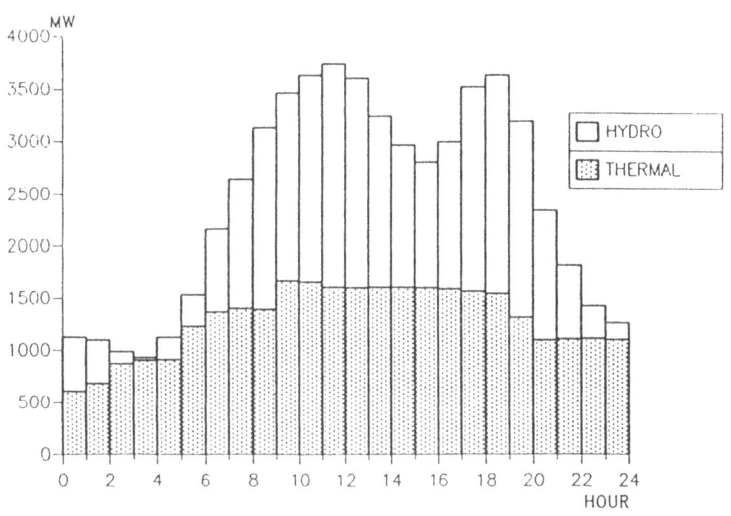

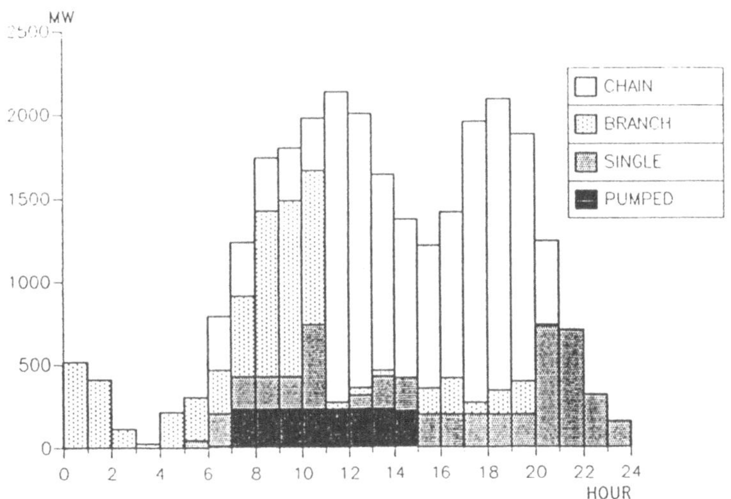

MATHEMATICAL MODELS FOR FLOW AND SEGREGATION OF PARTICULATE MATERIALS

S. A. HALVORSEN
Elkem a/s Research
P.O. Box 40, Vaagsbygd
N-4602 Kristiansand
Norway

ABSTRACT. A project on developing and applying mathematical models for granular flow and segregation is presented. The current models are described, emphasizing areas for further mathematical research.

1. Introduction

"Flow and Segregation of Particulate Materials" is a 3-year's research programme with the objective to develop and test mathematical models, measuring equipment and procedures for improved design of silos and other industrial units involving flow of particulate materials. The main project sponsors are The Nordic Fund for Technology and Industrial Development, The Royal Norwegian Council for Scientific and Industrial Research and Elkem a/s.

The project constitutes phase 2 of a Nordic cooperation between:
- Powder Science and Technology Research A/S, Porsgrunn, Norway
- Chalmers University of Technology (CTH), Department of Structural Mechanics, Göteborg, Sweden.
- Technical University of Denmark, Department of Structural Engineering, and Danish Building Research Institute, Lyngby, Denmark.
- Elkem a/s Research, Kristiansand, Norway (not involved in phase 1).

2. Powder flow

2.1. BASIC PROPERTIES

When a sufficiently large number of particles are put together, the behaviour of the bulk will be dominated by the collective behaviour of the huge number of particles and not by the properties of single, individual particles. Such a collection of particles is called a powder.

When discharging a flat-bottomed bin or a bin with a shallow hopper section, the powder flow will be concentrated to a channel stretching from the outlet and upwards (funnel flow). The powder surface normally forms a funnel and there will be a surface powder flow feeding the flow channel [1].

If the hopper section of a bin is sufficiently steep, the powder will flow over the whole cross-section throughout the whole bin (mass flow). In the upper vertical part of the bin, the powder will move downwards like a solid body. In the converging hopper, the powder will be plastically deformed and the velocity will be higher in the center than along the walls. The steepness of the hopper walls necessary to give mass flow, is determined by the internal powder friction and the friction between the powder and the hopper wall [1].

Mass flow ensures smooth flow through a silo on a first in first out basis, while funnel flow is often associated with flow problems. For industrial purposes it is therefore important to design equipment which guarantees mass flow. Today there exist procedures for two cases, silos with conical or wedge shaped hoppers. The current project aims at developing design tools for arbitrary 2-dimensional shapes (plane or axial symmetry).

It is generally accepted that slip occurs within a powder if the (internal) tangential stress reaches the value:

$$\tau = c + \sigma \tan \phi \tag{1}$$

where σ is the normal and τ the tangential component of the traction, c is called the cohesion and ϕ is the angle of internal friction. Beyond this yield condition (the Coulomb-Mohr criterion) the theory of the flow of granular materials is controversial [2].

2.2. MATHEMATICAL MODELS

2.2.1. *Finite element models.* A complex finite element model requiring mainframe computations has been developed by Prof. Eibl's group at the University of Karlsruhe [3]. For the development within the Nordic project we have aimed at a less complex model, suitable for computations on a workstation. Our main interest is the flow field, especially determining the necessary conditions for mass flow.

The current model has been developed at Chalmers University of Technology, CTH [4,5]. The equations for dynamic equilibrium can be formulated as:

$$\rho \left(\frac{\partial v}{\partial t} + v \cdot \nabla v \right) - \nabla \cdot \sigma(d) = \rho f \tag{2}$$

where v is the velocity vector, ρ the density, f the body force (gravity), t the time and $\sigma(d)$ the stress tensor as function of the rate of deformation tensor.

As a regularization, a highly viscous fluid behavior is assumed in regions where the powder is not yielding. Computationally the stress tensor can then first be computed as if the powder were a viscous fluid. Then, if the yield criterion is violated, the stress is projected onto the yield surface (in stress space), and the accelerations can be found from the (spatially) discretized version of (2). The Drucker-Prager cone, a generalization of the Coulomb-Mohr yield criterion, has been chosen for the yield surface.

The viscosity in the model does not correspond to a physical property. Regions where the model predicts low velocities are interpreted as no-flow regions.

Equation (2) together with the continuity equation have been spatially discretized (for 2-dimensional flow) using the finite element method on triangular elements. First the explicit forward Euler method was applied for the time stepping, resulting in extremely small time-steps (10^{-8} - 10^{-6} s). Now the implicit backwards Euler method has been implemented. The time-steps could then be increased with a factor of about 100, convergence problems (probably caused by strong nonlinearities) prohibiting further increase.

Computational time for a typical simulation is now in the order of 10-20 hours on an Apollo DN 4500 workstation. For a modern workstation (e.g. HP-type) we assume that the computational time will get down to a few hours. This is still too much for extensive parameter studies and the development proceeds at CTH in order to make the code more efficient. Advice from researchers who have worked with similar problems is requested.

2.2.2. *A double shearing model.* The problem of particle flow and segregation was presented at The European Study Groups with Industry (Southampton 1991). Here Prof. Spencer's Double Shearing Theory [2] was proposed as an alternative to the model developed at CTH. Prof. Spencer's theory assumes that when flow occurs, it occurs by simultaneously superimposed shearing on planes of maximum shear stresses. Equations for velocities and stresses have been developed for 2-dimensional stationary flow (plane or axial symmetry).

The equations are hyperbolic and can be solved by the method of characteristics. A computer code has been developed for flows within a hopper [6]. The computations start at the outlet boundary and continue upwards along the characteristics. For computations on a complete silo one would also need to take into account boundary conditions on the silo top. The conditions from the top and the bottom determine *different* stress fields, and a procedure for handling some type of discontinuity between the two fields is necessary. Major code modifications would also be necessary in order to adapt the code to general geometries [7].

3. Particle segregation

3.1. BASIC PROPERTIES

It is well known that mixtures of solid particles can segregate while being handled, due to differences in particle size (most important), density, shape or resilience [8].

On pouring a heap, flow occurs only in a top layer, a few particle diameters deep. Within the flowing layer the less flowable particles (small, dense, angular, less resilient) separate downwards by a screening mechanism and flow immediately on top of the static heap. These lower flowability particles will travel the shorter distance and concentrate close to the heap center. Small particles will leave the flowing layer, filling voids in the static region below [9].

Percolation where small particles move downwards relative to bigger ones, will occur within a powder whenever the mass of particles is disturbed [8, 10].

Many industrial processes are fed continuously with granular materials. Keeping the feeding tubes always filled, a stationary heap is maintained below each feeding point. The heap is renewed at the rate of material consumption in the process. Segregation will occur within top flowing layers, as for heap segregation, but now there will be percolation within the bulk immediately below, creating new voids and largely enhancing size segregation [11].

3.2. MATHEMATICAL MODELS

The heap segregation process has been described by Prof. Shinohara (Hokkaido University) by using a system of four hyperbolic differential equations [9]. The equations are solved numerically by some explicit method. Cooperation with Shinohara has been established, and his computer codes for plane or axial symmetric heap segregation are available for the project. Shinohara et. al. has further proposed a simple model for percolation in mass flow hoppers (applicable for some special cases) [10].

As Shinohara's models would benefit from mathematical analysis and we also wanted a model for the continuous case (combined heap segregation and percolation), the problem was presented at The European Study Groups with Industry (Southampton 1991). Here, Fitt and Wilmott proposed a cellular automaton model for the combined case [12]:

Within a 2-dimensional rectangular grid particles move to fill up voids. Voids in a cell are filled from the three neighboring cells immediately above, with greater probabilities for small particles to move within the bulk. The assigned probabilities depend on a "pressure", increasing big particle mobility closer to the surface. When updating the automaton, some cells in the bottom rows are first emptied (the discharge cells). Then the voids in these cells are filled, and the process continues row by row upwards.

Particle segregation is clearly demonstrated by the automaton [12]. The author has, however, limited confidence in the computed results as the basic segregation knowledge is not well enough represented by the current automaton rules.

4. Further work

Within the Nordic project we will continue developing the flow model from CTH, with emphasis on improving computational speed. A further analysis of the powder flow basis, including comparing the CTH approach and Prof. Spencer's Double Shearing Theory, highlighting similarities and differences, would be beneficial. So far we have not had sufficient resources to give priority to this task.

For segregation, cooperation with Shinohara continues. The work together with Fitt and Wilmott will also proceed, in order to establish more reliable automaton rules. If the latter approach succeeds, we get a cellular automaton model that can be applied for simulating heap formation, pure percolation, and the combined case of segregation during continuous flow.

5. References

[1]   Jenike, A.W. (1964) 'Storage and Flow of Solids', Bulletin 123, Utah University.

[2]   Spencer, A.M.J. (1991) 'Flow of Granular Materials', Lecture notes presented at The European Study Group with Industry & Training Course in Mathematical Modelling, Dep. of Mathematics, University of Southampton

[3]   Häussler, U. (1984) 'Velocity and Stress Fields at the Discharge of Silos', PhD dissertation, University of Karlsruhe

[4]   Runeson, K. and Nilsson, L. (1986) 'Finite Element Modelling of the Gravitational Flow of Granular Material', Int. J. Bulk Solids Handling, Vol. 6, No. 5, pp. 877-884

[5]   Runesson, K. and Klisinski, M. (1990) 'Simulations of the Motion of Bulk Materials in Silos - General Aspects and a Simple Flow Model', Department of Structural Mechanics, Chalmers University of Technology, Göteborg

[6]   Bradley, N.J. (1991) 'Gravity Flow of Granular Materials', PhD dissertation, University of Nottingham

[7]   Bradley, N.J. Private communications

[8]   Williams, J.C. , (1976) 'The Segregation of Particulate Materials. A Review', Powder Technology, Vol. 15, pp. 245-251.

[9]   Shinohara, K. (1987) 'General Mechanism of Particle Segregation during Filling Hoppers', Proceedings of 9th CHISA Congress, Section H: Particulate Solids, H.3.5, Prague,

[10]   Shinohara, K., Shoji, K. and Tanaka, T. (1970) 'Mechanism of Segregation and Blending of Particles Flowing out of Mass-flow Hoppers', Ind. Eng. Chem. Process Des. Develop., Vol. 9, No. 2, pp. 174-180.

[11]   Halvorsen, S.A. and Remnes, S. E. and Hyldmo, P. (1988) 'Segregation Patterns in a Smelting Furnace - Model Experiment', Electric Furnace Conference Proceedings, Vol. 46, Pittsburgh, pp. 153-160.

[12]   Fitt, A.D and Wilmott, P (1992) 'A Cellular Automaton Model for Segregation of a Two Species Granular Flow', to appear in Phys. Rev. A.

PAN – A PROBLEM SOLVING ENVIRONMENT FOR NUMERICAL ANALYSIS

T. HERCHENRÖDER, N. KÖCKLER, D. WENDT
Universität–Gesamthochschule–Paderborn, FB 17 – Mathematik/Informatik
Warburger Straße 100
D–4790 Paderborn (Germany)

ABSTRACT. A **P**roblem solving environment for **A**lgorithms in **N**umerical mathematics **PAN** is being developed in cooperation with NAG Ltd., Oxford, and with financial support from the EC program COMETT II. PAN is an open system running on UNIX machines under the X Window System with the possibility of individual user design. It provides an integrated graphical user interface for the development and management of programs and data. Parts of the system are the hypertext system HyTEX, a tutorial on numerical mathematics and about 30 elaborated Fortran programs using NAG libraries which are able to solve all basic numerical problems.

1 Introduction

Most students of mathematics or engineering will have to produce software and to work with software libraries during their vocational life. The development of Workstations into well organized graphical window systems makes it possible nowadays to work in a comfortable computing environment. While editing a file in one window you can read manual pages in another and look at graphics in a third. So why should you leaf through books on numerical mathematics and thumb an eight-volume user manual while learning numerical mathematics or solving one of the basic numerical problems?

2 The Main Targets of PAN

The PAN system integrates the following three main ideas to meet the requirements of a problem solving environment, or PSE (see [1]):

- A hypertext system controlling the tutorial, written as an extended LaTeX document
- A Fortran program sampler covering all basic numerical problems
- A management system for user and system programs, which controls HyTEX and also contains some important tools

3 The HyTEX Tutorial

The tutorial is based on the English translation [4] of [3]. Prepared for HyTEX, it will consist of about 600 pages of LaTeX text and graphics; see Figure 1, which shows a small page as an example. The whole tutorial is organized as a node system enabling the user

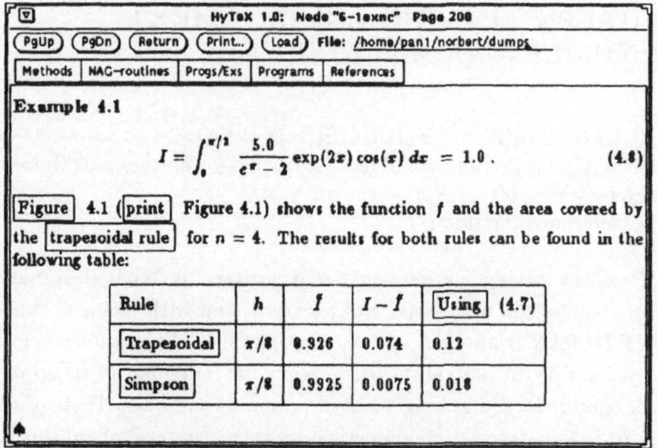

$$I = \int_0^{\pi/2} \frac{5.0}{e^\pi - 2} \exp(2x)\cos(x)\, dx \ = 1.0 \ . \qquad (4.8)$$

Rule	h	f	$I - f$	Using (4.7)
Trapezoidal	$\pi/8$	0.926	0.074	0.12
Simpson	$\pi/8$	0.9925	0.0075	0.018

Figure 1: The HyTEX Window

- to read the text in near print quality, including graphics and mathematical formulae
- to control the tutorial by means of a cascade menu
- to have quick access to different nodes throughout the tutorial
- thereby to learn numerical analysis with a mixture of
    - controlled selected reading
    - running examples and applications with different characteristics
    - developing programs in a comfortable programming environment

Within the text and the head of the window you see keywords marked by bounding boxes. A click on one of these keywords causes the display of the corresponding node within the tutorial.

The tutorial is divided into nine chapters dealing with the basic numerical subjects: linear and nonlinear systems, linear optimization, interpolation, approximation and integration, matrix eigenvalue problems and the solution of initial and boundary value problems for ordinary differential equations. For each of these subjects the text contains a short review of the theory, the algorithms, descriptions of the programs and when necessary decision trees for their use, manual pages of the NAG routines used, short examples and worked-out applications. The examples and applications are clarified through graphics whenever this is useful.

4 Fortran Program Sampler

The sampler contains about 30 elaborated Fortran programs connected to the tutorial and covering all basic numerical problems. For some of the programs, mainly those for interpolation, approximation and differential equations there are graphical versions as well.

The use of all programs is made easy through tools which are described in the next section; in particular there are input templates which allow a simple creation of input files controlled by the user and not by the Fortran program and its restrictions. These templates easily allow multiple test runs and also the use of some preassigned matrix types or example functions. The templates take advantage of a graphical user interface when generating the

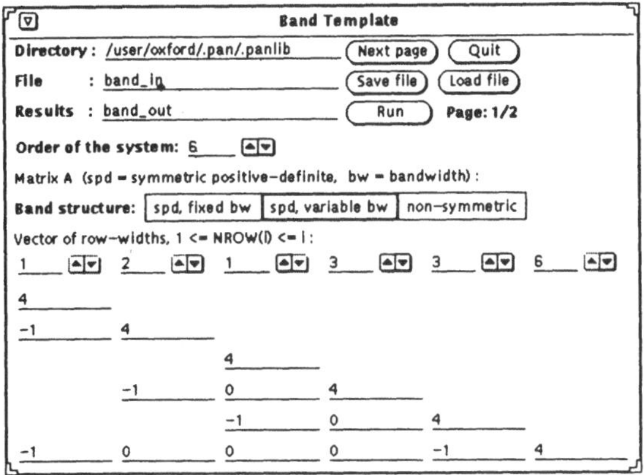

Figure 2: The Window of the Band Template

input data. As an example, Figure 2 shows the first page of the template for the program BAND, which solves linear systems with banded matrices of three different band forms.

5   The Management System

The hierarchical tree structure controlling the tutorial and the programs is embedded into a management system (see Figure 3) which provides the user with the ability to

- retrieve complex information
- run numerical programs
- use tools for program modification and development

The performance of these tasks should be highly adaptable to the user's needs – not only within the field of numerical analysis – and should be available to a wide range of users. Thus we have tried to build a portable, configurable and open system.

The basic features of the PAN management system are:

- dynamic control of HyTeX through a cascade menu
- on-line help system
- file management
- compilation and run control
- selection and definition of options for the available compilers
- tree management for creating an individual PSE
- management of the UNIX- and X11-environments used
- utilities for debugging, making screendumps etc.

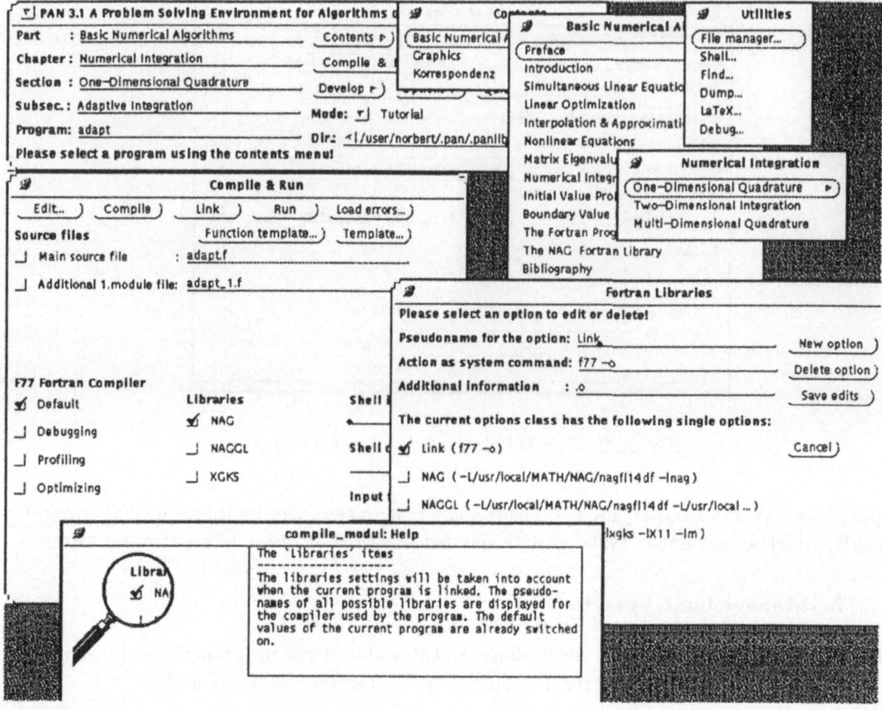

Figure 3: A Screen Full of PAN

6 Conclusion

Up to now the problem solving environment PAN has been used as an environment for numerical mathematics, only. But its design easily enables its use as an environment for different subjects, for example for statistical programs and continuously updated reports on the drug situation in a country. The system can deliver a skeleton which can be easily filled with program and text files to become a PSE.

References

[1] Ford, B., Iles, R.M.J., The What and Why of Problem Solving Environments for Scientific Computing, in: Ford, B.; Chatelin, F., eds., *Problem Solving Environments for Scientific Computing* (North Holland, Amsterdam, 1987)

[2] Grandine, T.A., The Numerical Methods Programming Projects Book, (Oxford University Press, Oxford, 1990)

[3] Köckler, N., Numerische Algorithmen in Softwaresystemen, (Teubner, Stuttgart, 1990)

[4] Köckler, N., Numerical Algorithms and Software Libraries, (Oxford University Press, Oxford) in preparation

IMPROVING THE PERFORMANCE OF A DYNAMOMETER WHICH MEASURES AXIAL FORCE AND TORQUE DURING DRILLING OF KEVLAR 49 COMPOSITES

R.Howard-Hildige, W.P.Hennessy & M.Hillery
University of Limerick

ABSTRACT A dynamometer has been constructed for the measurement of axial drill force and torque during the drilling of Kevlar 49 composites. Initial tests of the dynamometer showed a significant time delay on the torque output channel.
Consideration of the equations of motion of the force transducer elements has enabled an analysis of the dynamic response of the instrument. This has led to improved transducer design and allows prediction of the dynamic interaction between the transducer elements and other components of the overall system. In particular the likelihood of severe dynamometer output attenuation due to operation in the vicinity of structural resonance of large relative amplitude.

Acknowledgements   This work is supported by Mohawk Europa and Eolas

Introduction

Current practise for multicomponent dynamometer calibration involves determination of the output signal corresponding to known loading upon each measurement direction. This is achieved by applying single or combinations of loads and determining interactions between each measurement direction under static conditions. Subsequently dynamometers are often used under dynamic loading conditions. The drilling dynamometer shown in fig. 1 is typical of this application. It was made for the study of forces and moments arising during drilling of Kevlar 49 composites [1],[2].

Fig.1 Drilling Dynamometer

Dynamic Analysis

As a first approxiamation a lumped parameter model was assumed for the prototype axial force/torque transducer unit shown in fig.2(a) which uses strain gauged cantilevers as measuring elements. A force balance along the axis of the drill and a moment balance about the same axis yields equations 1 and 2.

$$F_D = m\frac{d^2x}{dt^2} + \mu\frac{dx}{dt} + kx \quad \text{Eq.1} \qquad T_D = J\frac{d^2\theta}{dt^2} + \mu_T\frac{d\theta}{dt} + k_T\theta \quad \text{Eq.2}$$

Fig.2(a) Prototype Transducer

Fig.2(b) Revised Transducer

Consideration of the coefficients of equations 1 and 2 enabled improvement of the original design by reducing non linearities due to friction as described by Meirovitch [3] by locating the spindle using needle roller bearings. The delay between axial force and torque graphs shown in fig.3 was reduced by reduction of the polar moment of inertia, the improvement is shown in fig.4. Figs 5 and 6 show the effect in the axial direction of adding viscous damping to reduce oscillations on the output signal. The revised transducer unit is shown in fig.2(b).

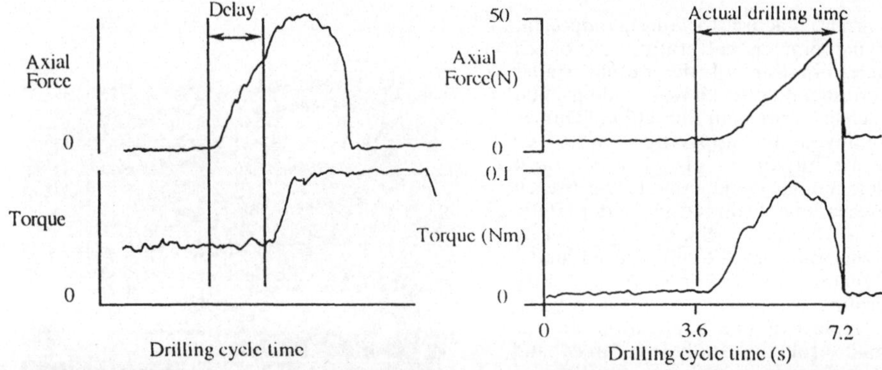

Fig.3 Output from the prototype transducer unit

Fig.4 Output from revised transducer unit

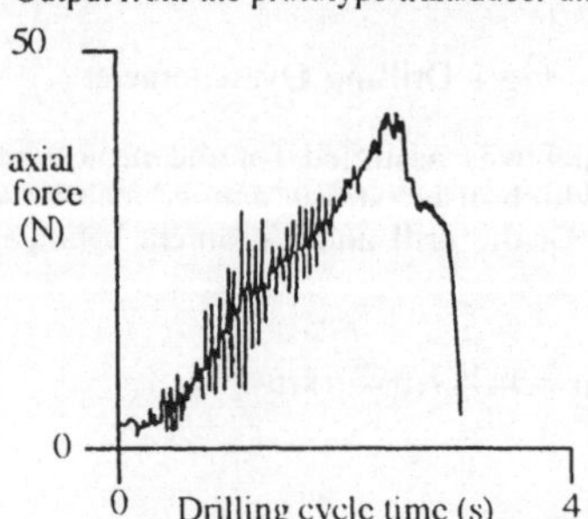

Fig.5 Output curve from the axial
force transducer before damping

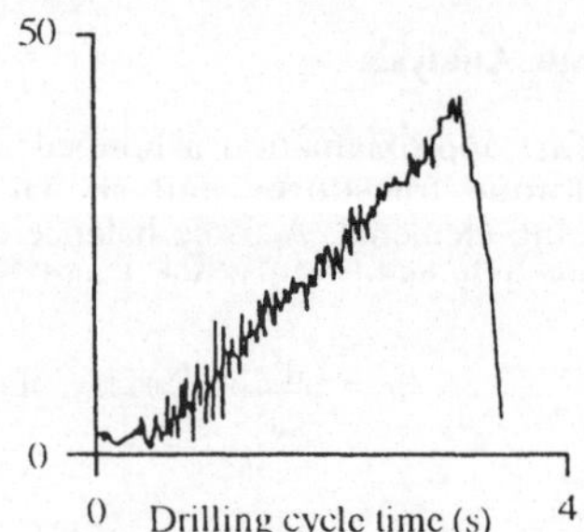

Fig.6 Output curves from the axial
force transducer after damping

Concentrating on the axial direction as the form of equations 1 and 2 are the same, taking Laplace transforms and rearranging gives

$$\frac{\overline{X}}{\overline{F}_D} = \frac{1/m}{s^2 + \mu\frac{s}{m} + \frac{k}{m}} \qquad Eq.3$$

Damping factor and natural frequency are readily found [3] from fig.7 which is the response to an applied pulse, and comparison with the standard quadratic form [4] enable the coefficients of equation 3 to be found when the moving mass of the transducer is weighed.

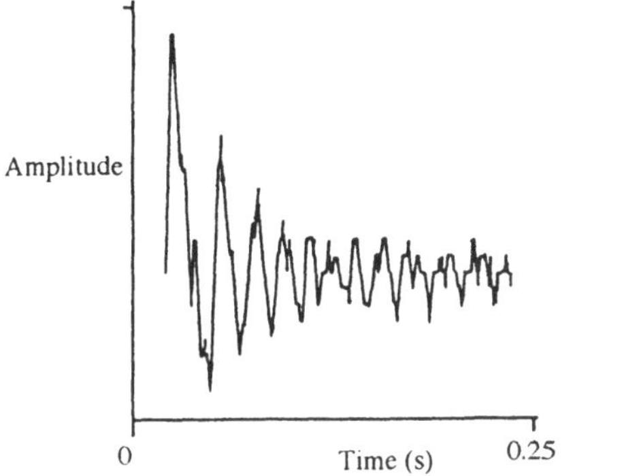

Fig.7 Response of transducer to a pulse   Fig.8 Schematic of system

Dynamic Interaction of the Transducer Elements with their surrounds

Kegg [5] found that interactions between the structural dynamics and the rest of the machining system cannot be ignored during a cutting process. As the drill in the dynamometer operates in the range 2000-30000 r.p.m. the input force to the elements must be considered in the frequency domain. Considering the flexible element of mass m, stiffness k and damping μ, mounted in a flexible casing of mass m_1 and stiffness k_1 and negligible structural damping as shown schematcally in fig.8. (Actually the casing may have any number of degrees of freedom but the argument still holds in the vicinity of an individual resonance).
The transformed equations of motion are:

$$ms^2 + \mu s(\overline{x}-\overline{x}_1) + k(\overline{x}-\overline{x}_1) = \overline{F}_D \qquad Eq.4 \qquad m_1 s^2 \overline{x}_1 - k(\overline{x}-\overline{x}_1) - \mu s(\overline{x}-\overline{x}_1) + k_1 \overline{x}_1 = 0 \qquad Eq.5$$

Solving for Eq.4 and Eq.5 gives

$$\frac{\overline{x}_1}{\overline{F}_D} = \frac{\mu s+k}{(m_1 s^2+k+\mu s+k_1)(ms^2+\mu s+k)-(\mu s+k)^2} = \frac{\mu s+k}{\lambda} \qquad Eq.6 \qquad \frac{\overline{x}}{\overline{F}_D} = \frac{m_1 s^2+k+\mu s+k_1}{\lambda} \qquad Eq.7$$

The strain measuired by the dynamometer is $x-x_1$ the compression of the spring k

$$\frac{\overline{x}-\overline{x}_1}{\overline{F}_D} = \frac{m_1 s^2+k_1}{\lambda} \qquad Eq.8$$

Equation 8 shows that the output of the dynamometer will approach zero when $m_1 s^2 = k_1$ i.e. at the i.e. at the natural frequency of the casing. In practise this attenuation can be minimised by keeping the moving mass of the dynamometer as low as possible.

To vary the drill angular velocity the dynamometer employs a series of interchangeable, fixed speed pneumatic motors. A knowledge of dynamometer attenuation characteristics is therefore useful prior to motor purchase, decomposing the response of the casing to an impact using a Fast Fourier transform analyser may provide this information.

Conclusions

The equations of motion of the dynamometer elements were obtained by treating the moving mass of the elements as a single lumped parameter. The subsequent analysis showed the polar moment of inertia of the measuring table to be the major cause of an appreciable time delay on the torque output channel. Table spindle friction was also found to introduce non linearities in the equations of motion. A revised design with lower inertia and needle roller bearings has improved dynamometer performance.

The analysis also led to reduction of undesirable oscillation of output traces by incorporating a dashpot in the revised design.

Consideration of the equations of motion has enabled the prediction of dynamometer output attenuation in the vicinity of casing resonance.

References

1   Hennessy, W.P. (1991) M,Tech Dissertation, University of Limerick.

2   Hennessy, W.P. and Hillery, M. (1991) The Design and Manufacture of a Drill Dynamometer for Drilling Kevlar Reinforced Composites. E.M.I. Conference, Jordanstown, pp803-816.

3   Meirovitch, L. (1986) Elements of Vibration Analysis 2nd Ed., McGraw Hill, pp30-34.

4   Raven, F.H. (1987) Automatic Control Engineering 4th Ed., McGraw Hill, p203.

5   Kegg, R.L. (1965) Cutting Dynamics in Machine Tool Chatter., Trans A.S.M.E. Journal of Engineering in Industry, pp464-470.

Nomenclature

F_D	axial drilling force
J	polar moment of inertia of transducer elements
k	stiffness of axial force transducer element
k_T	stiffness of torque transducer element
k_1	stiffness of dynamometer casing
m	mass of transducer elements
m_1	mass of dynamometer casing
s	Laplace operator
t	time
T_D	drilling torque
x	axial displacement of transducer elements
x_1	displacement of dynamometer casing
μ	coefficient of viscous damping in axial direction
μ_T	coefficient of viscous damping in torsion
θ	angular displacement
ω_n	natural frequency

A Quasi-likelihood Markov model for the hardenability of steel

Kyösti Huhtala
University of Jyväskylä, Dept. of Statistics
PL 35, SF-40351 Jyväskylä, Finland

1. Introduction

To evaluate the hardenability of steel the *Jominy test* can be used. A steel bar of fixed dimensions is heated up to about $850\ {}^\circ C$ and then quenched from one end by water spray. Its hardness y is then measured at increasing distances d from this quenched end giving observations $(d, y(d))$, which form the Jominy curve. For each Jominy curve these observations can be fitted by descending curve $\mu = \mu(d; \boldsymbol{\eta})$ with shape parameters $\boldsymbol{\eta} = (\eta_1, \ldots, \eta_p)^T$ which are assumed to vary along the alloying elements (C, Mn, Cr etc.).

A model based on stochastic independency of measurements $(d, y(d))$ is proposed by Huhtala (1991). We assume normally distributed measurements $y(d)$ with mean $\mu(d)$ and variance σ^2 and that the shape parameters are linear combinations of the covariates like the amounts of the alloying elements, i.e. $\boldsymbol{\eta} = \tilde{\mathbf{X}}\boldsymbol{\beta}$, where $\tilde{\mathbf{X}} = \in R^{p \times S}$ is the matrix of covariates and $\boldsymbol{\beta} \in \mathbf{R}^{S \times 1}$. For a sample of M Jominy curves with $n_i,\ i = 1, \ldots, M$, measurements each the maximum likelihood (ML-) estimate for the coefficient vector $\boldsymbol{\beta}$ can be computed using iteratively reweighted least-squares (IRLS) estimator. The updating equations are the normal equations of the linear regression model $\mathbf{Z} = \tilde{\mathbf{X}}\boldsymbol{\beta}^{(k+1)} + \boldsymbol{\zeta}$ with weights $\partial\boldsymbol{\mu}^{(k)}/\partial\boldsymbol{\eta}\,\mathbf{V}^{-1}(\partial\boldsymbol{\mu}^{(k)}/\partial\boldsymbol{\eta})^T$, where $\mathbf{Z}$ is the *working dependent variable* $\mathbf{Z} = \boldsymbol{\eta}^{(k)} + ((\partial\boldsymbol{\mu}^{(k)}/\partial\boldsymbol{\eta})^+)^T(\mathbf{y} - \boldsymbol{\mu}^{(k)})$. Here $\partial\boldsymbol{\mu}/\partial\boldsymbol{\eta} = \text{blockdiag}(\partial\mu_r/\partial\boldsymbol{\eta}_r,\ r = 1, \ldots, N) \in \mathbf{R}^{Np \times N}$ is the matrix of partial derivatives of the expected value or *link* function μ for each of the $N = \sum_{i=1}^{M} n_i$ observations, matrix $\mathbf{V}$ is the diagonal variance matrix of observations and $\mathbf{A}^+$ is the Moore-Penrose inverse of $\mathbf{A}$.

A set of Jominy test results for low alloyed special steels was modelled using $\mu(d, \boldsymbol{\eta}) = (\eta_1 \exp(\eta_2 + \eta_3 \log d) + \eta_4)/(1 + \exp(\eta_2 + \eta_3 \log d))$ and $S = 32$ covariates, which were chosen according to their capability to improve the fit significantly determined using approximate F-statistic of Jørgensen (1983). With $M = 48$ Jominy tests with $N = 639$ measurements we found the residual sum of squares to be $D = 1523.1$ reflecting quite satisfactory overall fit. However, if we inspect

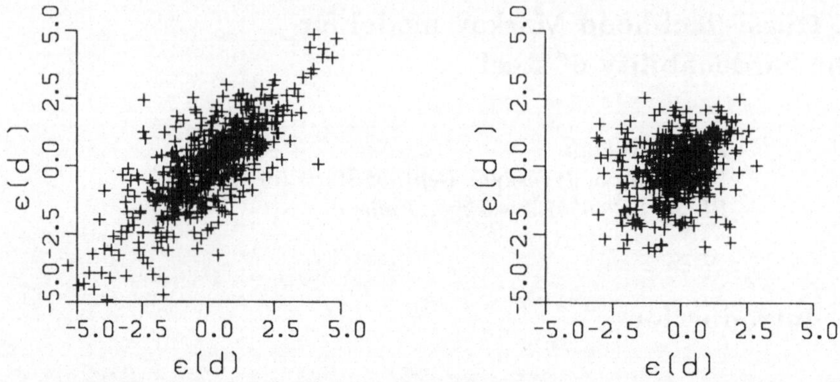

Figure 1: Plots of the lagged residuals against the original residuals for a) the independence model and b) the Markov model.

the plot of $\epsilon(d_-) = y(d_-) - \hat{\mu}(d_-)$ against $\epsilon(d)$ (Figure 1a), when d_- stands for the previous measurement distance, we find how they are strongly correlated.

This correlation may result from the link function being not flexible enough, inapplicable distributional assumption or the possible inaccuracy in the first measurement distance caused by scaling and decarburization during the heating.

2. Model and estimation

Applying the results of Zeger and Qaqish (1988) the conditional expected value of the Jominy hardness at the Jominy distance d depends on the previous measurement according to

$$\tilde{\mu}(d) = E(y(d) \mid d_-, y(d_-)) = \mu(d, \tilde{\eta}(d))$$
$$\tilde{\eta}(d) = \eta + \mathbf{f}(d_-, y(d_-), \eta, \boldsymbol{\beta})\boldsymbol{\gamma} \qquad (1)$$
$$\eta(d) = \tilde{\mathbf{X}}\boldsymbol{\beta},$$

where $\mathbf{f}$ is a $p \times L$-valued function of the previous measurement, and $\tilde{\boldsymbol{\beta}} = (\boldsymbol{\beta}^T, \boldsymbol{\gamma}^T)^T$ is an $S + L$-vector of coefficients to be estimated. The conditional variance is some known function of $\tilde{\mu}(d)$, i.e. $var(y(d) \mid d_-, y(d_-)) = v(\tilde{\mu}(d)) \cdot \sigma^2$.

Since the model is a direct extension of that of Section 1 we can compute the estimates for $\tilde{\boldsymbol{\beta}}$ using the same IRLS-algorithm. However, if the history function $\mathbf{f}$ contains also the $\boldsymbol{\beta}$-coefficients, as in Section 3, we need to use the following double iteration scheme:

(i) Compute $\eta(d)$, $\tilde{\eta}(d)$ and $\mathbf{f}$ using the current estimates of $\boldsymbol{\beta}$ and $\boldsymbol{\gamma}$.

193

(ii) Solve the linear regression problem of Section 1 with $\tilde{\boldsymbol{\beta}}$ in place of $\boldsymbol{\beta}$.

(iii) If not converged, return to (i).

The dispersion coefficient σ^2 is not usually involved in the estimating equations of $\tilde{\boldsymbol{\beta}}$ and thus not need to be estimated along $\tilde{\boldsymbol{\beta}}$. When the estimate $\hat{\tilde{\boldsymbol{\beta}}}$ has been computed, the dispersion coefficient can be estimated with $\hat{\sigma}^2 = (N - S - L - M)^{-1}D$, where $D = \sum_{i=1}^{M} \sum_{j=2}^{n_i}(y(d_{ij}) - \hat{\tilde{\mu}}(d_{ij}))^2 / \sqrt{v(\hat{\tilde{\mu}}(d_{ij}))}$.

3. A model for low alloyed special steel

A first attempt to correct the model of Section 1 was to use a crude quess $\mathbf{f} = \mathrm{diag}(y(d_-) - \mu(d_-, \boldsymbol{\eta}), \ k = 1, \ldots, 4)$ as the history function. This is a simplified version of the first order Taylor-approximation of the generalized inverse of link μ at the point $(d_-, y(d_-), (\boldsymbol{\eta}^{(k)})^T)$ with respect to $\boldsymbol{\eta}$ (Nashed, 1978).

The normality assumption made implies the assumption of the constant variance function, i.e. $v(\tilde{\mu}) = 1$. The fitting of this model gave $D \approx 613$ meaning a great improvement in the overall fit. The residuals were still a bit correlated so the link was extended to

$$\tilde{\mu}(d, \tilde{\boldsymbol{\eta}}) = (\tilde{\eta}_1 \exp(\tilde{\eta}_2 + \tilde{\eta}_3 \log^{\tilde{\eta}_5} d) + \tilde{\eta}_4)/(1 + \exp(\tilde{\eta}_2 + \tilde{\eta}_3 \log^{\tilde{\eta}_5} d)) \qquad (2)$$

and the history function to

$$\mathbf{f} = \mathrm{diag}((y(d_-) - \mu(d_-, \boldsymbol{\eta}))0.2/(d - d_-), \ k = 1, \ldots, 5). \qquad (3)$$

Here the constant 0.2 was used to scale the elements of $\mathbf{f}$ for more stable convergence. This removed the correlation of the residuals practically completely (see Figure 1b).

Since the model uses the first measurement of each curve for the conditioning, we used a simple linear regression model with carbon and its square as the only covariates fitted for the $M = 48$ first measurements to get starting values for prediction. The estimated coefficients were $\hat{y}(d_1) = 31.65 + 68.92 \cdot C - 78.23 \cdot (C - 0.25)^2$ and the estimated residual variance for this regression was $\hat{\phi}^2 \approx 2.56$.

The validity of the model was inspected using a sample of $M = 125$ tests independent of those used for the estimation with total number of $N = 1637$ measurements. The predictions followed quite well the observations for qualities close to those used for the estimation and for some qualities far away from these the results were disastrous. Adding some interaction terms bettered the prediction properties a bit and lowered the overall measure of fit to $D = 541.3$ giving $\hat{\sigma}^2 = 0.98$ with $S = 34$ and $L = 5$. The sum of the squared prediction errors for the original sample was about 348496 and for the independent sample about 45356. The poor result for the original sample follows from one disastrous prediction which constitutes over 95% of the sum of squared prediction errors. It seems that model can be used quite safely for those qualities same as or near those used for the estimation.

The final estimates are available from the author.

4. Concluding remarks

Zeger & Qaqish (1988) have shown that the QL-estimator is asymptotically normally distributed under quite moderate conditions at least in case of only one series. This has been used as a guide to select the covariates besides some metallurgical knowledge, of course.

Another method to evaluate the significance of the covariates would be the Monte Carlo techniques. Since constant variance function implies conditional normality of $y(d)$ conditioned on $y(d_-)$ We formed simulated Jominy curves for the Monte Carlo procedure using the normality assumption of measurements implied by the constant variance function. The estimates were then computed using simulated data and the procedure was repeated a large number of times. We monitored the variance of each estimate $\hat{\hat{\beta}}_i$, $i = 1, \ldots, S + L$, computed the sample variance of the corresponding Monte Carlo estimate and noticed no significant change in them after 200 replications. For the coefficients of η_1 the variances were almost the same as their asymptotic counterparts, for the coefficients of η_4 they were about twice as large and for the others the difference varied from near equality to multiples of about three. However, only 7 coefficients would have been considered insignificant, because their absolute values were less than two times the square root of their Monte Carlo variances. Removing all corresponding covariates from the model caused only a small increase in D so a smaller model is possible, but its prediction properties were not at all better than those of the larger model.

References

Huhtala, K. (1991), Modelling the hardenability of steel: A response curve approach, *Jyväskylä studies in computer science, economics and statistics, 15.* *University of Jyväskylä.*

Jørgensen, B. (1983), Maximum likelihood estimation and large-sample inference for generalized linear and nonlinear regression models, *Biometrika* **70**, 19 – 28.

Nashed, M. Z. (1978), Generalized inverse mapping theorems and related applications of generalized inverses in nonlinear analysis, in *Nonlinear equations in abstract spaces,* ed. V. Lakshmikantham, New York: Academic Press, 217 – 252.

Zeger, S.L. & Qaqish, B. (1988), Markov regression models for time series: a quasi-likelihood approach, *Biometrics* **44**, 1019 – 1031.

IMPLEMENTATION OF AN ELECTRICITY LOAD FORECASTING SYSTEM

O. HYDE and P.F. HODNETT
Department of Mathematics and Statistics
University of Limerick
Limerick, Ireland

ABSTRACT. The Electricity Supply Board (ESB) requires forecasts of system demand or load for up to 10 days in advance. A model has been developed which identifies a 'normal' or weather-insensitive load component and a weather-sensitive load component. Linear regression analysis of past load and weather data is used to identify the normal load model. The weather-sensitive component of the load is forecasted using the parameters of the regression analysis. The system is at present being implemented at the ESB. Here output from the forecast is presented, together with information on updating and improving the model.

1. Basic Model

Total load is assumed to be a linear combination of four separate components. The total load of the system at the kth hour, is written in the form

$$Y(k) = N(k) + W(k) + S(k) + r(k) \tag{1}$$

where $Y(k)$ represents total load, $N(k)$ represents the normal (weather-insensitive) load component, $W(k)$ represents the weather-sensitive load component, $S(k)$ represents the load component due to special events and $r(k)$ the random load component (see Hyde and Hodnett (1991)). The load forecasting model has been developed using real load and weather data for the years 1982 to 1988, and tested using real data for 1989.

A year is divided into six bimonthly periods of similar weather patterns, with separate normal load and weather load models being developed for each period. It is assumed that a stable relationship between weather and load exists for each of the bimonthly periods: January-February, March-April, May-June, July-August, September-October, November-December.

The normal load model consists of standardized normal load values for each hour of the day on ordinary weekdays, public holidays and church holidays, for each bimonthly period. Load data is standardized before the regression is performed to remove the effects of load growth from the data. Load growth for the present year is then taken into account when destandardizing the forecast total load.

Regression analysis on the data leads to the following weather load model (at k^{th} hour)

$$W(k) = \alpha_1 T(k) + \alpha_2 T(k\text{-}24) + \alpha_3 Sn(k) + \alpha_4 V(k) + \alpha_5 H(k) \tag{2}$$

where T represents dry bulb temperature in °C, Sn sunshine duration in hours, V wind velocity in knots, H relative humidity (%) and α_i ($i = 1$ to 5) are the corresponding parameters for these weather variables and may have value zero.

Load fluctuations due to special events such as strikes, school closures or elections are not handled separately by the system but are left to the discretion of the dispatching engineer for adjustments to load forecasts.

2. Some Results

As the load forecasting system has been in operation for only a short time, with load forecasts being generated daily using weather forecasts, a detailed analysis of the model output cannot at present be performed. Present results show a mean absolute forecasting error of 2.0% and a standard deviation of 1.9% of total load. Monday, 25th November 1991 has been taken as an example day, with forecasted loads using predicted weather data and actual loads for that day being shown in Table 1. Results for the same day are displayed in Table 2, with forecasted loads now generated using actual weather data as recorded by the Irish Meteorological Service. For the results shown forecasted loads using predicted weather have a mean absolute error of 37MW and a standard deviation of 41MW, while for forecasted loads using actual weather there is a mean absolute error of 30MW and a standard deviation of 38MW. It is reassuring to note that load forecasts in Table 1 using forecast weather data are close to the load forecasts in Table 2 using actual weather data.

3. Updating The Model

A deterioration in the model performance is expected over the long term as changes in weather patterns, the economic climate and social trends begin to influence the model. To ensure that the load forecasting system is adaptable to these and other changes, the model should be regularly updated through recalculation of the weather load model and its parameters, and subsequent recalculation of the normal loads. It is envisaged that the model will need to be updated annually, to allow the previous year's data to be added to the historical load and weather data base thus ensuring the most recent data is used in the model.

A program to automatically update the model has been written. The user enters the start and finish years between which data is selected from the historical database. For each bimonthly period, the program refits the weather load model. The best (in the statistical sense) weather variables are chosen as predictors. Points are dropped that appear to have a disproportionate influence on the estimated model (here called critical influential points), whether (a) due to improperly recorded data or (b) due to data not reflecting the expected relationship between load and weather arising from other causes such as special events. The values of the weather parameters are then estimated for that bimonthly period. Finally, the normal loads are computed after restoring all removed points to the database.

Predictors are selected for the weather load model using the partial F-test statistic and Mallow's C_p statistic (see Kleinbaum et al. (1988), Walpole and Myers (1989)). The full model is fitted, with the predictor which has the minimum partial F-test statistic tested to see if its true coefficient value is zero. The predictor is not dropped from the model unless it fails both the F-test and a second test involving the C_p value for that predictor. If the predictor is dropped, the reduced model is refitted and the procedure continues until one of the two tests is passed.

Having selected the form of the weather load model, the final values of the weather parameters are not calculated until all influential points having an undue influence on the choice of these values are removed. The statistics used in this procedure are the adjusted coefficient of determination, denoted by R^2adj, and the scaled change in fit resulting from deletion of the ith point, denoted by $DFITS_i$ (see Belsley et al. (1980), Weisberg (1985), Bollen and Jackman (1990)). The point with the highest absolute value of $DFITS_i$ is examined for removal. If the $DFITS_i$ value exceeds a certain cutoff level and the removal of the point leads to an improved model fit (as measured by an increase in R^2adj), then the point is dropped. The weather load model is then refitted and the procedure continues until all critical influential points are removed. The software and databases are stored on a Digital VAX 11/780 mainframe computer, which has ample memory space for the large databases involved. Runtime for the updating algorithm is a matter of minutes, with program output being immediately usable by the load forecasting system without any interruptions to the system or additional work required by the user.

4. Improving The System

Output from the system indicates that while a reasonable level of accuracy is achieved at most times in the load forecasts, there are periods during which the present system fails to produce an acceptable forecast. These include the Christmas holiday period, which is dependent on the day of the week on which Christmas day falls, other holiday periods and summer/winter time changes. Plots of the load at these times show a changed load shape, indicating that a separate normal load model is required. It is also possible that a change in the weather load model is another factor. It is proposed to adopt a rule based approach for treating these times. The resultant rule based program will then be used in conjunction with the other programs of the load forecasting system to produce accurate load forecasts at all times.

References

Belsley, D.A., Kuh, E. and Welsch R.E. (1980) Regression Diagnostics: Identifying Influential Data and Sources of Collinearity, John Wiley & Sons, New York.

Bollen, K.A. and Jackman, R.W. (1990) 'Regression Diagnostics: An Expository Treatment of Outliers and Influential Cases', in J. Fox and J.S. Long (eds.), Modern Methods of Data Analysis, Sage Publications, London, pp. 257-291.

Hyde, O. and Hodnett, P.F. (1991) 'An Electricity Load Forecasting System', in M. Heilio (ed.), Proceedings of the Fifth European Conference on Mathematics in Industry, B.G. Teubner, Stuttgart, pp. 99-103.

Kleinbaum, D.G., Kupper, L.L. and Muller, K.E. (1988) Applied Regression Analysis and Other Multivariate Methods, PWS-KENT Publishing Co., Boston, Chapter 16.

Walpole, R.E. and Myers, R.H. (1989) Probability and Statistics for Engineers and Scientists, MacMillan, Chapter 10.

Weisberg, S. (1985) Applied Regression Analysis, John Wiley & Sons, New York, Chapters 5 & 6.

TABLE 1. A comparison of actual loads and forecasted loads (in MW) using predicted weather data, with forecasting errors, for Monday 25th November 1991.

Hour	Load	Forecast	Error	Hour	Load	Forecast	Error
0	1741	1706	-35	12	2168	2248	80
1	1540	1507	-33	13	2196	2285	89
2	1427	1378	-49	14	2055	2111	56
3	1329	1329	0	15	2109	2179	70
4	1284	1278	-6	16	2189	2224	35
5	1234	1257	23	17	2379	2422	43
6	1253	1267	14	18	2445	2464	19
7	1457	1401	-56	19	2207	2236	29
8	1844	1782	-62	20	2229	2208	-21
9	2050	2025	-25	21	2104	2148	44
10	2126	2158	32	22	2007	2019	12
11	2161	2177	16	23	1932	1975	43

TABLE 2. A comparison of actual loads and forecasted loads (in MW) using actual weather data, with forecasting errors, for Monday 25th November 1991.

Hour	Load	Forecast	Error	Hour	Load	Forecast	Error
0	1741	1726	-15	12	2168	2224	56
1	1540	1526	-14	13	2196	2243	47
2	1427	1398	-29	14	2055	2109	54
3	1329	1322	-7	15	2109	2159	50
4	1284	1283	-1	16	2189	2224	35
5	1234	1237	3	17	2379	2411	32
6	1253	1240	-13	18	2445	2463	18
7	1457	1380	-77	19	2207	2235	28
8	1844	1755	-89	20	2229	2196	-33
9	2050	2000	-50	21	2104	2140	36
10	2126	2134	8	22	2007	2022	15
11	2161	2169	8	23	1932	1937	5

An Extinction Criterion for a Catalytic Wall Reactor

A.F. Jones and J.A. King-Hele
Mathematics Department
University of Manchester
Manchester. M13 9PL

ABSTRACT. When a catalytic wall reactor is operating properly the boundary walls of the reactor are at a high temperature and a strong temperature dependent exothermic reaction at the walls heats and reacts the inlet gas. However if the inlet gas temperature is reduced below a certain value T_I^*, the blow-out temperature, steady state combustion cannot take place. Combustion ceases first in the inlet region, but eventually combustion ceases in the whole reactor. In this paper we formulate the equations which govern steady heat and mass transfer in the boundary layer in the inlet region of the reactor. At high activation energies an order of magnitude analysis can be performed to yield an equation for the blow-out temperature T_I^* involving a single order one constant β^* a parameter which can be found by solving the governing equations. The solution for T_I^* is however insensitive to the value of β^*.

1. Introduction.

Catalytic wall reactors are used in a diverse range of applications including gas turbines, industrial boilers, furnaces as well as in automobile applications. Review articles have been written by Irandoust [1] and Prasad [2]. In such a reactor, for example the monolith reactor, a methane/air mixture flows past a large number of thin interlocking ceramic plates coated with a catalyst. The plates form many geometrical similar channels of small rectangular cross-section through which the fuel mixture passes. As gas enters the reactor, boundary layers are formed on both sides of every plate in the rectangular array. These boundary layers have a double functions since the gas comes in cold and has to be heated up to the temperature of the reactor, and the methane comes in at its highest concentration and rapid reaction quickly reduces this to a very low value at the wall. As we pass further into the reactor the boundary layers thicken and eventually fill the channels between the plates. However blow-out is caused by processes which occur in the inlet region where the boundary layers are thin. Moreover the boundary layers on adjacent plates do not interact with one another except in small corner regions. We conclude that for our purpose it is sufficient to consider the flow over a single semi-infinite flat plate. If the ceramic or metal plates have thickness 2ℓ, then by symmetry we can consider a half plate if thickness ℓ with no heat transport across its lower side and a fluid boundary layer on its upper side.

This problem has been considered previously by Linán and Trevino [3], who considered both the ignition and extinction problems for flow over a short thin plate. They formulated the problem in terms of a pair of coupled integral equations which they solved by a formal asymptotic method finally obtaining an extinction (or blow-out) criterion in a rather complicated form. It can be show that their criterion agrees with that derived here apart from a minor numerical factor, but we believe that the approach adopted here leads to a better physical understanding of the physical processes involved, and can therefore be extended more readily to situations where the integral equation approach cannot be used.

2. The Governing Equations.

We consider the steady flow of an air/fuel mixture past the upper surface of a semi-infinite plate $y = 0$, $x \geq 0$ where (y,x) are cartesian coordinates normal and parallel to the plate respectively. We assume that the inlet gas flow is uniform so that there is a Blasius boundary layer on the plate [4]. It is known that good estimates of the heat and mass transfer to a plate can be determined by using the near wall velocity field rather than the full Blasius solution (Lighthill [5]). The near wall velocity field (u,v) is given by

$$u(x,y) = \frac{\gamma y U}{(vx/U)^{1/2}}, \qquad v(x,y) = \frac{\gamma v y^2}{4(vx/U)^{3/2}}, \qquad (1)$$

where U is the inlet gas velocity, v is the constant viscosity and $\gamma \simeq 0.33$.

The fuel concentration $c(x,y)$ and gas temperature $T_g(x,y)$ satisfy the boundary layer equations

$$u\frac{\partial c}{\partial x} + v\frac{\partial c}{\partial y} = D\frac{\partial^2 c}{\partial y^2}, \qquad u\frac{\partial T_g}{\partial x} + v\frac{\partial T_g}{\partial y} = \alpha\frac{\partial^2 T_g}{\partial y^2}, \qquad (2)$$

where α and D are the constant thermal and mass coefficients respectively. We shall assume that the plate is thin so that the plate temperature is constant across any normal section. Then the heat balance equation for the plate is

$$\ell k_w \frac{d^2 T_w}{dx^2} = - k_g \frac{\partial T_g}{\partial y} - QD\frac{\partial c}{\partial y}, \quad \text{on} \quad y = 0, \qquad (3)$$

where $T_w(x)$ is the wall temperature, k_g and k_w are the constant thermal conductivities of the gas and wall respectively and Q is the heat of reaction.

The rate of diffusion of gas to the wall equals its rate of reaction at the wall so that for first order Arrhenius kinetics

$$D\frac{\partial c}{\partial y} = Ac \, \exp\left(\frac{-E}{RT_w}\right), \quad \text{on} \quad y = 0, \qquad (4)$$

where A is the pre-exponential factor, E is the activation energy and R is the gas constant.

These equations are to be solved subject to the boundary conditions $T_w(x) = T_g(x,0)$ on $x > 0$, $T_g = T_I$, $c = c_I$ at $x = 0$, and $T_g \rightarrow T_I$, $c \rightarrow c_I$, as $y \rightarrow \infty$, where T_I and c_I are the inlet temperature and methane concentrations respectively. Finally we shall assume that no heat is lost from the front end of the plate so that $dT_w/dx = 0$ at $x = 0$.

In the special case when the rate of reaction is infinitely fast i.e. $A \rightarrow \infty$ in (4), these model equations have an exact solution in which $c = 0$ at the wall and the wall temperature is constant and equal to

$$T_\infty = T_I + \Delta T, \quad \Delta T = \frac{Q\alpha c_I}{k_g}\lambda^2, \qquad (5)$$

where $\lambda = (D/\alpha)^{1/3}$. The corresponding solution for $c(x,y)$ and $T_g(x,y)$ is of similarity type, but its precise form is not of interest here. We note that if T_I is reduced then T_∞ is reduced by the same amount since ΔT is a constant.

3. The Blow-out Mechanism and an Order of Magnitude Analysis.

The way in which blow-out occurs may be explained briefly as follows. In a properly operating device the plate temperature $T_w(x)$ will be high but slightly less than T_∞ since the reaction rate is finite. The heat flux from the plate to the incoming gas will be slightly larger than the heat production near $x = 0$ and in a steady state this heat deficit will be supplied by heat conduction along the wall towards $x = 0$. Thus $T_w(x)$ will be smallest at $x = 0$ and will rise to T_∞ as x increases.

Suppose now that the inlet gas temperature T_I is slowly reduced. Then from (5) we see that T_∞ falls by the same amount. If the rate of heat production is a sensitive function of temperature this will lead to an increase in the heat deficit in the plate near $x = 0$ so that $T_\infty - T_w(0)$ will increase. Now for Arrhenius kinetics the rate of heat production falls by an order of magnitude if the temperature falls by RT_∞^2/E. We would therefore expect that when $T_\infty - T_w(0)$ has this magnitude there will be such a serious reduction in the heat production so that it is not possible for the heat deficit near $x = 0$ to be supplied by heat conduction in the plate. Consequently the temperatures near $x = 0$ will continue to fall until $T_w(0) = T_I$ at which stage combustion there has essentially ceased. We conclude that blow-out will occur when

$$\delta T \equiv T_\infty - T_w(0) \simeq \frac{RT_\infty^2}{E} . \tag{6}$$

The right hand side of (6) may be regarded as known, so we need to estimate δT from the model equations.

There are two physically distinct regions near $x = 0$ with very different x-length scales. In the first region, which we call the reaction region, combustion is limited by the chemical kinetics. In this small region the concentration of gas at the wall falls rapidly from c_I to zero and the gas and plate temperatures are almost constants. The x-length scale of this region L_r is readily determined from (2) and (4) and we find that

$$L_r = \frac{DU}{A^2} \left(\frac{D\gamma^2}{v} \right)^{1/3} \exp\left(\frac{2E}{RT_\infty} \right) . \tag{7}$$

We note that L_r is a sensitive function of T_∞. It is readily shown that the net heat loss from the plate in this region is of order

$$k_g \Delta T \left(\frac{UL_r}{\alpha} \right)^{1/2} \left(\frac{\gamma^2 \alpha}{v} \right)^{1/6} . \tag{8}$$

In the second, much thicker, region the temperature in the plate rises from $T_w(0)$ to T_∞. In this region, which we call the conduction region, there is a net heat flux along the plate towards the reaction region at $x = 0$. It can be shown that the x-length scale of this region is

$$L_C = \left[\frac{\ell^2 \alpha \, k_w^2}{U \, k_g^2} \right]^{1/3} \left[\frac{\nu}{\gamma^2 \alpha} \right]^{1/9} , \tag{9}$$

so that the heat flow towards the tip is $\ell k_w \delta T / L_C$. Equating this to the heat loss in the reaction region we find, after some rearrangment, that $\delta T = (L_r/L_C)^{\frac{1}{2}} \Delta T$. Hence from (6) we expect blow out to occur when

$$\left[\frac{L_r}{L_C} \right]^{1/2} \Delta T \simeq \frac{RT_\infty^2}{E} . \tag{10}$$

If we define a dimension parameter β by

$$\beta = \left[\frac{L_r}{L_C} \right]^{1/2} \frac{E \Delta T}{RT_\infty^2} , \tag{11}$$

we see that blow-out will occur when β is an order one number. We have solved the model equations using a combination of analytical and numerical methods and have shown that the equations have a steady state if $\beta < \beta^* = 1.08$, but this analysis will be given elsewhere.

The order of magnitude equation (10) may be regarded as an equation for T_∞. The key point is that L_r is a very sensitive function of T_∞ so that (10) defines a rather precise value of T_∞ for any order-one value of β. The corresponding blow-out inlet gas temperature T_I^* is then determined from (5).

4. Summary

The essential features of blow-out in the entry region of a monolith catalytic wall reactor can be determined by solving the equations of heat and mass transfer on a semi-infinite plate. Blow-out occurs as a consequence of the sensitivity of the rate of reaction to temperature. A blow-out criterion has been derived by an order of magnitude argument. This allows the blow-out inlet temperature to be determined rather precisely if the activation energy is high.

References

[1] Prasad, R. (1984). *Catalytic Combustion, Cat. Rev. Sci. Eng.* **26**, 1-57.

[2] Irandoust, S., Anderson, B. (1988). Monolithic Catalysts for Nonautomobile Applications. *Catal. Rev. Sci. Eng.,* **30**, 341-392.

[3] Linän, A., Trevino, C. (1984). Ignition and Extinction of Catalytic Reactions on a Flat Plate, *Comb. Sci. Tech.* **38**, 113-128.

[4] Schlichting, H. (1979). Boundary Layer Theory, *McGraw-Hill*, New York.

[5] Lighthill, M.J. (1950). Contributions to the Theory of Heat Transfer through a Laminar Boundary Layer, *Proc. Roy. Soc.* **A202**, 359-377.

EXTRACTION OF TRANSISTOR MODEL PARAMETERS

EWALD H. LINDNER   ULRICH WEINERT

1. Introduction

The field of application of semiconductor devices is still rapidly developing. Simulation is intensively used for their design and for designing integrated circuits. The transistor model parameters form the interface between the transistor modelled and the simulation package for an integrated circuit. Transistor models describe the performance of a transistor by means of a (set of) nonlinear equation(s). The parameters in the first transistor models were purely physical parameters (threshold voltage e.g.). Due to more and more complex models the number of parameters increased and the chance of knowing (part of) them in advance decreased. Hence, parameter extraction became a tool in semiconductor device simulation. The data used for extracting these parameters may origin mainly from two sources. Firstly, from the numerical solution of PDEs modelling a semiconductor device (see [7, 9]). Secondly, from experiments in some laboratory (see [1]). Especially in the second case all data may be subject to errors. Whereas electrical engineers tend to repeat their experiments with higher accuracy if their results are not promising we suggest to use the method of orthogonal regression instead of regression which is used by [1, 2, 11] and others. Here both the independent and the dependent variables are considered not to be known exactly.

The transistor model equations have to be considered as a black box so that the user may easily switch transistor models. Therefore, all derivatives are approximated by finite differences with self controlling stepsize.

2. A Method for Orthogonal Box Constrained Regression

Let $x \in I\!R^{N(x)}$ and $y \in I\!R^{N(y)}$ denote the vectors of state variables for the given model equations $y = f(x, \Theta)$ where $\Theta \in I\!R^{N(\Theta)}$ is a vector of parameters characterizing the system. Θ_L, and Θ_U, are the vectors of lower, and upper, bounds on Θ, resp. Let $\tilde{x}_i, \tilde{y}_i$ denote the measured values, and x_i^* and Θ^* the exact, but unknown vectors of independent variables and parameters, resp., for $i = 1, \cdots, M$. Let be $r(x_i, \Theta) \; : \; = \; f(x_i, \Theta) - \tilde{y}_i$ the i-th residual. Let the random vectors $(r(x_i, \Theta)^T, x_i^T - \tilde{x}_i^{\,T})^T$ be multivariately normally distributed with zero mean and covariance matrices Σ_i and be independent. For the unconstrained case the estimates $\hat{x}_1, \cdots, \hat{x}_M, \hat{\Theta}$ for $x_1^*, \cdots, x_M^*, \Theta^*$ are chosen to minimize the expression

$$(1) \qquad \frac{1}{2} \sum_{i=1}^{M} \begin{pmatrix} r(x_i, \Theta) - \tilde{y}_i \\ x_i - \tilde{x}_i \end{pmatrix}^T \Sigma_i^{-1} \begin{pmatrix} r(x_i, \Theta) - \tilde{y}_i \\ x_i - \tilde{x}_i \end{pmatrix} = : S(x_1, \cdots, x_M, \Theta)$$

For our case we use the identic objective and add the box constraints $\Theta_L \leq \Theta \leq \Theta_U$. We follow Schwetlick/Tiller [8] for solving this nonlinear least squares (LS) problem and decompose $\Sigma_i = R_i^T R_i$ by Cholesky factorization. (1) can be reformulated as a(n Euclidean) Minimum Norm Problem by using the following notations:

204

$$R_i = \begin{pmatrix} R_{i3} & R_{i2} \\ 0 & R_{i1} \end{pmatrix} \begin{matrix} \text{triangular} \\ \text{regular} \end{matrix}, \quad u := \begin{pmatrix} x_1 \\ \vdots \\ x_M \end{pmatrix}, \quad z := \begin{pmatrix} u \\ \Theta \end{pmatrix}, F(z) := \begin{pmatrix} F_1(z) \\ F_2(z) \end{pmatrix}$$

$$F_1(z) := \begin{pmatrix} R_{11}(x_1 - \tilde{x}_1) \\ \vdots \\ R_{M1}(x_M - \tilde{x}_M) \end{pmatrix}, F_2(z) := \begin{pmatrix} R_{12}(x_1 - \tilde{x}_1) + R_{13}r(x_1, \Theta) \\ \vdots \\ R_{M2}(x_M - \tilde{x}_M) + R_{M3}r(x_M, \Theta) \end{pmatrix}$$

and obtain the nonlinear box contrained LS problem

$$(2) \qquad S(z) = \frac{1}{2}\|F(z)\|_2^2 \longrightarrow \min_{x_1,\cdots,x_M,\Theta_L \leq \Theta \leq \Theta_U}$$

Let J_F be the Jacobian of F. The basic step of the Gauß-Newton method for solving (2) has the form $z^{(k+1)} = z^{(k)} - \Delta z^{(k)}$ for $k = 0,1,\cdots$ where the correction $\Delta z^{(k)}$ is the solution of the linear LS problem

$$(3) \qquad \|J_F(z^{(k)})\Delta z - F(z^{(k)})\|_2^2 \longrightarrow \min_{\Delta z} \quad \text{subject to } \Delta\Theta_L^{(k)} \leq \Delta\Theta \leq \Delta\Theta_U^{(k)}$$

where $\Delta\Theta_L^{(k)} := \Theta^{(k)} - \Theta_U, \Delta\Theta_U^{(k)} = \Theta^{(k)} - \Theta_L$.
Schwetlick/Tiller [8] derived an efficient and stable method of solving (3) for the unconstrained case. For brevity we omit all indices and arguments. As

$$J = \begin{pmatrix} J_{11} & 0 \\ J_{21} & J_{22} \end{pmatrix} \qquad\qquad J_{11} = diag(R_{11},\cdots,R_{M1}) \quad \text{blockdiagonal}$$

$$J_{21} = diag(J_{211},\cdots,J_{21M}) \qquad\qquad J_{22} = \begin{pmatrix} J_{221} \\ \vdots \\ J_{22M} \end{pmatrix} \text{ with } J_{22i} = R_{i3}r_\Theta(x_i,\Theta)$$
$$\text{blockdiagonal}$$
$$\text{with } J_{21i} = R_{i2} + R_{i3}r_x(x_i,\Theta)$$

we may determine an orthogonal matrix Q so that $QJ =: K = \begin{pmatrix} K_{11} & K_{12} \\ 0 & K_{22} \end{pmatrix}$, $K_{11} = diag(K_{111},\cdots,K_{11M})$ blockdiagonal with K_{11i} upper triangular. $QF =: g = \begin{pmatrix} g_1 \\ g_2 \end{pmatrix}$ is partitioned accordingly. By this orthogonal transformation we gain a decoupled problem as

$$(4) \qquad \|J\Delta z - F\|_2^2 = \|K_{11}\Delta u + K_{12}\Delta\Theta - g_1\|_2^2 + \|K_{22}\Delta\Theta - g_2\|_2^2$$

which can be solved very effectively due to the structure of K_{11}.
In general we will use the partially regularized LS problem

$$(5) \qquad (1 + \lambda^2)\|K_{11}\Delta u + K_{12}\Delta\Theta - \frac{g_1}{1+\lambda^2}\|_2^2 + \|\begin{pmatrix} K_{22} \\ \lambda I_{n(\Theta)} \end{pmatrix}\Delta\Theta - \begin{pmatrix} g_2 \\ 0 \end{pmatrix}\|_2^2 \rightarrow \min$$

subject to $\Delta\Theta_L \leq \Delta\Theta \leq \Delta\Theta_U$

which is solved by the standard LS technique combined with the Active Set Method (see e.g. [4]) for inequality constrained optimization problems. So we obtain the following partially regularized Marquardt-like method which we derived from [8] (called variant A):

Step 0:   Choose $z^{(0)} := (u^{(0)T}, \Theta^{(0)T})^T, \alpha \in]0,1[, \delta \in]0,\tfrac{1}{2}[;$

             Set $\gamma^{(-1)} := 1, k := 0$

Step 1:   Calculate $F(z^{(k)}), J_F(z^{(k)}), \nabla S(z^{(k)}) = J_F^T(z^{(k)})F(z^{(k)}),$

             $\mu_L^{(k)}, \mu_U^{(k)}, \mu_{Lj}^{(k)} \mu_{Uj}^{(k)} = 0$ for $j = 1, \cdots, N(\Theta), \nabla\mathcal{L}(z^{(k)}, \mu_L^{(k)}, \mu_U^{(k)})$

             If $\|\nabla\mathcal{L}\|_2$ sufficiently small and $\mu_L^{(k)} \geq 0, \mu_U^{(k)} \geq 0$, stop.

Step 2:   Choose $\gamma^{(k)} \in [\min(\gamma^{(k-1)}/\alpha, 1), 1]$

Step 3:   Compute Q so that $QJ_F(z^{(k)}) = K$

Step 4:   If $(\gamma^{(k)} = 1$ and $\mathrm{rank}(K_{22}) < N(\Theta))\ \gamma^{(k)} := \alpha\gamma^{(k)}$

             $\lambda^{(k)} := \sqrt{(1 - \gamma^{(k)})/\gamma^{(k)}}$

             Solve (5) for $\Delta\Theta(\lambda^{(k)}), \Delta u(\lambda^{(k)})$

             Set $\Delta z^{(k)}(\lambda^{(k)}) := (\Delta\Theta^T, \Delta u^T)^T(\lambda^{(k)})$

Step 5: $G_k(\gamma^{(k)}) := (S(z^{(k)}) - S(z^{(k)} - \Delta z^{(k)}(\lambda^{(k)})))/\nabla S(z^{(k)})^T \Delta z^{(k)}(\lambda^{(k)})$

             If $(G_k(\gamma^{(k)}) \geq \delta)$ then

                $z^{(k+1)} := z^{(k)} - \Delta z^{(k)}(\lambda^{(k)})$; $k := k + 1$; goto Step 1

             else

                Calculate an improved estimate $\tilde{\gamma} \in [0, \gamma^{(k)}]$ for $G_k(\gamma) = \tfrac{1}{2}$

                by the improved Pegasus method [5]; $\gamma^{(k)} := \tilde{\gamma}$; goto Step 4

             end if

Here $\mathcal{L}(z, \mu_L, \mu_U) := S(z) + \sum_{j=1}^{N(\Theta)} [(-\Theta_j + \Theta_{Lj})\mu_{Lj} + (\Theta_j - \Theta_{Uj})\mu_{Uj}]$, μ_L and μ_U denote the Lagrangian, and the Lagrange multiplier with respect to the lower, and upper, bounds, resp.

Additionally we suggest to modify steps 4 and 5 (denoted as variant B):

Step 4B: Choose the regularization parameter $\lambda^{(k)}$ proportional

             to $\|K_{22}\|_2\|\nabla\mathcal{L}(z^{(k)})\|_2/\|\nabla\mathcal{L}(z^{(0)})\|_2$

             Solve (5) for $\Delta\Theta(\lambda^{(k)}), \Delta u(\lambda^{(k)})$

             Set $\Delta z^{(k)}(\lambda^{(k)}) := (\Delta\Theta^T, \Delta u^T)^T(\lambda^{(k)})$

Step 5B: $G_k(\gamma^{(k)}) := (S(z^{(k)}) - S(z^{(k)} - \gamma^{(k)}\Delta z^{(k)}(\lambda^{(k)})))/\nabla S(z^{(k)})^T \Delta z^{(k)}\gamma^{(k)}$

             If $(G_k(\gamma^{(k)}) < \delta)$ perform some (e.g. 3) parabola interpolation steps

                (partial line search, see e.g. [3, 6]) for improving $\gamma^{(k)}$.

             $z^{(k+1)} := z^{(k)} - \gamma^{(k)}\Delta z^{(k)}$; $k := k + 1$; goto step 1

In variant A the step size $\gamma^{(k)}$ is always reduced at least to $\alpha(< 1)$ for the rank deficient case which leads to linear convergence only which is avoided in variant B.

3. An n-MOSFET Example

The metal-oxide-semiconductor field-effect transistor was described by SPICE Lev. 3.0 (Berkely) (cp. [10]), $x = (Uds, Ugs, Uss)^T \in I\!\!R^3$ represents the drain, gate and substrate voltage [V], $y = (Cds) \in I\!\!R^1$ is the drain current [A]. 72 data points were given for $x \in [0,5] \times [0,5] \times [0,-5], y \in [0, 0.045], T_{len} = 5.5\mu m, T_{wid} = 609\mu m, \Theta \in I\!\!R^{15}$. A priori 2 redundant parameters were known. So $N(\Theta)$ is effectively 13. We used $\|\nabla\mathcal{L}\|_2 \leq 10^{-4}$ as stopping criterion. All calculations were performed on an IBM 3084-Q96.

For variant B we needed 22 iterations and 29 sec CPU-time, whereas variant A needed more than twice as much. The objective, and the norm of the Lagrangian, were reduced by 6, and 11, orders of magnitude, resp. for both variants. The maximum error in y was 8.99 % which seems to be reasonably low compared to [2] whereas the estimated error in x was lower than 10^{-7}. In addition this orthogonal regression method took just 5% more CPU-time than the regression method used by the second authors formerly.

Acknowledgement

The scientific and financial support granted by Siemens AG Munich, FRG, for this project is gratefully acknowledged by the first author.

References

[1] B. Ankele, P. O'Leary, W. Hölzl: An Optimized SPICE Parameter Extraction Strategy for Mixed Analogue and Digital Circuit Simulation, paper presented at the 8th International Custom Microelectronics Conference, London 1988, pp. 1-8

[2] P. Conway, C. Cahill, W.A. Lane, S.U. Lidholm: Extraction of MOSFET Parameters Using the Simplex Direct Search Optimization Method, IEEE CAD-4, 1985, pp. 694-698

[3] R. Fletcher: Practical Methods of Optimization Vol. 1: Unconstrained Optimization, John Wiley & Sons, Chichester, 1980

[4] R. Fletcher: Practical Methods of Optimization Vol. 2: Constrained Optimization, John Wiley & Sons, Chichester, 1981

[5] R.F. King: An Improved Pegasus Method for Root Finding, BIT 13, 1973, pp. 423-427

[6] David G. Luenberger: Introduction to linear and nonlinear programming, Addison-Wesley, Reading, 1973

[7] Peter A. Markowich: The Stationary Semiconductor Device Equations, Springer-Verlag, Wien, 1986

[8] Hubert Schwetlick, Volker Tiller: Numerical Methods for Estimating Parameters in Nonlinear Models With Errors in the Variables, Technometrics 27, 1985, pp. 17-24

[9] Siegfried Selberherr: Analysis and Simulation of Semiconductor Devices, Springer-Verlag, Wien, 1984

[10] A. Vladimirescu e.a.: SPICE Version 2G User's guide, Dept. Elec. Eng. and Computer Sci., University of California, Berkely, CA, Aug. 1981

[11] K. Yokoyama, M. Tomizawa, T. Takeda, A. Yoshii: Optimization of Process and Device Characteristics for MOSFETs by Using the BFGS Method, Solid State Electronics 27, 1984, pp. 545-551

Addresses:

Ewald H. Lindner
Institut für Mathematik
Johannes Kepler Universität Linz
A-4040 Linz/Auhof, Altenbergerstr. 69
AUSTRIA

Ulrich Weinert
ZFE SPT 32, Siemens AG
Otto-Hahn-Ring 6
D-W-8000 München 83
FRG

Mathematical Modelling of Modular
Lower Limb Prostheses

E.A.Magnissalis and S.E.Solomonidis

Bioengineering Unit
University of Strathclyde
Glasgow, Scotland, U.K.

Abstract. One type of modular lower limb prosthesis consists of a number of components, connected by means of adaptors, responsible for altering the intersegmental inclinations. Equations for the function of the adaptors were developed to allow the prediction of the imposed tilts. Use of frames of reference provided a model on which 3-D transforms could operate. Thus, the geometrical configuration of prostheses could be derived by simple measurements on the adaptors and by the use of the associated computer program.

1. Introduction

The term alignment in prosthetics refers to the relative geometrical position and orientation of the components of a prosthesis (artificial leg). The alignment has been found [Solomonidis (1980)] to be very important in the mechanical behaviour of the prosthesis and has been expressed as a set of parameters relating the components to a universal reference. Frames of reference have been established for these components and thus, alignment assessment is possible [Zahedi et al (1986)]. When an amputee is fitted with an artificial leg, adjustments are performed by the use of adaptors and a suitable alignment is finally achieved. Subsequent measurements of the prosthesis in a special rig lead to the calculation of the alignment parameters. The objective of the work presented in this paper was to derive a method for the assessment of alignment following every individual adjustment, without need for additional measurements.

2. Description of Prostheses

The prostheses dealt with are above-knee (AK) modular prostheses made by Otto Bock Orthopadische GmbH, Germany. Such prostheses consist of the prosthetic foot, shank, knee and the socket into which the residual limb is fitted. Each component can be given an appropriate frame of reference and the alignment parameters are defined with respect to the foot reference as shown in figure 1.

The Otto Bock prostheses are provided with special set-adaptors between the components, which allow angular adjustments in two perpendicular planes (figure 2). The machined cup-shaped surface of one component (1) swivels onto the spherical surface (dome) of the other component (2) and can be locked in any orientation by means of set-screws, which are tightened against the pyramid-shaped protrusion of the dome; thus, changing the relative inclination of any two adjacent components.

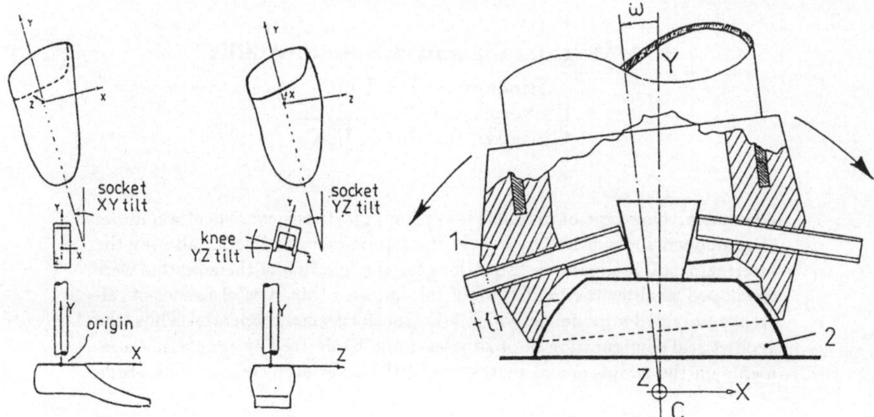

<table>
<tr><td>Figure 1. AK prosthesis.</td><td>Figure 2. Set-adaptor.</td></tr>
</table>

3. Model for the Adaptors

Measurements of the geometry of the adaptors allowed the determination of the position of the centre C of the spherical surface. Due to the spherical shape of the surface, any adjustment is considered as a rotation about the X or the Z axes of the adaptor frame (figure 2) and the value of any angular change is related to the protrusion length of the corresponding set-screw, which is directly measurable.

Calibration of the adaptor allowed the compilation of a linear relationship between the imposed angular change ω (in degrees) and the protrusion screw length l (in mm). The obtained equation is :

$$\omega = 30.59 - 1.92l \tag{1}$$

This linear model exhibited an r^2-value of 99.9% and predicted ω within a range of ± 0.12 degree, for a confidence level of 95% [Magnissalis (1992)].

4. Mathematical Description of Alignment Changes

The frames of reference of the prosthetic knee and socket as well as of the adaptors can be described with respect to a universal reference by means of matrices. If the frames under consideration are denoted by F and the universal reference frame is denoted by U, then frames F can be described by matrices T_F^U of the form :

$$T_F^U = \begin{bmatrix} X_x & Y_x & Z_x & F_{ORG_x} \\ X_y & Y_y & Z_y & F_{ORG_y} \\ X_z & Y_z & Z_z & F_{ORG_z} \\ 0 & 0 & 0 & 1 \end{bmatrix} \tag{2}$$

where $X_{x,y,z}$, $Y_{x,y,z}$ and $Z_{x,y,z}$ are the direction cosines of frame F and $F_{ORG_{x,y,z}}$ are the coordinates of the origin of frame F, with respect to the reference frame U.

Frame U is chosen to be the prosthetic foot frame (figure 1). Suppose that : 1) an adjustment is made about the X or the Z axis of an adaptor, 2) the origin of the adaptor frame is, with respect to frame U at a point (P_x , P_y, P_z) and 3) the rotational adjustment has a value ω. The position of the adaptor as well as the rotational adjustment can be expressed in the form of matrices P and R

$$P = \begin{bmatrix} 1 & 0 & 0 & P_x \\ 0 & 1 & 0 & P_y \\ 0 & 0 & 1 & P_z \\ 0 & 0 & 0 & 1 \end{bmatrix} \tag{3}$$

$$R = \begin{bmatrix} K_x K_x v + c & K_x K_y v - K_z s & K_x K_z v + K_y s & 0 \\ K_x K_y v + K_z s & K_y K_y v + c & K_y K_z v - K_x s & 0 \\ K_x K_z v - K_y s & K_y K_z v + K_x s & K_z K_z v + c & 0 \\ 0 & 0 & 0 & 1 \end{bmatrix} \tag{4}$$

where K_x, K_y, K_z are the direction cosines of the axis of rotation with respect to the frame U and s, c and v stand for $sin(\omega)$, $cos(\omega)$ and $[1 - cos(\omega)]$ respectively.

In order to determine *the new position and orientation* of a frame F, after a particular adjustment, two auxiliary frames F' and U' are defined. Both frames have the orientation of frame U and they share the same origin (P_x, P_y, P_z). Frame F can be described with respect to frame F' by the following matrix :

$$T_F^{F'} = \begin{bmatrix} X_x & Y_x & Z_x & F_{ORG_x} - P_x \\ X_y & Y_y & Z_y & F_{ORG_y} - P_y \\ X_z & Y_z & Z_z & F_{ORG_z} - P_z \\ 0 & 0 & 0 & 1 \end{bmatrix} \tag{5}$$

Thus, the new frame F can be obtained by: a) description of frame F with respect to frame F' (equation 5), b) description of frame F' with respect to U', after rotation of the former by ω (equation 4) and c) description of frame U' frame U (equation 3). The above operations can be expressed as :

$$newT_F^U = T_{U'}^U T_{F'}^{U'} T_F^{F'}$$
$$newT_F^U = PRT_F^{F'} \tag{6}$$

where multiplications are performed from right to left. The developed program, considering a known initial position for all frames, performs the above operations after each adjustment and provides the resulting frame matrices and the corresponding alignment parameters.

5. Results

The values of the alignment parameters obtained using the 3-D model were compared to values obtained using the rig mentioned in section 1, for several tests. A typical set of results is given in table 1.

6. Discussion

The results obtained showed that the model provides the values of the alignment parameters with a practically satisfactory approximation. The main advantage of the developed model is that the alignment parameters can be derived immediately and without removing the prosthesis from the amputee. Furthermore, the reported method consitutes a basis for the development of versatile software, which could assist the clinical work and decision making.

Table 1. Comparative Results for Alignment Parameters.

Alignment Parameters	3-D Model	Rig
Knee: x-coord.(mm)	−7	−8
Knee: y-coord.(mm)	462	462
Knee: z-coord.(mm)	0	1
Knee Axis: tilt (degrees)	0	0
Socket: x-coord.(mm)	−17	−14
Socket: y-coord.(mm)	731	731
Socket: z-coord.(mm)	−19	−20
Socket Axis: XY tilt (degrees)	13	12
Socket Axis: YZ tilt (degrees)	−5	−6

Acknowledgements
The first author would like to thank the Science and Engineering Research Council who provided some financial support throughout the period of the project work.

References
Magnissalis, E.A. (1992) Studies of prosthetic loading by means of pylon tranducers, PhD thesis (in process of completion), University of Strathclyde, Glasgow.
Solomonidis, S.E. (editor) (1980) Modular artificial limbs / Second report on a programme of clinical and laboratory evaluation / Above-Knee systems', Scottish Home and Health Department.
Zahedi, M.S., Spence, W.D., Solomonidis, S.E., Paul, J.P. (1986) 'Alignment of lower limb prostheses', Journal of Rehabilitation Research and Development 23(2), 2-19.

OPTIMUM SHAPE OF A COOLING FIN

K. G. Mahmoud[1]

Mathematics Institute, Johannes Kepler University Linz, Austria

ABSTRACT

Optimum shape design solves a class of problems involving continuous mechanical components where the optimum shape (the shape of the boundary and the surfaces of the components) is determined. In this work, a computer code has been developed for analysis and design/optimization of 2D fin subjected to both convective and radiative heat dissipation. The procedure proposed involves numerical solution of field equations, using finite element approximation, and optimization of parameters defining fin geometry using mathematical optimization techniques. The finite element mesh or grid is reinterpolated automatically due to the change in the fin shape during the problem solution.

INTRODUCTION

A new strong impulse came to the area of optimal design with the introduction of finite element methods (FEM) and mathematical optimization procedures. In 1960 Schmit [1] proposed for the first time coupling of FEM with mathematical optimization procedures in order to optimize a structure. The engineering problem of our concern here is the optimal shape design of extended heat transfer surfaces (fins). Fin optimization problem, referred to in the literature as the optimum volume, is to find the shape of the fin which would minimize the fin volume for a given amount of heat dissipation, or alternatively, to maximize the heat dissipation for a given volume.

Most of the previous work in fin optimization employs one-dimensional lumping [2-3] to allow for the approximation of unidirectional heat flow in thin fins. For purely convective fins, Schmidt [4] proposed the first optimal fin solution by assuming that the temperature profile is a linear function of distance from the root of the fin. Duffin [5] used variational approach to prove the assumption of Schmidt. Wilkins [6] showed that for a minimum mass fin a linear temperature profile was only true for the case of pure convection. Razelos and Imre [7] require that the annular fin has a trapezoidal profile with only convective heat dissipation. However, most of literature used thin fin approximation where the fin has a small width-to-height ratio. It is shown in refs. 8&9, this is not true in most of real fins.

This paper presents an approach for shape optimization of a 2D fin subjected to both convective and radiative heat dissipation. This optimization approach considers fin profile that is amendable to automatic machining, fabrication limits for the fin thickness at the tip, and restricted dimensions of the fin base.

STATEMENT OF THE PROBLEM

Consider a 2D fin of uniform density, with symmetric profile, attached to a surface with a temperature T_o. The profile of the fin is Y(x), with thermal conductivity, k, convective heat transfer coefficient, h_c, and both surfaces are exposed to an environment of temperature, T_c, and exchange heat by radiation with a surface at temperature T_r. The geometry of this arrangement is shown schematically in Fig. 1. The problem is consider to be steady state two-dimensional with constant heat generation rate, Qs, thermal conductivity, convective and radiative temperature, and convective heat transfer coefficient.

The heat transfer from the fin can be described by the equation of energy conservation. Dute to profile symmetry, the the upper half of the fin is considered.

---

[1]On leave form Faculty of Engineering, Assiut University, Assiut 71516, Egypt

$$\frac{\partial q_x}{\partial x} + \frac{\partial q_y}{\partial y} + Qs = 0, \quad V \in R^2 \qquad (1)$$

$$-q_x n_x - q_y n_y = q \quad on \quad \Gamma_1 \ \& \ \Gamma_2 (2)$$

$$T(x,y) = To \quad on \quad \Gamma_3 \qquad (3)$$

$$q_x n_x + q_y n_y = 0 \quad on \quad \Gamma_4 \qquad (4)$$

where $q = q_r + q_c$,

$$q_r = \epsilon\sigma\left(T^4(x,y) - T_r^4\right) \& q_c = h_c\left(T(x,y) - T_c\right)$$

$q_x \& q_y$, and $n_x \& n_y$ are the conductive heat transfer and direction cosine vectors in x and y directions respecitively.

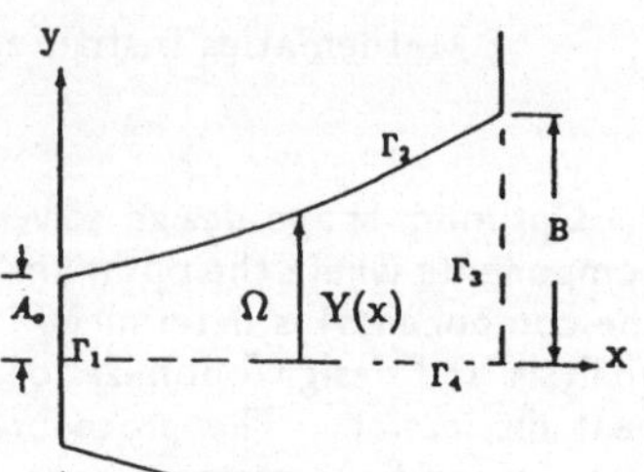
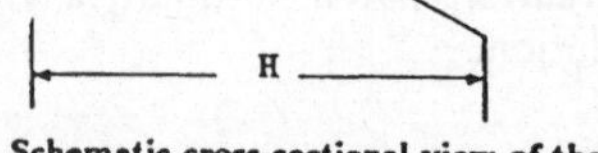

Fig. 1. Schematic cross-sectional view of the fin

Using Galerkin formulation together with Green-Gauss theorm to obtain a discrete representation of the energy equation (see ref. [10]). The resulting set of nonlinear algebric equations is of the form.

$$[K]^e\{T\}^e = \{R\}^e \qquad (5)$$

where $k_{ij} = \int_\Omega \left(\frac{\partial N_i}{\partial x}\frac{\partial N_j}{\partial x} + \frac{\partial N_i}{\partial y}\frac{\partial N_j}{\partial y}\right)d\Omega + \int_{\Gamma_1,\Gamma_2} N_i N_j(h_c + \epsilon\sigma T^3)d\Gamma,$

$r_i = \int_\Omega Qs N_i d\Omega + \int_{\Gamma_1,\Gamma_2} N_i(h_c T_c + \epsilon\sigma T_r^4)d\Gamma,$ and

ϵ is the surface emissivity and σ is the Stefan-Boltzmann constant.

Assembly of these elements equations to obtain the global equations, the resulting system can be solved to obtain the nodal temperatures vector. Having calculated the temperature at all the nodes, the heat flux, q, can be estimated. Then the optimization problem can be stated as follows:

Given the fin volume V_o

$$Maximize \quad Q \qquad (6)$$

$$s.t. \qquad a_o \le Y(x) \le B, \quad H \le H_o, and \quad V = V_o \qquad (7)$$

where $V = W_o A = W_o \int_0^H Y(x)dx$ and $Q = W_o \int_{\Gamma_1,\Gamma_2} qd\Gamma$.

The way used to describe the shape of the design problem is the key element in the process of obtaining the optimum shape. If the shape variables are not carefully selected, the reliability of the results is affected seriously. In this work the boundary shape of the fin is described by a piecewise polynomial.

In design optimization problems, the analysis and optimization are coupled by either nested or simultaneous approach. The traditional nested design problem is usually solved by repeatedly calculating the state variables and their derivatives with respect to the components of the design variables. Based on the state variables and their derivatives the constraints functions and their derivatives can be evaluated and a numerical optimization technique can use this information to improve the design variables. The simultaneous approach treats the state variables and the design variables equally as design variables and treats the governing equation in discretized form as nonlinear equality constraints.

However, the nested approach is used in this study and the algorithm can be summarized as follows: 1. Initialize the design variable vector, $\{p\}$, 2. Solve the energy conservation equation for the temperature using the finite element method

and calculate the heat dissipated, 3. Solve the optimization problem presented by Eqs. (6)& (7) for the design vector $\{p\}$, 4. Go to step 2 if convergance is not achieved, otherwise stop.
In recent years, many optimization techniques have been developed in mathematical programming. Belegundu and Arora [11] presented a comprehensive study of various mathematical programming methods used for structural optimization. In this work, Augmented Lagrangian Multiplier (ALM) technique [12] together with variable metric algorithm [13] are used to solve the optimization part.

RESULTS AND DISCUSSION

The probelm of optimal shape design of two-dimensional fin subjected to both convective and radiative heat transfer is presented. The approach is demonstrated by two examples. The first example is to find the shape of a trapezoidal fin and the second is to find the profile of parabolic fin. In the two cases the geometrical and physical properties are given in table 1.

$$h_c = 5.677 \times 10^{-4} \quad W/cm^2K \quad V_O = 615.6 \quad cm^3 \quad T_c = 289 \quad K \qquad \epsilon = 0.95$$
$$k = 0.3461 \quad W/cmK \qquad A_o = 0.125 \quad cm \quad T_r = 250 \quad K \quad Z_o = 30.4 \quad cm$$
$$\sigma = 5.669 \times 10^{-12} \quad W/cm^2K^4 \quad B = 50A_o \quad cm \quad T_o = 555 \quad K \qquad Q_e = 0 \quad W$$

Table 1. Fin geometrical and physical properties

Trapezoidal fin:

Using the data in Table 1, determine the optimum dimensions of a trapezoidal fin simultaneously dissipated heat by convection and radiation.
Using $a_o = .25$ cm, $a_1 = 0$ and $H = 5$ cm as initial guess and carrying out the optimization process, the results are: $Q = 268.5908$ W, $a_o = .125$ cm, $a_1 = 1.92821$ and $H = 3.1765$ cm. Figure 2 shows the finite element mesh after the optimum conditions have been achieved.

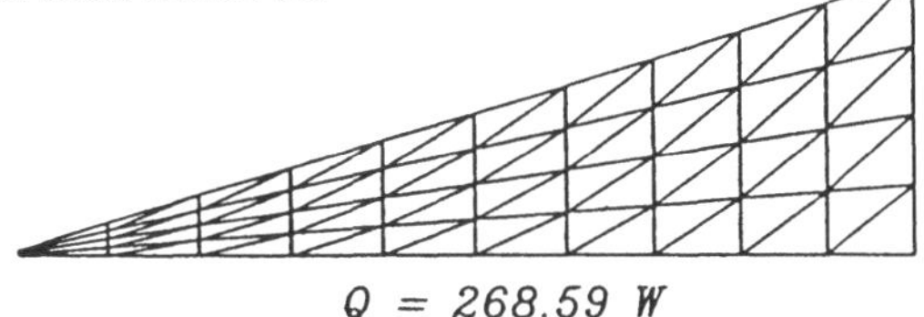

Fig. 2. Final grid of the trapezoidal fin

Fin with a parabolic profile:

It is required to find the optimum shape of a parabolic profile fin subject to both convection and radiation at the surface using the parameters given in table 1.
Set $a_o = .25$, $a_1 = 1.0$, $a_2 = 0.0$ and the initial height to 2.5 cm. After carrying out the numerical test the results are: $a_o = .125$ cm, $a_1 = 0.60$, $a_2 = 0.24032$, H $= 3.95216$ cm and $Q = 286.016$ W. Figure 3 shows the finite element grid of the optimum shape.
From the numerical tests, we found out that in both cases (trapezoidal and parabolic fins), there are three active constraints at the optimum conditions. These are the equality constraint and the restriction on the minimum and maximum heights

of the fin base. From these tests one can see also that for the same volume, the fin with quadratic profile dissipate more heat than trapezoidal fin.

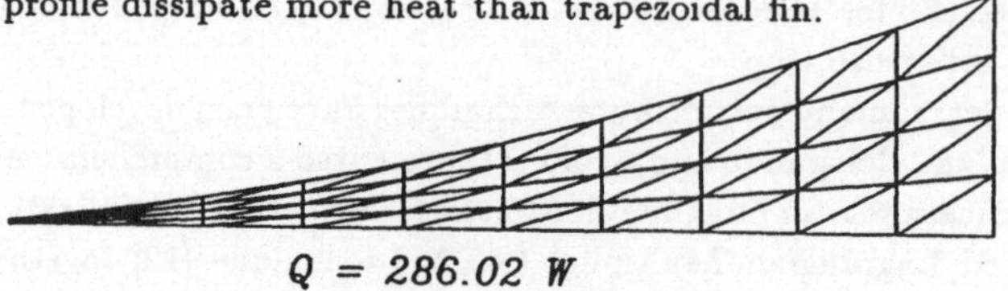

Fig. 3. Final grid of the parabolic fin

CONCLUSION

An optimal shape design approach for two-dimensional fin has been presented. The approach uses finite element approximation to calculate the state variables and ALM optimization technique to improve the design variables. Some numerical tests have been carried out and it has been found that the constraint on the maximum fin base height is always active. It has been found also that for the same volume the heat dissipated from the parabolic profile is higher than that dissipated from trapezoidal fin.

REFERENCES

1. Schmit, L. A., Structural Design by Systematic Synthesis, Proceeding, 2nd Conference on Electronic Computation, ASCE New York, pp. 105-122, 1960.

2 Kern, D. Q., and Kraus, A. D., Extended Surface Heat Transfer, McGraw-Hill, New York, 1972.

3 Ozisik, M. N., Basic Heat Transfer, McGraw-Hill, New York, 1977.

4 Schmidt, E. , Die Warmeübertragung durch Rippen, Zeitschrift des VDI, Vol. 70, No. 26, pp. 885-889, No. 28, pp. 947-951, 1926.

5 Duffin, R. J. , A variation Proplem Related to Cooling Fins, Journal of Mathematics and Mechanics, Vol. 8, No. 1, pp. 47-56, 1959.

6 Wilkins, J. E., Minimum-Mass Thin Fins and Constant Temperature Gradients , SIAM Journal, Vol. 10, pp. 62-73, 1962.

7 Razelos, P., and Imre, K., The Optimum Dimensions of Circular Fins With Variable Thermal Parameters, ASME Journal of Heat Transfer, Vol. 102, pp. 420-425, 1980.

8 Suryanarayana, N. V., Two-Dimensional Effects on Heat Transfer Rates From an Array of Straight Fins, ASME Journal of Heat Transfer, Vol. 97, pp. 129-132, 1977.

9 Ingham, D. B., Heggs, P. J., Manzoor, M., and Stones, P. R., Two- Dimensional Analysis of Fin Assembly Heat Transfer: A comparison of Solution Techniques, In: Numerical Properties and Methodologies in Heat Transfer; Proc. of the Second National Sympsium, Ed. T. M. Shih, Hemisphere, New York , 1983.

10 Huebner, K. H. and E. A. Thornton, The Finite Element Method for Engineers, John Wiley and Sons, New York, 1982.

11 Belegundu, A. D., and Arora, J. S., A Study of Mathematical Programming Methods For Structural Optimization, Part I : Theory, Int. J. Numr. Methods Engineering, Vol 21, pp. 1583-1599, 1985.

12 Vanderplaats, G. N., Numerical Optimization Techniques, For Engineering Design With Applications, McGraw-Hill, New York, 1984.

13 Huang, H. Y., Unified Approach to Quadratically Convergent Algorithms for Function Minimization, J. Opt. Theory Appl., Vol. 5, pp. 405-423, 1970.

THE OPTIMAL COLLAR

J. Molenaar

Institute for Mathematics Consulting
Eindhoven University of Technology
P.O. Box 513
5600 MB Eindhoven
The Netherlands

1. Introduction

We deal with the design of a part of certain packaging machines. In these machines the packaging material (paper or plastic sheet) is unrolled from a horizontal cylinder and folded against the inner side of a vertical, hollow cylinder. During this folding process, the sheet passes over a curved surface, the so-called 'collar' or 'shoulder'. In Fig. 1 the geometry is drawn. After a piece of sheet has been positioned inside the vertical cylinder, it is sealed at the bottom and the side and filled by dropping the product to be packed from above into the newly formed bag. Then, the bag is drawn downwards, sealed at the top and cut off. This technique allows for packaging at high speed (hundreds of bags per minute), but is sensitive to disturbances. The curvature of the shoulder should be such, that the sheet is nowhere stretched or torn. In the literature a mathematical description of possible surfaces is given by Mot [1] and Culpin [2]. The former author constructs the shoulder out of pieces of a plane and a cone.

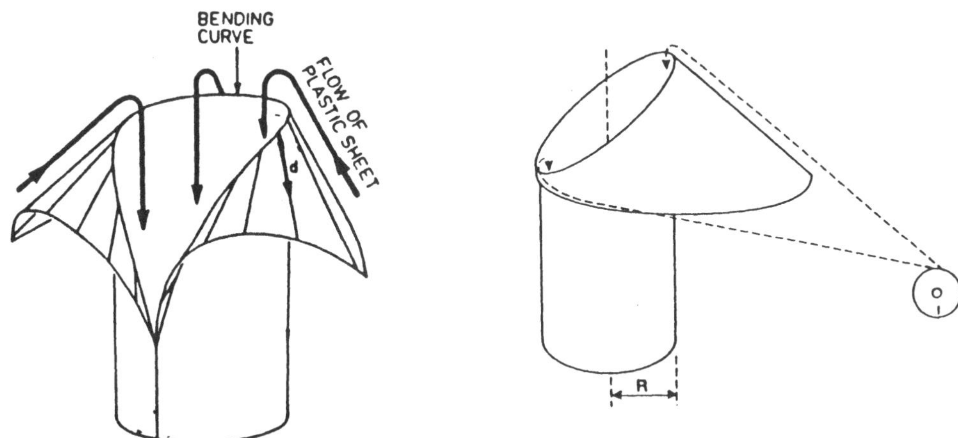

Fig. 1. Schematical sideviews of the 'collar'.

Each collar is designed for a specific bag and a specific packaging machine. Therefore, the customer must have the possibility to prescribe parameters such as the radius, the height, and the angle between the collar and the vertical cylinder at the top and at the lowest point. An extra condition is that the back edge should be flat, because the packaging sheet leaves the last roll in a horizontal, flat position. This last condition cannot be fulfilled if the theory in [2] is used. In the present paper we show how this problem can be solved.

2. Determination of a shoulder representation

In this section we summarize the essential ingredients of the geometrical description of the collar. The collar must be isometric with the plane, i.e., it can be mapped unto a part of the plane such that all distances (and thus angles) are preserved. A cylinder and a cone are for example isometric with the plane. We call the three-dimensional intersection of the shoulder and the vertical cylinder the bending curve BC. Under an isometric mapping BC transforms into a planar bending curve $\overline{BC}$. In the following, we shall denote quantities referring to the plane by an overbar. Because BC is obtained by wrapping $\overline{BC}$ around a given cylinder, the relation between BC and $\overline{BC}$ is explicitly known. We note that BC is the intersection of two surfaces, which are both isometric with the plane. It is known from differential geometry (see e.g. Forsyth [3]), that there exist precisely two surfaces which contain a given three-dimensional curve and are isometric with the plane.

We obtain an appropriate representation for the collar in the following way. Let a point of BC be represented by a three-dimensional vector $\mathbf{r}(s)$ with parameter arclength s. We assume $\mathbf{r}(s)$ to be twice differentiable and choose $\mathbf{r}(0)$ as the highest point of the shoulder. We notice, that in terms of differential geometry the collar (and also the vertical cylinder) are developable surfaces. Through each point of BC a straight line passes, which is completely contained in the collar. These lines "generate" the surface and the surface is specified by giving their directions. See e.g. Weatherburn [4], and Struik [5]. We specify the straight line passing through the point $\mathbf{r}(s)$ on BC by the unit vector $\mathbf{d}(s)$. A point $\mathbf{P}$ on the collar is then parametrized by

$$\mathbf{P}(s, u) = \mathbf{r}(s) + u\,\mathbf{d}(s)\,.$$

Let us introduce a local, orthonormal coordinate system $(\mathbf{t}, \mathbf{n}, \mathbf{b})$ at $\mathbf{r}(s)$, with $\mathbf{t}$ the tangent, $\mathbf{n}$ the normal, and $\mathbf{b}$ the binormal. The generator $\mathbf{d}$ is characterized by two angles α and φ with respect to these coordinates:

$$\mathbf{d}(s) = \cos\,\alpha(s)\,\mathbf{t}(s) + \sin\,\alpha(s)\,(\cos\,\varphi(s)\,\mathbf{n}(s) + \sin\,\varphi(s)\,\mathbf{b}(s))\,.$$

A point of $\overline{BC}$ is represented by a two-dimensional vector $\bar{\mathbf{r}}(s)$. Note that both $\mathbf{r}(s)$ and $\bar{\mathbf{r}}(s)$ have arclength s as parameter, because s is preserved under an isometric mapping. Analogous to the representation of $\mathbf{d}$ we have

$$\bar{\mathbf{d}}(s) = \cos\,\alpha(s)\,\bar{\mathbf{t}}(s) + \sin\,\alpha(s)\,\bar{\mathbf{n}}(s)\,.$$

Here we use that the angle α is preserved under the isometric mapping. A point $\bar{\mathbf{P}}$ in the plane is parametrized by

$$\bar{\mathbf{P}}(s,u) = \bar{\mathbf{r}}(s) + u\,\bar{\mathbf{d}}(s) \ .$$

For each pair (s,u) $\mathbf{P}_s$ and $\mathbf{P}_u$ are linearly independent tangent vectors which form a basis for the tangent plane at $\mathbf{P}(s,u)$. The fact that under the isometric mapping all distances (and angles) are preserved can be expressed by the following three conditions:

(i) $\qquad |\mathbf{P}_u| = |\bar{\mathbf{P}}_u|$

(ii) $\qquad \mathbf{P}_s \cdot \mathbf{P}_u = \bar{\mathbf{P}}_s \cdot \bar{\mathbf{P}}_u$

(iii) $\qquad |\mathbf{P}_s| = |\bar{\mathbf{P}}_s| \ .$

From lengthy algebra we find

$$\cos \varphi = \frac{\bar{\kappa}}{\kappa}$$

with $\bar{\kappa}$ and κ the curvatures of the planar and three-dimensional bending curves $\overline{BC}$ and BC respectively, and

$$\tan \alpha = \frac{-\kappa \, \sin \varphi}{\varphi_s + \tau} \ , \quad 0 \leq \alpha < \pi \ ,$$

with τ the torsion of BC. These two equations determine φ and α if the BC (or $\overline{BC}$) is given. Note, that in general two values φ_1 and φ_2 are obtained with $0 \leq \varphi_1 \leq \pi/2$ and $\varphi_2 = 2\pi - \varphi_1$, and two corresponding values of α. In [6] it is shown that φ_2 corresponds with the cylinder and φ_1 with the collar.

3. The effect of a discontinuity in the bending curve

The sheet of packaging material is unrolled from a horizontal cylinder and thus initially completely flat. If the sheet is guided over the shoulder it will not exactly fit to it, because the axis of the horizontal cylinder is not parallel to any generator of the shoulder. The discrepancy can be decreased by enlarging the distance between the vertical and horizontal cylinders, but this appears to be highly disadvantageous to the stability of the process. In this report, we solve the problem by constructing a shoulder which is still isometric with the plane and contains a planar area. This can be managed if one introduces a discontinuity in the third derivative of the $\overline{BC}$ at its top. To show this, we assume the planar bending curve $\overline{BC}$, given as $z(\nu)$ (see Fig. 2), to be three times differentiable except for the top at $\nu = 0$, where a discontinuity in the third derivative is introduced. To retain symmetry, we choose $z_{\nu\nu\nu}(0+) = -z_{\nu\nu\nu}(0-)$ with 0+ and 0− indicating the right and left hand derivatives respectively. Other conditions on $z(\nu)$ are $z(-\nu) = z(\nu)$, $z(0) = h$ and $z(\pm\pi R) = 0$ with h the height of the shoulder and R the radius of the vertical cylinder. Furthermore, it is assumed that $z_{\nu\nu} < 0$, i.e. $\overline{BC}$ is concave downwards.

It is important to realize that the discontinuity introduced above affects neither the planar tangent $\bar{\mathbf{t}}$, normal $\bar{\mathbf{n}}$ and curvature $\bar{\kappa}$ along $\overline{BC}$, nor the tangent $\mathbf{t}$, normal $\mathbf{n}$, binormal $\mathbf{b}$ and curvature κ along BC. These quantities are determined by $z(\nu)$ and its first and second derivatives only, which are supposed to be continuous functions everywhere along $\overline{BC}$ and BC. However, the torsion τ exhibits a discontinuity at the top.

The techniques in Section 2 can be applied to the parts $\nu > 0$ and $\nu < 0$ of the bending curves separately. Only at the top we have to distinguish between $\nu = 0+$ and $\nu = 0-$ indicating from which side we approach it. Because $\bar{\kappa}$ and κ are everywhere continuous, we find (also at the top) one value for φ in the interval $0 \leq \varphi \leq \pi/2$. In [7] it is derived that

$$\varphi_{\bullet}(0\pm) = \tau(0\pm) = z_{\nu\nu\nu}(\pm 0) / \kappa^2(0)\, R \;.$$

At the top we thus have two values for α determined by

$$\tan \alpha(0\pm) = -\kappa^2(0) / 2\, z_{\nu\nu\nu}(0\pm) \;.$$

Here, we made use of the relation $\sin \varphi(0) = 1/\kappa(0)\, R$. Because $z_{\nu\nu\nu}(0+) = -z_{\nu\nu\nu}(0-)$ we have $\alpha(0-) = \pi - \alpha(0+)$. The situation is sketched in Fig. 2.

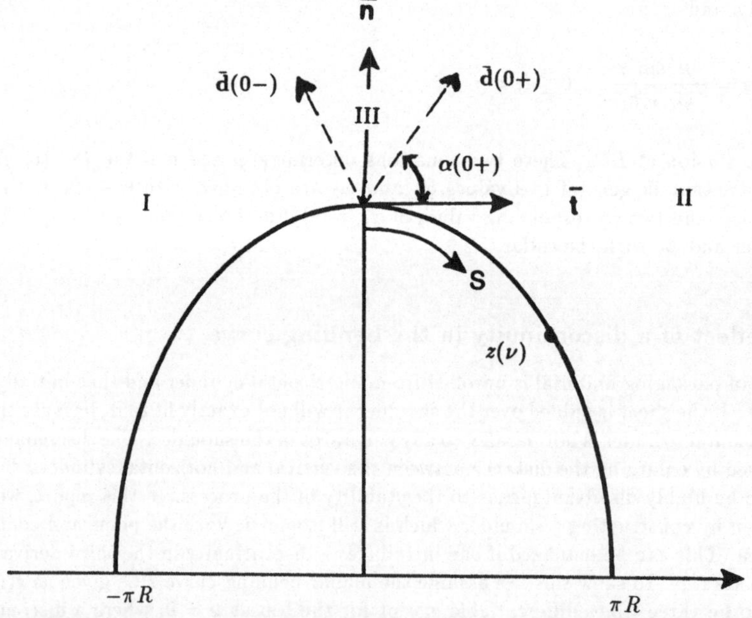

Fig. 2. Planar bending curve $z(\nu)$ with a discontinuity in the third derivative at $\nu = 0$. At the top two generators $\bar{\mathbf{d}}(0+)$ and $\bar{\mathbf{d}}(0-)$ exist. The plane above $\overline{BC}$ is divided into the parts I, II, and III, which are separately mapped onto the shoulder in space.

In Fig. 2 three regions I, II and III are indicated. By folding $\overline{BC}$ around a cylinder with radius R, the regions I and II are isometrically mapped onto the collar in space in the way described in [2] and [6]. Region III is kept flat. The question remains whether the triangular region III still fits between $\mathbf{d}(0+)$ and $\mathbf{d}(0-)$ after mapping. The answer to this question is positive and a proof is easily given. The angle between $\bar{\mathbf{d}}(0+)$ and $\bar{\mathbf{d}}(0-)$ is $\pi - 2\alpha(0+)$. We show that this is also the angle between $\mathbf{d}(0+)$ and $\mathbf{d}(0-)$ by using the representations:

$$\mathbf{d}(0\pm) = \cos \alpha(0\pm)\, \mathbf{t}(0) + \sin \alpha(0\pm)\, [\cos \varphi(0)\, \mathbf{n}(0) + \sin \varphi(0)\, \mathbf{b}(0)] \ .$$

The orthonormality of the triple $(\mathbf{t}, \mathbf{n}, \mathbf{b})$ then yields

$$\mathbf{d}(0+) \cdot \mathbf{d}(0-) = -\cos^2 \alpha(0+) + \sin^2 \alpha(0+) = -\cos 2\alpha(0+) \ .$$

Here, we use that $\cos \alpha(0-) = -\cos \alpha(0+)$ and $\sin \alpha(0-) = \sin \alpha(0+)$. Because $\mathbf{d}(0\pm)$ are unit vectors, we conclude that $\mathbf{d}(0+)$ and $\mathbf{d}(0-)$ indeed make an angle $\pi - 2\alpha(0+)$.

4. Determination of the planar bending curve

To fulfil all conditions mentioned in the introduction the planar bending curve $z(\nu)$ has to contain a number of adjustable parameters. In practice it suffices to use a curve of the form

$$z(\nu) = c_0 + c_2\, \nu^2 + c_3\, |\nu|^3 + c_4\, \nu^4 + c_6\, \nu^6 \ .$$

The coefficients are to be determined from the geometrical constraints. It is found immediately that c_0 corresponds with the collar height. The parameter c_2 is fixed by specification of the angle θ_0 between the collar and the vertical at the top. We have

$$c_2 = -\left(\frac{1 - \cos \theta_0}{1 + \cos \theta_0}\right)^{1/2} \Big/ (2R)$$

with R the collar radius. Expressions for c_3, c_4, and c_6 are not that simple and therefore omitted here. See also [7].

References

1. MOT, E., The "Shoulder problem" of Forming, Filling and Closing Machines for Pouches, Appl. Sci. Res. **27**, october 1972, pp. 1–13.
2. CULPIN, D., A Metal-Bending Problem, Math. Scientist **5**, 1980, pp. 121–127.
3. FORSYTH, A.R., Lectures on the Differential Geometry of Curves and Surfaces, Cambridge University Press, Cambridge, 1912.
4. WEATHERBURN, C.E., Differential Geometry of Three dimensions, Vol. I, Cambridge University Press, Cambridge, 1927.
5. STRUIK, D.J., Lectures on Classical Differential Geometry, Addison-Wesley, 1957.
6. MOLENAAR, J., Shoulder Design for Packaging Machines, Report IWDE 89–06, Eindhoven University of Technology, The Netherlands, 1989.
7. MOLENAAR, J., The Optimal Shape of Shoulders, Report IWDE 89–09, Eindhoven University of Technology, The Netherlands, 1989.

DRAG PREDICTION FOR AIRCRAFT AT TRANSONIC SPEED

Uno G. Närvert
Swedish Institute of Applied Mathematics
Chalmers Teknikpark
S-412 88 GÖTEBORG
Sweden

Yngve C.-J. Sedin
SAAB-SCANIA AB
Saab Aircraft Division
S-581 88 LINKÖPING
Sweden

ABSTRACT. We present a method to compute inviscid drag for aircraft at transonic speed. The formulation is based on small perturbation theory using a far field approach for the drag calculation. Computed results are validated against experimental data for different configurations.

1. Introduction

Aerodynamic drag is a key performance parameter for an aircraft having a direct impact on speed, range, and fuel consumption. Efficient and robust computational methods for drag prediction, easy to use also in an early, preproject phase, are therefore highly valuable tools.

Great progress has been made in Computational Fluid Dynamics (CFD) in recent years. Codes based on the Euler or Navier-Stokes equations are however still complicated and expensive to use for realistic three dimensional geometries. In particular, they require an efficient and flexible grid generator capable of producing body fitted grids about complex aircraft configurations.

A major simplification, allowing the use of Cartesian computational grids, is offered by formulations based on Transonic Small Perturbation (TSP) theory assuming slender configurations at small angles of attack. We consider a particular method of this type, previously introduced in [1].

In the present paper we address the problem of drag prediction based on the computed flow field using an appropriate drag formula. To find the pressure drag one could in principle integrate the computed pressure force over the surface of the aircraft. This procedure is however doubtful for a TSP method, being less accurate at wing leading edges, where nose suction is important.

We propose the use of a far field approach based on considerations given in [2]. The lift induced drag is computed using a quadrature formula introduced in [3] to solve a certain class of singular integral equations. The capability of the proposed method is illustrated by comparisons between computed and experimental values for several test cases.

2. Drag formula based on far field data

Inviscid flow is governed by a set of conservation laws for mass, momentum, and energy, here written in integral form:

$$\frac{\partial}{\partial t} \iiint_V \rho \, dV + \iint_S \rho (\vec{V} \cdot \hat{n}) dS = 0,$$

$$\frac{\partial}{\partial t} \iiint_V \rho \vec{V} dV + \iint_S \rho \vec{V}(\vec{V} \cdot \hat{n}) dS + \iint_S p \, d\vec{S} = 0,$$

$$\frac{\partial}{\partial t} \iiint_V \rho (e + \tfrac{1}{2}V^2) dV + \iint_S \rho (h + \tfrac{1}{2}V^2)(\vec{V} \cdot \hat{n}) dS = 0,$$

where t denotes time and ρ, p, $\vec{V}$, e, h denote density, pressure, velocity, internal energy per unit mass, and enthalpy per unit mass, respectively. Here V is an arbitrary control volume fixed in space with boundary surface $\partial V = S$ having outward pointing unit normal $\hat{n}$ and directed surface element $d\vec{S} = \hat{n}dS$. We assume the fluid to be a perfect gas with constant specific heat ratio γ.

To calculate the force $\vec{F}$ acting on an object with boundary surface B immensed in the fluid we could in principle solve the equations in a region V about the object and then integrate the resulting pressure over the surface to obtain

$$\vec{F} = \iint_B p \, d\vec{S}.$$

For a TSP formulation to be valid the configuration should be slender, giving small disturbances of the free stream flow. This assumption is not satisfied in the wing leading edge region, where the wing surface forms a large angle to the direction of the free stream and a stagnation point may well occur. Since this region gives a significant contribution to the drag, the above formula is not well suited to compute the drag component in small disturbance codes.

Let S now denote the outer boundary of the computational domain V, i.e., $\partial V = B \cup S$. Using the momentum equation we may then alternatively express $\vec{F}$ for stationary flow as

$$\vec{F} = -[\iint_S p \, d\vec{S} + \iint_S \rho \vec{V}(\vec{V} \cdot \hat{n})dS] = -[\iint_S (p - p_\infty)d\vec{S} + \iint_S \rho(\vec{V} - \vec{V}_\infty)(\vec{V} \cdot \hat{n})dS],$$

where the equation has further been cast into perturbation form using the identities $\iint_S d\vec{S} = 0$ and $\iint_S \rho(\vec{V} \cdot \hat{n})dS = 0$, with subscript ∞ denoting free stream data. This formulation is entirely based on far field data.

3. Small disturbance formulation

We consider three-dimensional transonic flow about an aircraft at angle of attack α relative to a freestream of velocity V_∞ and Mach number M_∞. The configuration is symmetric with respect to the xz-plane of a Cartesian coordinate system (x,y,z) having its x axis aligned with the longitudinal axis of the fuselage, cf. Figure 1.

We consider a stationary potential approximation to the equations above, with the velocity field decomposed into

$$\vec{V} = \vec{V}_\infty + V_\infty \nabla \phi = V_\infty[(1 + u)\hat{x} + v\hat{y} + (\alpha + w)\hat{z},]$$

where the perturbation potential ϕ satisfies the TSP equation, cf. e.g. [4],

$$T\phi \equiv (1 - M^2)\phi_{xx} + L\phi \equiv (1 - M_\infty^2 + M_\infty^2[3 - (2 - \gamma)M_\infty^2]\phi_x)\phi_{xx} + \phi_{yy} + \phi_{zz} = 0.$$

This equation is solved using domain decomposition, with inner and outer equations governed by the operators L and T, as described in [1]. To derive a formula for the drag coefficient C_D, normalized to unit reference surface, we follow the procedure described in detail in [2]. Expanding the far field formulation in Section 2 together with the energy equation and neglecting higher order terms assuming small perturbation velocities u, v, w and weak shocks, we arrive at

$$C_D = \iint_{S_T} (v^2 + w^2)dS - 2\iint_{S_L} u\phi_n dS - C\iint_{S_S} [u]^3 dS.$$

Here [u] denotes the jump of u across a surface of discontinuity and C is a known constant depending on the free stream Mach number. The surface integrals extend over the far downstream Trefftz' plane surface S_T, the lateral surface S_L, and all shock surfaces S_S, with $\phi_n \equiv \nabla\phi \cdot \hat{n}$.

The evaluation of the last two integrals is straightforward. Since a type dependent finite difference scheme is used, shock surfaces are detected as a part of the normal solution procedure.

The first integral, giving the lift induced drag $C_{D,i}$, can be rewritten as a line integral using Green's formula in Trefftz' plane. We arrive at the following expression, see also [4]:

$$C_{D,i} \equiv \iint_{S_T} (v^2 + w^2)\, dS = -\int_{\partial S_T} \phi\phi_n\, ds \approx -\int_{\partial B} \phi\phi_n ds - \int_{\partial W} \Gamma w ds,$$

where ∂B and ∂W are the contours of the fuselage sting and the wing wake(s), with outward pointing normal $\hat{n}$ and $\phi_n \equiv \nabla\phi \cdot \hat{n}$. Here $\Gamma = \Gamma(y)$, $|y| < b$, is the circulation given by the potential difference along the wing trailing edge of span 2b. Far downstream the downwash $-w$ is given by

$$-w(y) = -\frac{1}{2\pi} \int_{-b}^{b} \frac{\partial \Gamma}{\partial \eta} \frac{1}{\eta - y} d\eta \ .$$

4. Numerical quadrature for the lift induced drag integral

For an elliptic circulation distribution $\Gamma(y) = \Gamma_0[1 - (\frac{y}{b})^2]^{\frac{1}{2}}$, which is qualitatively representative, the downwash takes the form

$$-w(y) = -\int_{-1}^{1} \frac{\omega(t)}{t - \frac{y}{b}} dt \equiv \frac{\Gamma_0}{2\pi b} \int_{-1}^{1} \frac{t(1 + t)^\alpha (1 - t)^\beta}{t - \frac{y}{b}} dt \ , \qquad \alpha = \beta = -\tfrac{1}{2} \ .$$

In [4] a quadrature formula based on a tanh transformation is introduced for a class of integral equations having singular kernels of this type for $\alpha, \beta > -1$. The proposed quadrature rule gives

$$w(y) \approx h \sum_{n=-N^-}^{n=N^+} \frac{\omega(tanh(nh))cosh(s)}{cosh(nh)sinh(nh - s)} + \pi cot(\frac{\pi s}{h})\omega(tanh(s)) \ ,$$

where $tanh(s) = \frac{y}{b}$ and $h = \frac{\pi}{\sqrt{2(1+\alpha)N^+}}$ or $h = \frac{\pi}{\sqrt{2(1+\beta)N^-}}$.

5. Computational results

The method proposed above has been implemented in the small disturbance code SPICA developed at Saab. In this section we present some validating test runs for drag evaluation applied to a single wing, an isolated body, and a wing-body combination. Each test case typically requires 15 min CPU time on a Cray X-MP in one processor mode.

We first consider the M6 wing at free stream Mach number 0.84 and angle of attack 3.06°. This is a standard transonic test case for CFD codes with pressure distributions available from wind tunnel measurements. In Table 1 we compare lift and drag coefficients C_L and C_D computed with SPICA with published results taken from [5] for different Euler codes.

If cross terms are included in the TSP equation the value of C_D obtained with SPICA is increased to 0.0128. The agreement is good between the different codes except for the drag reported for the MBB code, which differs from the results of the other codes by a factor of two.

We next consider a bumpy and indented body with windtunnel data reported in [6]. Figure 2 shows a good agreement between calculated and experimental drag coefficient curves.

Our final test case is a wing-body combination tested at the Swedish Aeronautical Research Institute. The experimental curve for the drag coefficient is well reproduced by the calculations as shown in Figure 3.

6. Conclusions

We have described a far field formulation for inviscid drag prediction of aircraft. The method is tailored for small disturbance codes, where complicated generation of body fitted grids is avoided. For a number of test cases the proposed procedure has shown capable of producing load estimates, also for the drag component, that agree well with experimental data and results obtained with more sophisticated full potential and Euler codes.

Figures and tables

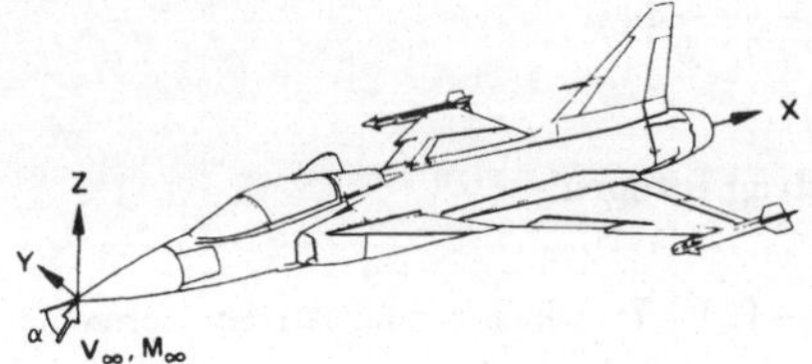

Fig. 1 Cartesian coordinate system.

Code	C_L	C_D
SPICA	0.278	0.0101
Dornier/Jameson	0.279	0.0141
FFA	0.283	0.0111
MBB	0.290	0.0251

Table 1. M6 wing results, $M_\infty = 0.84, \alpha = 3.06°$

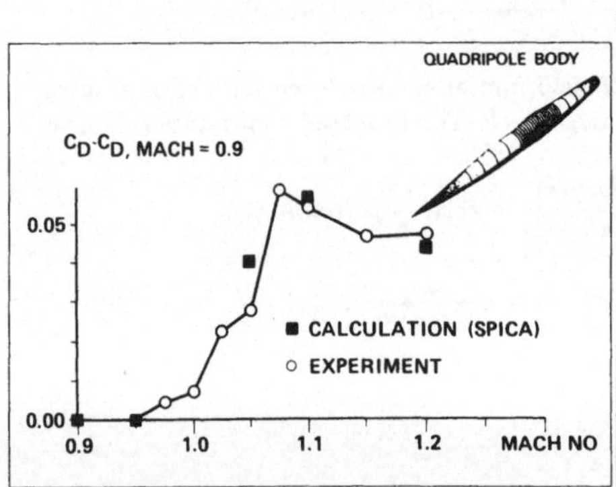

Fig. 2 Quadripole body drag comparison

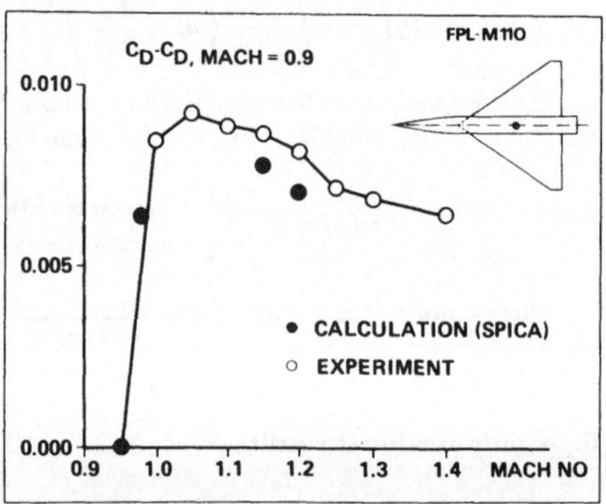

Fig. 3 Wing-body drag comparison (Fpl 110)

References

1. Nävert, U.G. and Sedin, Y.C.-J. (1988) 'Transonic Computations About Complex Configurations Using Coupled Inner and Outer Flow Equations, J.Aircraft Vol.25, No.8, pp.677-683.

2. Sedin, Y.C.-J. (1978), Qualitative Calculations of Transonic Drag-Rise Characteristics Using the Equivalence Rule, 11th ICAS Congress, Lisboa.

3. Tsamasphyros, G. and Androulidakis, P. (1987), The Tanh Transformation for the Solution of Singular Integral Equations, Int.J.Num.Meth.Eng., Vol.24, pp.543-556.

4. Ashley, H. and Landahl, M. (1965), Aerodynamics of Wings and Bodies, Addison-Wesley, Reading, Massachusetts.

5. AGARD Advisory Report 211 (1985), Test Cases for Inviscid Flow Field Methods.

6. Taylor, R.A. (1959), Pressure Distributions at Transonic Speeds for Bumpy and Indented Midsections of a Basic Parabolic-Arc Body, NASA Memo 1-22-59A.

PATHS OF REINFORCED PLASTIC FIBRES
AND STRUCTURES ON BLOOD VESSEL WALLS

MARTIN NITSCHKE
Rheinisch-Westfälische Technische Hochschule Aachen
Institut für Geometrie und Praktische Mathematik
Templergraben 55
W-5100 Aachen
Germany

ABSTRACT. In many technical applications surfaces (for example components of aircrafts and cars) are wound by reinforced plastic fibres constituting a system of different layers. Each of these layers should be "almost" non-skidding, i.e. every fibre path is "close" to a geodesic curve. Moreover, the fibre system should be as regular as possible. A list of references concerning this problem is given in [3]. A related problem occurs while modelling blood vessel walls and ventricular walls (e.g. [1] page 92ff, [2] page 38ff and 75ff, [4], [5], [6]). The fibre pattern reinforcing these walls *do* have similar properties as the patterns in the technical applications *should* have. The goal of this paper is to give a common mathematical model for describing *both* the reinforced plastic fibre paths and the structures on blood vessel walls. Therefore a system of parameter dependent ODEs is constructed and solved by numerical methods. The parameter controls properties like the skidding risk (of the plastic fibre path) and the regularity of the pattern.

1. Paths of reinforced plastic fibres

1.1. REQUIREMENTS

An ideal fibre pattern should have the follwing properties:
  a) The fibres do not slip away during the winding process.
  b) The surface is covered by a homogenous pattern.
  c) The winding angle is constant.
To illustrate the requirements a) and c) a skeleton sketch of an NC winding machine is given in figure 1.
  To fulfill c) we have to define the winding angle. If the mandrel is a surface of revolution, the winding angle is defined as angle between fibre path direction and meridian direction. For non-rotational surfaces we shall introduce generalized meridian curves in 1.3.2.

1.2. SURFACE REPRESENTATION

Mandrels which occur in technical applications are "almost" rotational symmetric. Therefore, these mandrels are parameterized by

$$\Phi(t,\varphi) := \mathbf{m}(t) + r(t,\varphi)(\cos(\varphi)\mathbf{n}(t) + \sin(\varphi)\mathbf{b}(t)).$$

The geometrical interpretation of this parametrization is given in figure 2.

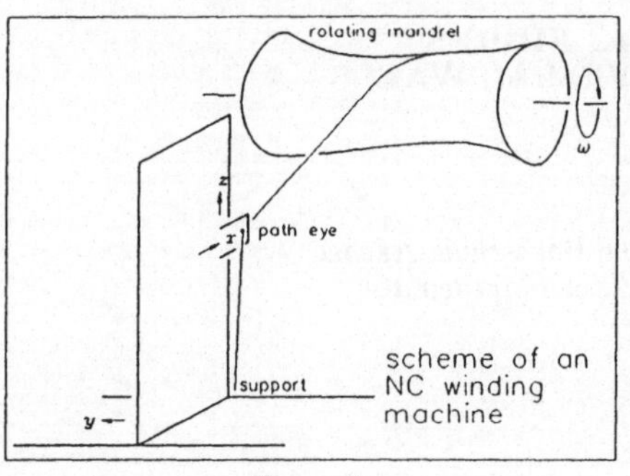

Figure 1

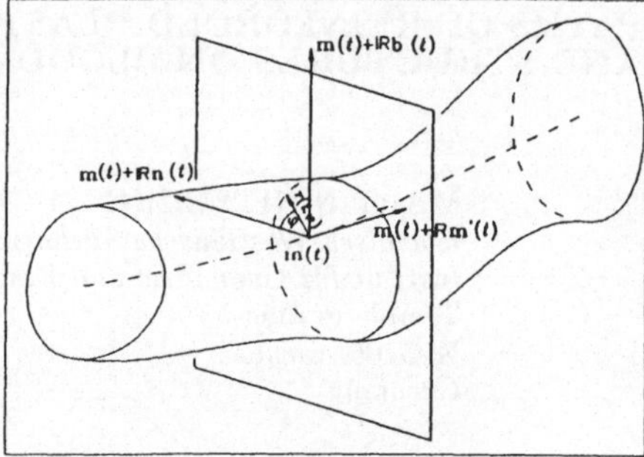

Figure 2

The *centreline* m (which can be seen as a generalized axis of revolution) is an arbitrary piecewise twice continous differentiable space curve. The vectors $n(t)$ and $b(t)$ form an orthonormal base of the plane perpendicular to the centreline at the point $m(t)$, i.e. $\left(\dfrac{m'(t)}{\|m'(t)\|}, n(t), b(t) \right)$ is an orthonormal base of $\mathbb{R}^3$. If the centreline is a Frenet curve, $n(t)$ can be choosen as normal vector field and $b(t)$ as binormal vector field. Furthermore, $n(t)$ and $b(t)$ depend piecewise twice continous differentiable on the centreline parameter t. The intersection curves between the mandrel and the planes perpendicular to the centreline are all star-shaped piecewise twice continous differentiable curves.

1.3. PATH COMPUTATION

To compute the fibre paths we give a mathematical description of the curves which fulfill the requirements a) and c):

Requirement	Mathematical description of curves
slip free	geodesics, i.e. $E(c) := \int \|c'\|^2 = \min!$
constant winding angle	$\left\langle \dfrac{c'}{\|c'\|}, \text{meridian direction} \right\rangle = const$

As an example a geodesic pattern and a constant angle pattern on a cone are given in the figures 3 and 4.

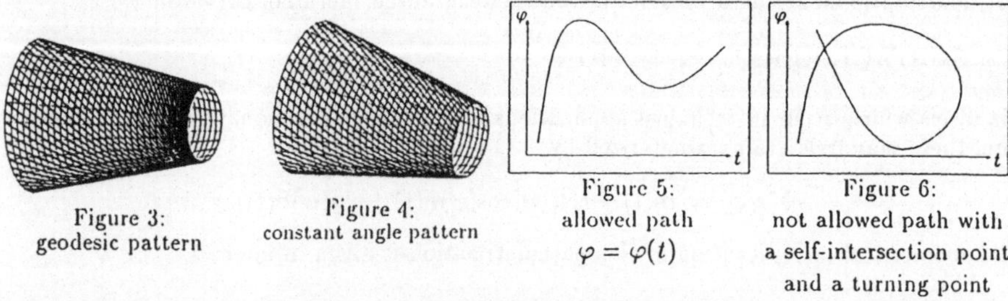

Figure 3:
geodesic pattern

Figure 4:
constant angle pattern

Figure 5:
allowed path
$\varphi = \varphi(t)$

Figure 6:
not allowed path with
a self-intersection point
and a turning point

The main problems concerning fibre path computation are the following:

1. It is necessary, that each fibre path goes from one end of the surface to the other and neither intersects itself nor turns around. This problem is easy to sove: The mandrel is parameterized by $\Phi(t,\varphi)$ and a path is allowed, iff it is represented by a function $\varphi = \varphi(t)$ in the (t,φ) parameter plane (see figures 5 and 6).

   In the geodesic case we replace the energy functional $E(c) = \int \|c'\|^2$ by the functional $E(\varphi) = \int \left\|\frac{d}{dt}\Phi(t,\varphi(t))\right\|^2$ or more general: we restrict the functional to be minimized to curves, for which φ can be represented as a function of the centreline parameter t.

2. We have to generalize the definition of "meridian direction" to non-rotational surfaces. For many technical applications, it is suitable to take the curves $\varphi = const$ of the parametrization $\Phi(t,\varphi)$ as generalized meridian curves. From a mathematical point of view, this is not satisfying, because these generalized meridian curves depend on the surface representation. The two families of principle lines of a surface depend only on its geometry and on a rotational symmetric surface one of these families consists of the meridian curves. Therefore, we define one of the families of principle lines as generalized meridian curves on a non-rotational symmetric surface.

3. Even on the simple surface shown in the figures 3 and 4, it is not possible to fullfil the requirements a) and c) by one pattern. Therefore, we make a compromise by computing the extremals of the *modified energy functional*

$$E_{w,\alpha}(c) := \int (1 - w)\,\|c'\|^2 + w\,\left(\left\langle \frac{c'}{\|c'\|}, \text{meridian direction} \right\rangle - \cos(\alpha)\right)^2.$$

   $\alpha \in [0;\pi]$ is the desired winding angle, and $w \in [0;1]$ is a compromise parameter: For $w = 0$ we obtain the usual energy functional and its extremals are geodesic curves; for $w = 1$ we obtain a functional, which extremals are constant angle curves; and for other values $w \in (0;1)$ the extremals of $E_{w,\alpha}$ are the wanted "compromise" curves. The influence of the compromise parameter w is shown on a test mandrel in figure 7. Each path of these patterns is computed by solving the Euler Lagrange differential equation of the modified energy functional. To construct a pattern from these paths, the inital conditions are varied.

4. As we see in figure 7 ($w = 0$), the geodesic pattern can be very irregular on non-rotational symmetric surfaces. By increasing w it becomes more regular, but we need more friction between mandrel and fibres to realize it. In some applictions, there is not such a great friction. Then we have to replace the inital conditions of the Euler Lagrange differential equation by a sequence of boundary conditions: We compute one fibre path and prescribe the distance between two neighbour paths and some discrete values of the centreline parameter t.

2. Structures on blood vessel walls

According to [4] blood vessel walls are fibre-reinforced, and it is a reasonable hypothesis that the fibres follow geodesic paths. Geodesics on *tubular surfaces* (these are mandrels with only *circular* intersections are discussed. Similar structures occur in the arrangement of the muscle fibres in the human/mammalian heart; see e.g. [2], [5], [6]. The muscle fibres are multi layer system of curves. If they are geodesics, there is again the problem of a full and regular covering. For an investigation we compute some geodesics on the model of the left human ventricle given in [1]. The result is shown in figure 8: The geodesics on this surface form a very regular pattern. So the technical requirements to be slip free, regular and to cover homogenous are fulfilled by only *one* curve family.

228

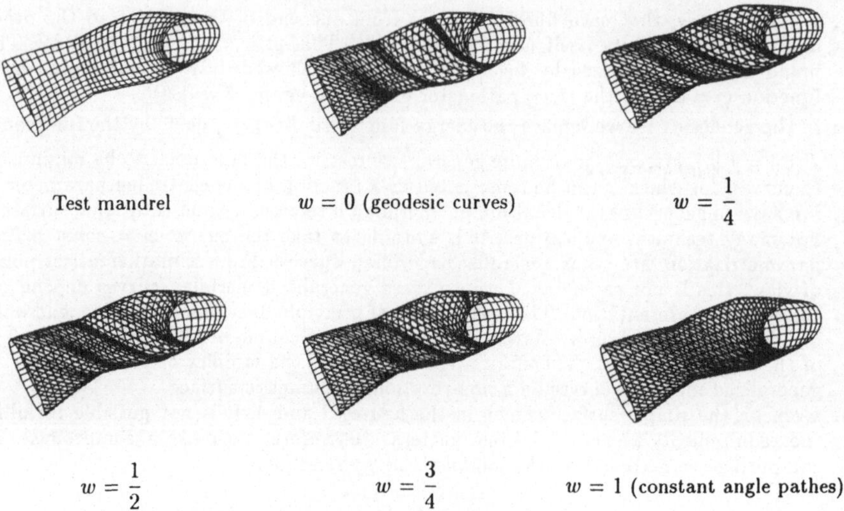

Test mandrel $\qquad$ $w = 0$ (geodesic curves) $\qquad$ $w = \dfrac{1}{4}$

$w = \dfrac{1}{2}$ $\qquad$ $w = \dfrac{3}{4}$ $\qquad$ $w = 1$ (constant angle pathes)

Figure 7: influence of the compromise parameter

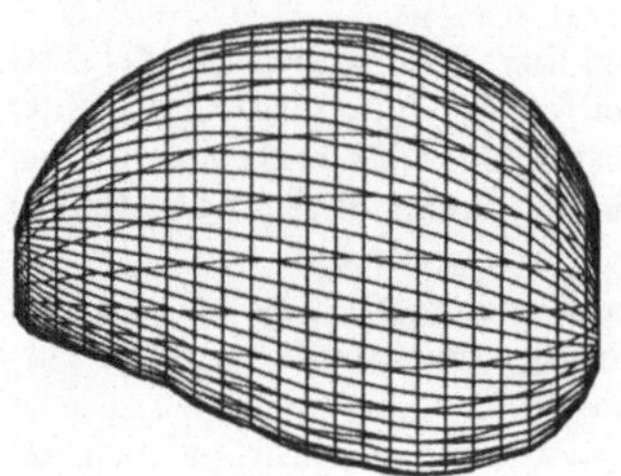

Figure 8: geodesics on the left human ventricle

REFERENCES

[1] P. T. HADINGHAM, *The Stress State in the Human Left Ventricle*, Advances in Cardiovascular Physics, Vol. 5: Cardiovascular Engineering, Part I: Modelling, D. N. Ghista, E. van Vollenhoven, W.-J. Yang, H. Reul, W. Bleifeld eds., Basel 1983.

[2] P. P. LUNKENHEIMER, A. LUNKENHEIMER, F. TORRENT-GUASP, *Kardiodynamik: Wege zur strukturgerechten Myokardfunktion*, Erlangen 1985.

[3] M. NITSCHKE, *Zur numerischen Berechnung von Faden- und Bandablagespuren auf nicht notwendig rotationssymmetrischen Wickelkernen*, Fortschr.-Ber. VDI Reihe 20 Nr. **31**, Düsseldorf 1990.

[4] J. SNEYD, C. S. PESKIN, *Computation of geodesic trajectories on tubular surfaces*, SIAM J. Sci. Stat. Comput., Volume **11**, No. 2, 230-241, March 1990.

[5] D. D. STREETER, JR., W. E. POWERS, M. A. ROSS, F. TORRENT-GUASP, *Three-dimensional fiber orientation in the mammalian left ventricular wall*, Cardiovascular Systems, J. Baan, A. Noordergraaf and J. Raines eds., Cambridge MA 1978, 73-84.

[6] C. E. THOMAS, *The muscular architecture of the ventricles of hog and dog hearts*, Amer. J. Anat., **101**(1957), 17-57.

Small drops, surface tension and contact angle

S. B. G. M. O'Brien

Philips Research Laboratories,

P. O. Box 80000, 5600 JA Eindhoven, The Netherlands.

Abstract: The shape of small sessile and pendant drops is investigated by means of a parameterisation of the Laplace capillary equation and the use of perturbation techniques. Matched asymptotic expansions are necessary for a full description of the drop shape. These solutions can then be used to determine contact angle or surface tension of a system from a measurement of two characteristic lengths of the drop. Solutions for double pendant drops are also presented.

1. Introduction

The shape of a static free liquid/gas interface is described by the well-known Laplace capillary equation which represents a force balance between hydrostatic pressure, gravity and surface tension effects. If attention is restricted to axisymmetric sessile drops, this reduces to a second order non-linear ordinary differential equation (o.d.e.), see Landau & Lifshitz (1987). Progress by analytic means is not possible but by directing attention to small drops, perturbation methods can be employed. Previous work by Shanahan (1982, 1984) and Chesters (1977) used the traditional formulation but O'Brien & van den Brule (1991) and O'Brien (1991) used an alternative formulation (1) which simplifies the zero order equations. Rienstra (1990) uses an arclength formulation. Higher order solutions are derived and the improved accuracy is investigated from the point of view of an experimentalist attempting to measure contact angle indirectly. It is also shown how the knowledge of the sessile solutions simplifies the associated more complicated pendant drop problem and we derive solutions for double pendant drop shapes.

2. Governing equations and analysis

The axisymmetric form of the capillary equation for a sessile drop (i.e. one resting on a horizontal surface) can be written as a system of coupled non-linear o.d.e.s:

$$\frac{dX}{d\phi} = \frac{X\cos\phi}{XY+XP-\sin\phi}; \quad \frac{dY}{d\phi} = \frac{X\sin\phi}{XY+XP-\sin\phi} \tag{1}$$

as for example in O'Brien & van den Brule (1991). X and Y are dimensionless (relative to the capillary length $a \equiv (\gamma/\rho g)^{1/2}$, where γ is surface tension, ρ is density and g is gravitational acceleration) and represent (r, z) coordinates as in fig.1. P is the pressure made dimensionless with $a/\rho g$ and ϕ is the inclination of the free surface. The relevant boundary conditions are:

$$X=0, \; Y=0 \text{ when } \phi=0; \quad X=R \text{ when } \phi=\pi/2. \tag{2}$$

The extra condition is used to determine the as yet unknown pressure P. $R=L/a$ is the dimensionless radius and we will restrict attention to the case where R is small.

3. Solutions

We now rescale as follows: $X=Rx$, $Y=Ry$, $P=p/R$ and seek perturbation solutions in the form: $x=x_0+R^2x_1+..$ etc. The following solutions are obtained:

$$x_0 = \sin\phi \qquad (3a)$$

$$y_0 = 1-\cos\phi \qquad (3b)$$

$$p_0=2 \qquad (3c)$$

$$x_1= 1/3 \cos^2\phi \tan \phi/2 \qquad (4a)$$

$$y_1=1/3 \cos\phi -1/3 \cos^2\phi + 2/3 \ln(\cos\phi/2) \qquad (4b)$$

$$p_1= -1/3 \qquad (4c)$$

Second order solutions are to be found in O'Brien & van den Brule (1991).

Near $\phi=\pi$, it is clear that (4) blow up and it is necessary to insert a boundary layer by rescaling as follows: $X=R^2\xi$; $Y=2R+R^3\zeta$, $P=p/R$; $\phi=\pi-R\phi$. The details are straightforward and are given in O'Brien (1991).

4. Determination of contact angle and surface tension

One difficulty which arises in measuring contact angle is in objectively determining the position of the three-phase line. If an analytic expression for the drop shape is known, then by measuring the (unambiguous) lengths L and H (see fig.1) and using the outer solutions of the last section, the contact angle can be estimated by solving the following algebraic equation for ϕ: $y(\phi,R)= h=H/R$ where $R= L/a$ and $y=y_0+R^2y_1+R^4y_2$. If the contact angle is acute, we can modify the procedure slightly. The problem here is that the half-width of the drop is not known but we can retain the same formulation by analytically continuing the drop and extending it in this way past the point where $\phi=\pi/2$.

Surface tension can also be determined in an analogous fashion without a knowledge of the contact angle provided the latter is greater than $\pi/2$. At the point where $\phi=\pi/2$, the effective contact angle is ϕ and the drop profile below this point is effectively disregarded. In figs. (2,3) we illustrate the accuracy of the zero, first and second order estimates of contact angle and surface tension using the above methods. Using the second order solutions allows one thus to work with larger drops than was previously possible. The advantage of using perturbation solutions instead of full numerical solutions is that the latter have to be continually recalculated for each set of data.

4. Pendant drops

The governing equation for this problem (a hanging drop) is identical to equation (1) with the sign of the XY term changed. Physically this indicates the fact that the drop is stretched out by gravity rather than flattened by its own weight in the sessile case. In this instance it is possible to identify several distinct regions ,O'Brien (1991), where different scales are required: (1) an outer region resembling the outer solution for the sessile drop, (2) an interior layer region centred around the point of inflection, (3) a neck region, (4) a top boundary layer region centred around a second point of inflection, (5) a second outer solution. Here we will restrict attention to the second outer solution. We define rescaled variables u,v as follows: X=Ru; Y=2R+Rv; P=p/r. The corresponding equations are then:

$$\frac{du}{d\phi} = \frac{u\,\cos\phi}{-R^2(2u+uv)+up-\sin\phi} \qquad \frac{dv}{d\phi} = \frac{u\,\sin\phi}{-R^2(2u+uv)+up-\sin\phi} \tag{5}$$

The lowest order solutions are of the same form as (3) while the first order corrections are given by:

$$u_1 = 4/3\ \sin\phi + (1/3\ \cos^3\phi\ -1)/\sin\phi \tag{6a}$$

$$v_1 = -4/3\ \cos\phi + 1/3\ \cos^2\phi\ +\ 2/3\ \ln(\ \sin\phi/2)-4/3\ \ln(\cos(\phi/2)+F \tag{6b}$$

where F is obtained by matching with the top boundary layer solution and is given by: $F=-8/3\ln R+ 4/3\ \ln 6 +1/3$.

We note that the maximum width of the upper drop is $R+ 1/3R^3$ i.e. the upper drop is slightly larger than the lower. We further note that eqs (6) blow up when ϕ approaches π radians. At this point it is necessary to insert another boundary layer and the process continues in a similar fashion to that adopted for the lower drop. In this way it is possible to define an infinite sequence of drops, each drop slightly larger than the proceeding one and joined to it by a thin neck (fig.4).

Finally we note that it is possible to also extend the sessile solutions past the point where $\phi=\pi$ using scales similar to those in the pendant case. In this way we can describe the sort of sessile profiles mentioned by Padday (1970). We give two results for the drop profiles in the "neck" portions of a sessile drop:

$$X= -2/3\ R^3\ \mathrm{cosec}\phi; \quad Y= 2R+2/3R^3\ln(\ \tan\phi/2) +2/3R^3(2\ln R-\ln 6) \tag{7}$$

References

1. Chesters A.K., 1977, J.Fluid Mech., **81**, 609.
2. Landau L.D., Lifshitz E.M., 1987, *Fluid Mechanics*, Pergamon.
3. O'Brien S.B.G.M., van den Brule, 1991, J. Chem. Soc: Faraday Trans. 1, **87**, 1579.
4. O'Brien S.B.G.M., 1991, J.Fluid Mech., to appear.
5. Padday J.F., 1971, Phil.Trans.R.Soc.Lond.**A269**, 265.
6. Rienstra S.W., 1990, J.Engng.Maths, **24**, 193.
7. Shanahan M.E.R., 1982, J.Chem.Soc., Faraday Trans 1, **78**, 2701.
8. Shanahan M.E.R., 1984, J.Chem.Soc., Faraday Trans 1, **80**, 37.

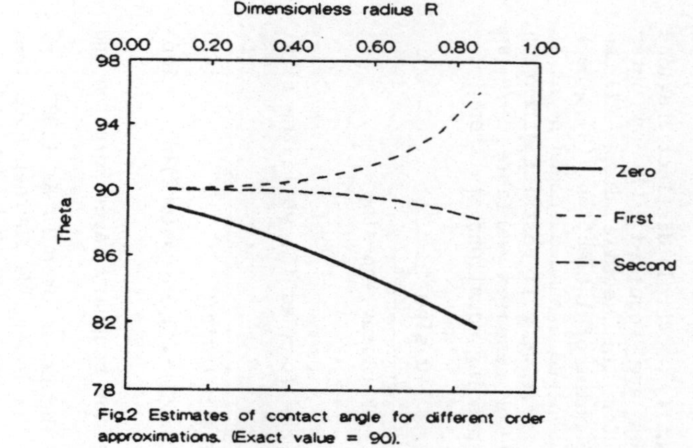

Fig.1 Schematic drawing of sessile drop showing coordinate system and characteristic lengths.

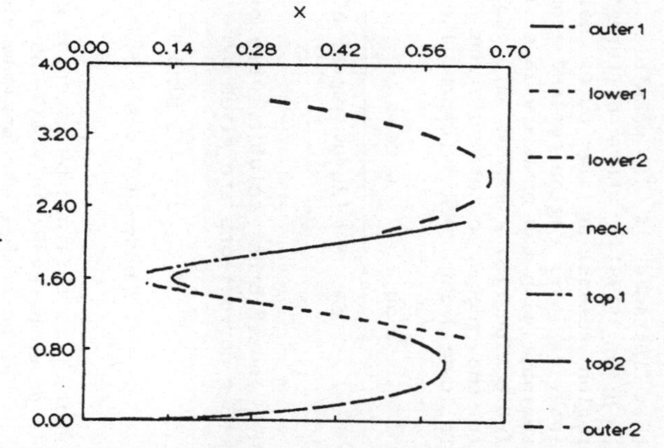

Fig.2 Estimates of contact angle for different order approximations. (Exact value = 90).

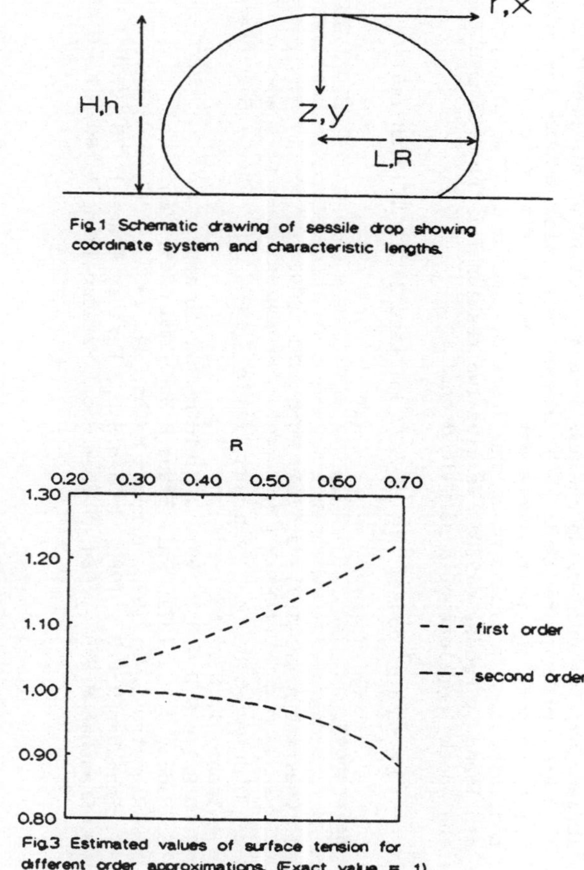

Fig.3 Estimated values of surface tension for different order approximations. (Exact value = 1).

Fig.4 Constituent solutions for the double pendant drop. (R=0.6).

OPTIMAL SCHEDULING OF HYDRO-THERMAL POWER SYSTEMS

P. Oliveira, S. McKee, C. Coles and J. Blair-Fish

Department of Mathematics
University of Strathclyde
Glasgow G1 1XH
Scotland, U.K.

Edinburgh University Computing Service
The King's Buildings
Edinburgh EH9 3JZ
Scotland, U.K.

ABSTRACT. This study is concerned with the optimal scheduling of an electricity power system consisting of both hydro and thermal units. Using Lagrangian relaxation the original primal problem may be written in dual formulation; the problem then admits decomposition into more tractable subproblems. This feature considerably reduces the computation time and has enabled an implementation to be performed on a transputer array.

1. Introduction

Power system scheduling involves decisions concerning which units should be run and what their level of output should be. These two decisions are usually referred to as the Unit Commitment and Economic Dispatch problems. Together these manifest themselves as a large scale mixed integer programming problem. By relaxing the coupling constraints the original primal problem is decomposed into separable problems, so that each subproblem involves only one individual unit subject to its own local (operating) restrictions. The local subproblems are parameterized by the Lagrange multipliers and the master problem maximizes a dual function, producing new estimates of the Lagrange multipliers while ensuring that the two global constraints are met. This allowed the implementation of the algorithm on the Edinburgh Concurrent Supercomputer.

2. The system

The system considered here is a small part of the total set of generating units of Scottish Hydro-Electric plc. This subsystem consists of five generating units: one nuclear, two thermal (oil and coal), one hydro unit run off a river and a pump-storage unit. The scheduling can be modelled as the mixed integer linear programming problem (P)

$$
F^* = \min_{\alpha_i^t, x_i^t, y_k^t, s_k^t, \mu_l^t, q_l^t, \nu_l^t, p_l^t} \sum_{t=1}^{T} \left\{ \sum_{i=1}^{N} f_i^I(\alpha_i^t, x_i^t) + \sum_{k=1}^{K} f_k^{II}(y_k^t, s_k^t) + \sum_{l=1}^{L} f_l^{III}(\mu_l^t, q_l^t, \nu_l^t, p_l^t) \right\} \quad (1)
$$

where f_i^I refers to the costs of starting, running and closing down the ith thermal unit, f_k^{II} to the running costs of the kth hydro unit and f_l^{III} to the running costs of the lth pump-storage unit [Oliveira, McKee and Coles (1991)] subject to the demand and reserve

234

constraints, which are global in the sense that they couple all the committed generating units

$$\sum_{i=1}^{N} x_i^t + \sum_{k=1}^{K} y_k^t + \sum_{l=1}^{L} (q_l^t - \Theta_l p_l^t) \geq d^t \tag{2}$$

$$\sum_{i=1}^{N} \bar{x}_i \alpha_i^t + \sum_{k=1}^{K} \bar{y}_k + \sum_{l=1}^{L} (\bar{q}_l \mu_l^t - \Theta_l p_l^t) \geq d^t + R \tag{3}$$

and the local constraints $h_i^I, h_k^{II}, h_l^{III}$ for $t = 1, 2, \ldots, T$, representing the operating restrictions of each unit

$$
\begin{array}{ccc}
h_i^I(\alpha_i^t, x_i^t) \leq 0 & h_k^{II}(y_k^t, s_k^t) \leq 0 & h_l^{III}(\mu_l^t, q_l^t, \nu_l^t, p_l^t) \leq 0 \\
\text{for } i = 1, 2, \ldots, N & \text{for } k = 1, 2, \ldots, K & \text{for } l = 1, 2, \ldots, L
\end{array}
\tag{4}
$$

Notational definitions are provided in the Appendix.

3. Lagrangian relaxation in power scheduling

In power scheduling the most suitable restrictions to be relaxed are the global ones, that is, the demand and reserve constraints [Bertsekas *et al.*(1983), and Bard (1988)]. The inclusion of these two constraints gives the following Lagrangian problem which in a condensed form, emphazising the problem decomposition, can be written as

$$
\begin{aligned}
\Phi(\lambda_1, \lambda_2) = &\sum_{i=1}^{N} \min_{\alpha_i^t, x_i^t} \sum_{t=1}^{T} \left(f_i^I(\alpha_i^t, x_i^t) - \lambda_1^t x_i^t - \lambda_2^t \bar{x}_i \alpha_i^t \right) \\
&+ \sum_{k=1}^{K} \min_{y_k^t, s_k^t} \sum_{t=1}^{T} \left(f_k^{II}(y_k^t, s_k^t) - \lambda_1^t y_k^t - \lambda_2^t \bar{y}_k \right) \\
&+ \sum_{l=1}^{L} \min_{\mu_l^t, q_l^t, \nu_l^t, p_l^t} \sum_{t=1}^{T} \left[f_l^{III}(\mu_l^t, q_l^t, \nu_l^t, p_l^t) - \lambda_1^t(q_l^t - \Theta_l p_l^t) - \lambda_2^t(\bar{q}_l \mu_l^t - \Theta_l p_l^t) \right] \\
&+ \sum_{t=1}^{T} \left[\lambda_1^t d^t + \lambda_2^t(d^t + R) \right]
\end{aligned}
\tag{5}
$$

Clearly, given the values of the Lagrange multipliers, the subproblems can be solved independently of each other. Consequently algorithmic parallelisation can be implemented in such a way that all subproblems are solved simultaneously; so the subproblems may be regarded as one stage in a two-stage sequential program consisting of this stage and the master problem. This will be efficient provided that the master program is not too time consuming compared with the subproblems and there are not great imbalances between the computational times of the subproblems. Thus, the dual problem (D) can be stated as

$$
\begin{aligned}
\Phi^* &= \max \Phi(\lambda_1, \lambda_2) \\
\text{s. t.} \quad &\lambda_1 \geq 0, \ \lambda_2 \geq 0
\end{aligned}
\tag{6}
$$

4. The subproblems and the master problem

A solution to (D) provides a lower bound on (P), $\Phi^* \leq F^*$. This is accomplished by dynamic programming, which is suitable for the type of constraints in the subproblems. For power scheduling the duality gap $F^* - \Phi^*$ has been shown to be strictly positive, and the relative duality gap $(F^* - \Phi^*)/\Phi^*$ decreases as the problem size increases [Bertsekas *et al.*(1983)]. Given feasible points for the primal and dual problems, it is possible to estimate the relative duality gap, which can be used as a termination criteria. The master problem controls the iterative process, verifies that the dual solution satisfies the reserve constraint, generates new multiplier values and stops the algorithm whenever the duality gap is less than a specified value. The maximization of Φ^* with respect to λ_1, λ_2 implies the successive computation of the multipliers λ_1, λ_2. It should be noted that $\Phi(\lambda_1, \lambda_2)$ is a concave function and not differentiable so the subgradient method was used. Given the solution to problem (D), the question arises as to how to construct a feasible solution to the original problem (P). Usually the solution obtained from the optimization of (D) will be nearly feasible for (P), and an heuristic was developed to generate a feasible solution which takes into account the information provided by the dual solution.

5. Results

The results obtained are listed in Table 1: the dual feasible solution, the primal feasible solution, the percentage difference between the dual and primal feasible solutions, the number of iterations, the serial cpu time (CPU) used (min:sec) and the parallel processing time (PPT) (min:sec). The parallel implementation was carried out on the Edinburgh Concurrent Supercomputer [Wallace (1991)]. In the present study each processor was assigned a subproblem corresponding to an individual generating unit with the master problem allocated to a master processor which controlled the iterative process by exchanging information between the different processors. No information exchange takes place between the subproblems. Message passing, that is information exchange, was accomplished using CS-tools [Blair-Fish *et al.* (1990)], a communication 'harness' for multiprocessor programming.

TABLE 1. Numerical Results - 1 to 7 days

Days	Primal †	Dual †	% Diff.	Iterations	CPU	PPT
1	$1.656E5$	$1.649E5$	0.4	9	$00:07$	$00:21$
2	$3.392E5$	$3.385E5$	0.2	11	$00:19$	$00:31$
3	$4.940E5$	$4.928E5$	0.2	16	$00:49$	$00:47$
4	$6.488E5$	$6.481E5$	0.1	23	$01:41$	$01:24$
5	$8.009E5$	$7.976E5$	0.4	20	$02:01$	$01:38$
6	$9.281E5$	$9.237E5$	0.5	42	$05:00$	$03:41$
7	$1.056E6$	$1.049E6$	0.7	150	$20:55$	$14:19$

†(Solution in £)

6. Conclusions

The implementation of Lagrangian relaxation using dynamic programming to optimise the individual units has considerably reduced the computational times over those taken by the branch and bound implementation of the problem [Oliveira, McKee, Coles (1991)]. Furthermore, the Lagrange multipliers may be perceived as shadow costs and in the context of the recent privatization of the Generating Boards in the UK, they are a valuable tool for costing the energy, whether from thermal, hydro or pump-storage. Significant reductions in computational time have been achieved by implementing the algorithm on a transputer array, although not as much as is theoretically possible. This is due to the existence of substantial imbalances between the subproblems; in particular, the hydro subproblem is by far the most time consuming. In certain cases those imbalances can be overcome by assigning several thermal units to one processor. In conclusion, this approach has demonstrated that a transputer array can be used to solve very large power systems with a corresponding decrease in the computational time.

Acknowledgements—The first author has been supported by the Junta Nacional de Investigação Científica e Tecnológica (JNICT), Portugal, research grant BD/210/90-RM.

References

Bard, J. (1988) 'Short-term scheduling of thermal-electric generators using Lagrangian relaxation', Operations Research 36, 756-766.

Bertsekas, D., Lauer, G., Sandell, N. and Posbergh, T. (1983) 'Optimal short-term scheduling of large-scale power systems', IEEE Trans. Aut. Control AC-28, 1-11.

Blair-Fish, J., Brown, S., MacDonald, N., Smith, N. and Stroud, N. (1990) 'Using CS tools on the Edinburgh Concurrent Supercomputer', EPCC UG91-02, Edinburgh University, Edinburgh.

Oliveira, P., McKee, S. and Coles, C. (1991) 'The optimal scheduling of hydro electric power generation', in M. Hellio (ed.), Proceedings of the 5th ECMI Conference, Teubner, Stuttgard, pp. 109-114.

Oliveira, P., McKee S. and Coles, C. (1991) 'Lagrangian relaxation and its application to the unit commitment/economic dispatch problem', Research Report 22, Department of Mathematics, University of Strathclyde.

Wallace, D. (1991) 'Supercomputing with transputers', Computing Syst. in Engineering 1, 131-141.

Appendix

i	a thermal unit with $i = 1, 2, \ldots, N$
k	a hydro unit with $k = 1, 2, \ldots, K$
l	a pump-storage unit with $l = 1, 2, \ldots, L$
t	a time period with $t = 1, 2, \ldots, T$
x_i^t	the power produced by the ith thermal unit during period t (MW)
$\bar{x}_i$	the maximum power output of the ith thermal unit (MW)
α_i^t	the commitment (integer) variable of the ith thermal unit
y_k^t	the discharge from the kth hydro unit (MW)
$\bar{y}_k$	the maximum discharge for the kth hydro unit (MW)
s_k^t	the spillage from the kth hydro unit (MW)
q_l^t	the discharge from the lth pump-storage unit (MW)
$\bar{q}_l$	the maximum discharge for the lth pump-storage unit (MW)
p_l^t	the pumped water to the lth pump-storage unit (MW)
μ_l^t	the commitment (integer) variable associated with generation of the lth pump-storage unit
ν_l^t	the commitment (integer) variable associated with pumping of the lth pump-storage unit
Θ_l	the inverse of the thermodynamic efficiency of the pumping process in the lth pump-storage unit
R	the constant reserve in every time period (MW)
d^t	the demand which the system has to meet (MW)

SINGULAR BEHAVIOUR OF EIGENVALUE LOCI ARISING FROM AN MHD INSTABILITY IN THE PRESENCE OF FLOW.

R.B. PARIS
Department of Mathematical and
Computer Sciences
Dundee Institute of Technology
Dundee DD1 1HG, UK.

A.D. WOOD and S. STEWART
School of Mathematical Sciences
Dublin City University
Dublin 9
Ireland.

ABSTRACT. The growth rate of the resistive tearing instability in the presence of equilibrium plasma flow and fluid viscosity corresponds to the eigenvalue of a certain ordinary differential equation problem in the resistive boundary layer. We consider the perturbation of such eigenvalues as the flow and viscosity are increased from zero.

1. Introduction

This work forms part of EC Fusion Contract 401-89-11 FUA IRL for a computational study of the effects of equilibrium plasma flow, fluid viscosity and a gravitational acceleration on the onset of resistive instabilities in magnetohydrodynamics. Such flows occur in many Tokamak machines where modes which have been predicted to be stable in the absence of flow are found to be unstable when the flow is switched on. The work has application in the long-term generation of electricity from nuclear fusion. It is a good example of the need for mathematical modelling in a situation where experiment is extremely expensive.

2. Physical Background

Physical details of the effects of equilibrium flow on the resistive tearing mode instability appear in a separate paper[3] by the present authors. We offer here only a broad outline of the derivation of the eigenvalue problem in the resistive boundary layer. Furth, Killeen and Rosenbluth[2] in 1963 studied resistive instabilities in a plane current layer with zero-order flow ignored. In that and subsequent papers the problem was simplified by scaling variables in such a way that resistive diffusion across the boundary layer occurs on a much longer time scale than the growth of the resistive modes. In the present work, which is based on the papers of Paris and Sy[4] and Bondeson and Persson[1], we consider the effect of a sheared plasma flow in the current layer along the equilibrium magnetic field, and order the variables in such a way that the equilibrium flow significantly modifies the linear growth of the modes. This ordering is relevant to tokamak plasmas and has the advantage that the effect of flow need only be considered in the resistive boundary layer. The external region of infinite conductivity is not modified.

Our starting point is the set of resistive MHD equations for an incompressible plasma, which we linearise and manipulate into a pair of coupled equations in the velocity and magnetic field. By assuming a complex exponential variation in all perturbed quantities and applying the standard transformation to dimensionless variables of [2], we arrive at a pair of coupled ordinary differential equations of second order in the scaled components of velocity and magnetic field. In the high conductivity limit as the magnetic Reynolds number tends to infinity, a boundary layer appears, centred on the resonant plane. The dynamic and dissipative effects are concentrated in this narrow boundary layer and we, in turn, concentrate attention on it. All we require from the outer region of infinite conductivity is the discontinuity in the logarithmic derivative of the magnetic field perturbation across the resonant plane. We take this as a given quantity Δ' dependent on the global structure of the problem.

In the boundary layer, after various ordering arguments and a further rescaling described in [1,4], we arrive at the problem

$$(P + iR\theta)\frac{d^2 H}{d\theta^2} - \theta^2 H + \frac{G}{F'^2}\frac{H - iR}{P + iR\theta} - N\frac{d^4 H}{d\theta^4} = P\theta - \left(\frac{F''}{F'}\right)\Delta \tag{1}$$

$$\Delta' = \Delta \int_{-\infty}^{\infty} (P + \theta H)d\theta \tag{2}$$

where θ is the stretched variable across the boundary layer and, following [1], we have written the complex variable $\Delta' - i\pi F''/F'$ in polar form

$$\Delta' - \frac{i\pi F''}{F'} := \Delta e^{i\chi}, \quad \Delta > 0, \chi \in (-\pi, \pi]. \tag{3}$$

The new scaled variables P, R, H, G and N are related to the physical quantities growth rate, flow velocity, velocity perturbation, gravitational forces and viscosity. For the tearing mode we take the gravitational term $G = 0$. The location of the boundary layer is determined by $F = 0$, where F is the component of the equilibrium magnetic field in the direction of the wave vector and F' and F'' denote the numbers $F'(0)$ and $F''(0)$. Equation (2) is the matching condition between the outer region and the resistive boundary layer.

3. Analysis By Fourier Transformation

Let L denote the formal differential operator

$$Lh := \frac{d^2 h}{dk^2} + R\frac{d}{dk}(k^2 h) - (k^2 P + k^4 N)h.$$

Under the Fourier transform $h(k) = \int_{-\infty}^{\infty} H(\theta)e^{-ik\theta}d\theta$, (1) becomes

$$Lh = 2\pi i P\delta'(k) - \left(\frac{F''}{\Delta F'}\right)2\pi\delta(k).$$

If we replace these delta functions by the corresponding jump conditions, we obtain a pair of boundary value problems on $(-\infty, 0)$ and $(0, \infty)$, namely

$$Lh = 0, \quad -\infty < k < 0 \quad \text{and} \quad 0 < k < \infty \tag{4}$$

$$h(k) \to 0 \quad \text{as} \quad k \to \pm\infty$$

$$h(0+) - h(0-) \;=\; 2\pi i P \tag{5}$$

$$h'(0+) - h'(0-) \;=\; -2\pi \frac{F''}{\Delta F'}. \tag{6}$$

When we transform condition (2) we find that

$$\Delta' = \frac{1}{2} i \Delta \{ h'(0+) + h'(0-) \}.$$

Also, from (3) and (6),

$$h'(0\pm) = -i \exp[\pm i\chi].$$

Substitution in (5) yields the relation for the eigenvalue P

$$2\pi P = \frac{h(0-)}{h'(0-)} e^{-i\chi} - \frac{h(0+)}{h'(0+)} e^{i\chi}. \tag{7}$$

If we set $h(k) = y(k)\exp[-\frac{1}{6}k^3 R]$, and observe that under this change of dependent variable the ratios $h'(0)/h(0)$ and $y'(0)/y(0)$ remain equal, we obtain the problems

$$y'' + [kR - k^2 P - k^4(N + \frac{1}{4}R^2)]y = 0, \quad -\infty < k < 0, \quad 0 < k < \infty \tag{8}$$

$$y(0) = 1, \quad y(\pm\infty) = 0 \tag{9}$$

where the eigenvalues P are the zeros of the function

$$f(P) := 2\pi P + \frac{e^{i\chi}}{y'(0+, P)} - \frac{e^{-i\chi}}{y'(0-, P)}. \tag{10}$$

Notice that the only quantity from the differential equation (8) required in the eigenvalue relation (10) is the slope of the solution $y(k, P)$ on either side of the origin.

4. Perturbation Of Eigenvalues

The number χ is given, through (3), by the global equilibrium structure but the flow parameter R arising from neutral beam injection and viscosity parameter N may be varied. We study the locus of the eigenvalue $P \equiv P(R, N, \chi)$ in $\mathbf{C}$ as R and N increase from zero for values of χ/π between 0 and 1. The paths obtained for $-1 < \chi/\pi < 0$ are simply complex conjugates. We consider first the inviscid limit $N = 0$. When $R = 0$ it is known from [2] that the eigenvalue $P(\chi)$ lies on the non-negative real axis for $0 \le \chi/\pi \le 0.5$ and for $0.5 < \chi/\pi \le 1$ we obtain a pair of complex conjugate eigenvalues located on the rays $\arg P = \pm 4\pi/5$. As $R \to \infty$, the location of the eigenvalue is known from the asymptotics of [1] and [4].

For intermediate values of R we adopt the following numerical scheme. Fix R, χ and choose $k_\infty > 0$ and a tolerance $\varepsilon > 0$. Set $P = P_0$, our first guess at the eigenvalue. For $n = 0, 1, 2, \cdots$, until $|P_{n+1} - P_n| < \varepsilon$ solve (by finite differences) the differential equation (8) with $P = P_n$ and $y(0) = 1, y(k_\infty) = 0$. Compute $y'(0\pm)$ and solve the eigenvalue relation (10) (by a generalised Newton-Raphson method) to obtain a new estimate $P = P_{n+1}$.

We check the validity of our choice of k_∞ by solving the problem again with the condition $y(2k_\infty) = 0$ and comparing the solutions. For most values of P and R it is sufficient to take $k_\infty = 3$. The difficult case is when $R << 1$ and Re $P < 0$. The small R asymptotic analysis of the differential equation shows that there is a domain of growing oscillation of the solution on the positive k-axis before exponential decay takes over. It is necessary to choose k_∞ beyond this point, which may also be found from a numerical plot of the eigenfunction. For small R and small negative P the choice $k_\infty = 50$ was needed, greatly increasing the computation. For very small values of R it was found easier to solve the boundary layer equations (1), (2) by a direct numerical method, rather than applying the Fourier transform, although the latter was always more efficient for reasonably large values of R, when good agreement with the asymptotics was obtained.

It may be seen from Figure 1 that, for $0 \leq \chi/\pi \leq 0.5$, the growth rate Re P increases with R and when $0.5 < \chi/\pi < 1$ new eigenvalues emerge from $P = 0$, initially with Re$P < 0$ (stable) but moving over into Re$P > 0$ (unstable) as R increases. This discontinuous behaviour of the spectrum is not unexpected, as the differential equation (8) is singularly perturbed for small R. In Figure 2 we see that the stable eigenvalues originally on the ray $\arg P = 4\pi/5$ simply move towards and out along the negative real axis with increasing R. The physical interpretation of Figure 1 is that shear flow along the field lines destablises the tearing mode, with important consequences for neutral injection in tokamaks.

As soon as we introduce viscosity effects, the spectrum undergoes another qualitative change. The loci emerging from $P = 0$ for $0.5 < \chi/\pi < 1$ are no longer present. These originate instead from the eigenvalues which lie on the ray $\arg P = 4\pi/5$ when $R = N = 0$, although this ray is itself distorted as N increases, being pulled down gradually on to the negative real axis. This is shown in Figure 3, which has the not unexpected physical interpretation that increasing viscosity improves stability, although it remains true for any viscosity that the tearing mode is destablised by sufficiently increasing flow.

References

1 Bondeson, A. and Persson, M. (1986) 'Resistive tearing modes in the presence of equilibrium flows' Phys. Fluids $\underline{29}$, 2997-3007.

2 Furth, H.P., Killeen, J. and Rosenbluth, M.N. (1963) 'Finite-resistive instabilities of a sheet pinch', Phys. Fluids $\underline{6}$, 459-484.

3 Paris, R.B., Stewart, S. and Wood A.D. (1992) 'Effects of equilibrium flow on the resistive tearing mode instability' submitted to Phys. Fluids.

4 Paris, R.B. and Sy, W.N-C. (1983) 'Influence of equilibrium shear flow along the magnetic field on the resistive tearing instability' Phys. Fluids $\underline{26}$, 2966-2975.

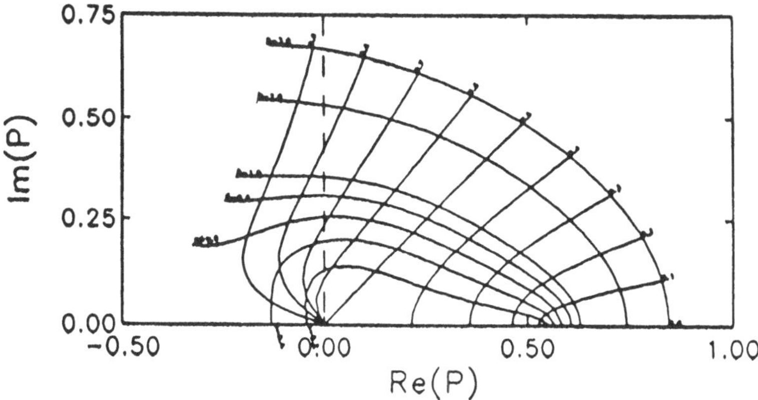

Figure 1. Loci of the eigenvalue P originating on the non-negative real axis as R increases from 0 to 3 for $0 \leq \chi/\pi \leq 0.9$ $(N = 0, G = 0)$.

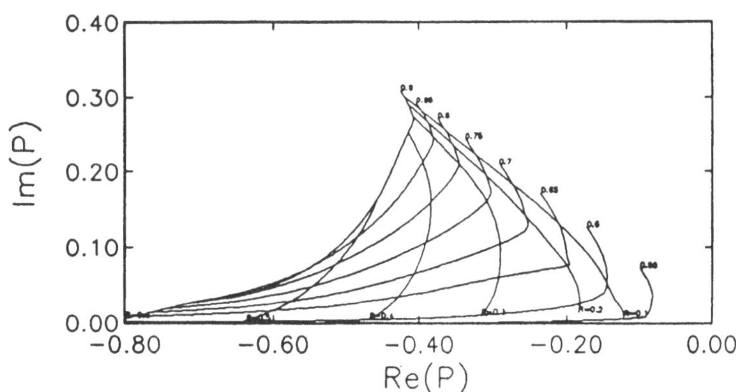

Figure 2. Loci of the eigenvalue P originating on the ray $\arg P = 4\pi/5$ as R increases from 0 to 0.6 for $0.55 \leq \chi/\pi \leq 0.9$ $(N = 0, G = 0)$.

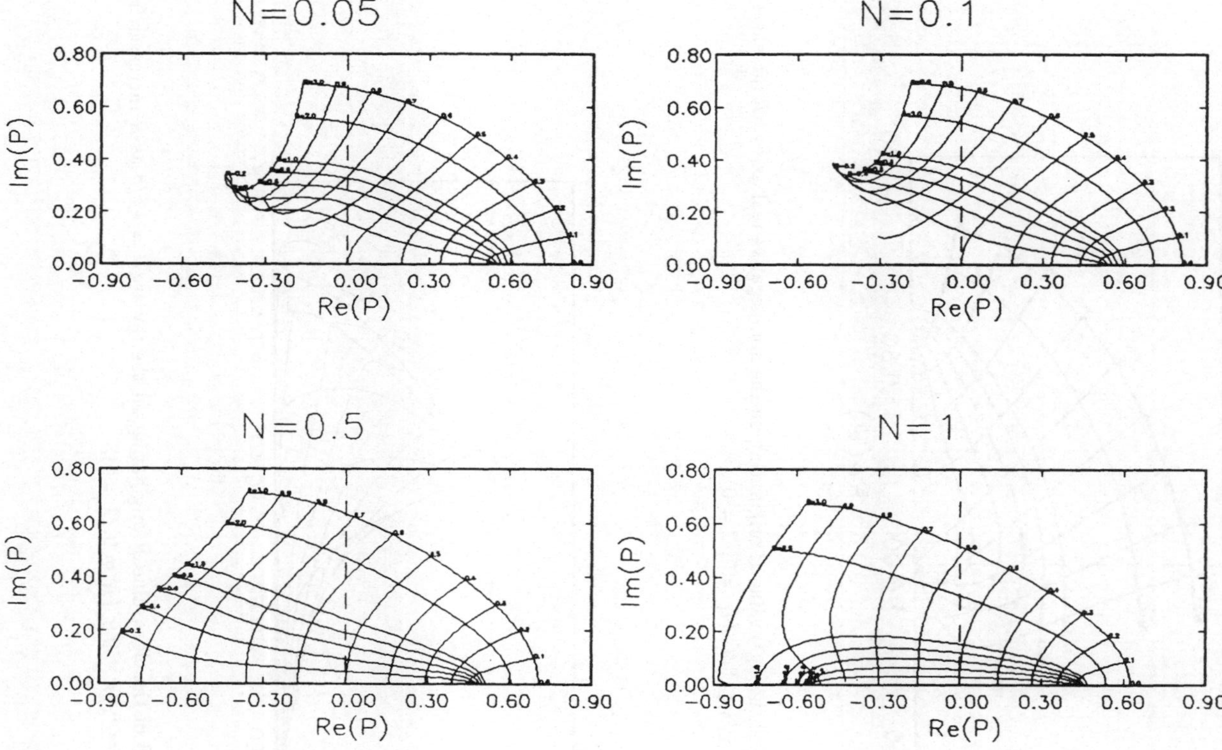

Figure 3.   Loci of the eigenvalue P as R increases from 0 to 3 for the viscosity parameter $N = 0.05, 0.1, 0.5, 1.0$ with $0 \leq \chi/\pi \leq 1$ $(G = 0)$.

A GENERALIZED CUSUM SCHEME FOR AUTOCORRELATED DATA

M. Pereira-Leite and R.J. Rowlands
Department of Chemical and Process Engineering
University of Newcastle-upon-Tyne, United Kingdom
August 1991

ABSTRACT

The need to improve and maintain productivity and product quality in industry has led to the development and application of a set of statistical techniques which are grouped under the term Statistical Process Control (SPC). In particular statistical control charts are used to monitor the performance of the process in real time. Traditionally, independence between sucessive measurements is assumed but data from process industries are usually autocorrelated, often similar to a first or second order autoregressive stochastic process. This paper[1] presents a monitoring scheme to cope with this kind of data. The major objective of the work reported here consists in extending the minimax-optimal cumulative sum procedure to the case of dependent data. The performance of the extended scheme is assessed using data from a noisy autoregressive model of order one. Comparisons with the standard CuSum procedure are made.

1. INTRODUCTION

Statistical Process Control (SPC) is now recognized as having a very important role to play in modern industry and it should be seen as an objective statistical analysis of process variation and its causes. Statistical methods can be used in order to: obtain evidence of what a process is doing and what it is likely to do; provide an assessment of the quality levels the process is currently capable of meeting; tell when to look for trouble and when not to; provide clues as to where trouble is likely to occur; help towards an understanding of the operation of the system and so help in making improvements to the process or product.

All industrial processes display variation and there are many different reasons for it, such as:

1. variational noise (this is the variation we observe between product manufactured under the same conditions and specifications);

2. causes external to the process (ambient temperature, humidity, etc);

3. process causes (build-up of waste products, ageing of a catalyst, etc);

4. special causes of variation (a poor quality of batch of raw material, incorrect setting of equipment, etc).

The procedure adopted in SPC is to try to identify and separate these kinds of variation. The idea of a *state of statistical control* was first introduced by Shewhart [1]. When a chosen characteristic of a process is in statistical control, its behaviour is predictable in the sense that it is governed only by common, every-day variation that is an intrinsic characteristic of the system of production. If, on the other hand, variation outside this pattern, which may be due to special causes that can be discovered and corrected, is present then the process is not in statistical control.

In order to decide between these alternatives, some form of statistical appraisal is necessary and the most well-known graphical form is probably the statistical control chart. There are several types of control charts, namely Shewhart charts, moving average charts, exponentially weighted moving average charts and cumulative sum (CuSum) charts. Their great value is that they show when to set up procedures to look for special causes of variation (which may involve stopping the process) and when to leave the process alone.

The objective of this paper is to present a new efficient monitoring procedure based in cumulative sums. The proposed Generalized CuSum scheme is designed to cope with data following autoregressive moving average models of complex processes. The technique proposed is an extension of the minimax-optimal cumulative sum procedure based on the likelihood ratio and it is used to detect changes in the overall mean of the observations.

---

[1]This research has been partly supported by JNICT - Portugal

2. AUTOREGRESSIVE MOVING AVERAGE MODELS

The design of most control charts is based on the assumption that the observations to be monitored are independently and distributed about some mean (μ) with constant variance (σ^2). If the data are not independent, but serially correlated, these charting procedures are not totally invalidated but their performance is seriously degraded.

In practice, the assumption of independence between sucessive sample results is only a very crude approximation. For many modern industrial processes, the "natural variation" observed is not pure noise, but includes an inherent dynamical component. The process dynamics cause successive sample results to be correlated. In such situations a stationary time series model that includes the process dynamics will give a better description of process behaviour in the in-control state than the simpler model with independent observations.

Disturbances arising in chemical production processes can often be well represented by linear stochastic models of autoregressive moving average (ARMA) type. These models were first introduced by Box and Jenkins [2]. The general form of a zero-mean ARMA model of order (p,q) is given by

$$x_t - \phi_1 x_{t-1} - \cdots - \phi_p x_{t-p} = a_t - \theta_1 a_{t-1} - \cdots - \theta_q a_{t-q}$$

where $\phi_1, \cdots, \phi_p, \theta_1, \cdots, \theta_q$ are the model parameters and a_t is a sequence of independent normally distributed random disturbance terms with mean zero and variance σ_a^2. Using the backshift operator B defined by

$$Bz_t = z_{t-1} \qquad\qquad B^j z_t = z_{t-j}$$

this expression reduces to

$$\Phi(B)x_t = \Theta(B)a_t$$

where $\Phi(B) = 1 - \phi_1 B - \cdots - \phi_p B^p$ is termed the autoregressive AR(p) operator and $\Theta(B) = 1 + \theta_1 B + \cdots + \theta_q B^q$ is the moving average MA(q) operator.

In this paper we investigate the problem of detecting a change in the mean μ of an ARMA model of order (1,1):

$$(x_t - \mu) = \phi_1(x_{t-1} - \mu) + a_t - \theta_1 a_{t-1}$$

An equivalent "state space" representation is[2]

$$\begin{cases} x_t = \mu + b_t + \xi_t & (\textit{measurement equation}) \\ b_t = \rho b_{t-1} + \eta_t & (\textit{transition equation}) \end{cases}$$

where $\xi_t \sim N(0, \sigma_\xi)$ and $\eta_t \sim N(0, \sigma_\eta)$ are two independent sequences of mutually independent normal disturbances. In the state space model the state variable b_t evolves over time according to the *transition equation*. This variable is not directly observable but is related to the variable x_t by the *measurement equation*. The overall mean of the series is μ, which remains constant until it is changed at an unknown time m by some special cause. Our problem consists in finding a procedure that will detect the change with minimum average delay.

Page [3] introduced a procedure to detect changes in distributions based on cumulative sums of independent observations. Supposing that $x_1, \cdots, x_{m-1}$ have distribution function F_0 and $x_m, x_{m+1}, \cdots$ have distribution function $F_1 \neq F_0$, Page pointed out that his scheme is equivalent to performing a sequential probability ratio test on sucessive new observations until a decision in favour of F_1 is reached. Moustakides [4] proved that this cumulative sum scheme is minimax optimal.

3. GENERALIZED CUMULATIVE SUM SCHEME

The minimax-optimal CuSum procedure for independent data signals a change from state μ_0 (the target) to state μ_1 on the first occasion that $Q_n \geq h > 1$, where

$$Q_n = \max\left\{1, Q_{n-1}\frac{f(x_n;\mu_1)}{f(x_n;\mu_0)}\right\} = \max\left\{1, \max_{1 \leq r \leq n} \prod_{j=r}^n \frac{f(x_n;\mu_1)}{f(x_n;\mu_0)}\right\} \quad (n = 1, 2, \cdots; Q_0 = 1)$$

---

[2]The two representations are equivalent provided $(1 - \phi_1^2)\sigma_\xi^2 + \sigma_\eta^2 = (1 - 2\theta_1\phi_1 + \theta_1^2)\sigma_a^2$, $\sigma_\eta^2 = (\phi_1 - \theta_1)(1 - \phi_1\theta_1)\sigma_a^2$ and $\rho = \phi_1$.

and $f(x; \mu)$ is the probability density function (pdf) of the observations for state μ. The value of the constant h is chosen to control the false alarm rate of the procedure.

A natural generalisation of Q_n for dependent data is

$$Q_0 = 1$$
$$Q_n = \max\left\{1, \max_{1 \le r \le n} \frac{f(x_n, \cdots, x_r | x_{r-1}, \cdots, x_0; \mu_1)}{f(x_n, \cdots, x_r | x_{r-1}, \cdots, x_0; \mu_0)}\right\} \quad (n = 1, 2, \cdots)$$

where $f(x_n, \cdots, x_r | x_{r-1}, \cdots, x_0; \mu)$ denotes the joint conditional pdf of $x_n, \cdots, x_r$ given $x_{r-1}, \cdots, x_0$ for state μ. We want to implement the procedure on-line but the evaluation of these conditional densities is time consuming. Thus a rapid approximation for Q_n is required.

For our assumed model, the joint pdf of $x_n, \cdots, x_r$ and $\xi_n, \cdots, \xi_r$ given $x_{r-1}, \cdots, x_0$ and $\xi_{r-1}, \cdots, \xi_0$ for state μ is

$$\prod_{j=r}^{n} f_\gamma(x_j - \rho x_{j-1} - \xi_j + \rho \xi_{j-1}) f_\xi(\xi_j)$$

where $f_\gamma(x)$ denotes the pdf of $\gamma_j = (1 - \rho)\mu + \eta_j$ and $f_\xi(x)$ denotes the pdf of ξ_j. This suggests that we replace Q_n by Q_n' where

$$Q_n' = \max\left\{1, \max_{1 \le r \le n} \prod_{j=r}^{n} \frac{f_\gamma(x_j - \rho x_{j-1} - \hat{\xi}_j + \rho \hat{\xi}_{j-1}; \mu_1)}{f_\gamma(x_j - \rho x_{j-1} - \hat{\xi}_j + \rho \hat{\xi}_{j-1}; \mu_0)}\right\}$$

and $\hat{\xi}_j$ is an estimate of the unobservable disturbance ξ_j. The estimate used here is given by

$$\hat{\xi}_j = X_j - \hat{X}_j \qquad \hat{X}_j = p X_{j-1} + (1 - p)\hat{X}_{j-1}$$

with $\hat{X}_0 = 0$ and $p = 0.1$.

3.1. PERFORMANCE EVALUATION BY SIMULATION

The performance of monitoring procedures is usually assessed by considering the *average run length* (ARL). The run length T is defined as the reaction time when a shift from the target occurs, or as the time to the first false alarm if there is no shift, i.e.:

$$T = \begin{cases} N - m + 1 & if \quad m < \infty \\ N & if \quad m = \infty, \end{cases}$$

with N denoting the time of the signal. Thus ARL=$E(T)$. The ARL should be large when the process is at the target level but small when the process shifts to an undesirable level.

The ARLs for the proposed GCuSum scheme were estimated using computer simulation. A program generating observations following the ARMA(1,1) model and implementing Q_n' with $\mu_0 = 0$ and $\mu_1 = 1$ was written in FORTRAN77. The values of the normal deviates for the ξs and ηs were obtained using the NAG routines G05DDF and G05CBF. The ARL for an out-of-control situation was computed assuming that the mean of the observations had already shifted from the target value at the time the procedure was started. Thus, after generating 30 zero-mean observations, an increase $\delta(= .5, (.5), 5)$ was added to the value of the mean and the scheme started to operate. The values chosen for the standard deviation of the ηs and xs were respectively .4 and 1 and each simulation consisted of 2500 runs.

Several simulations were made for different values of the correlation coefficient ρ. In order to obtain the same ARL=300 when the process is in a state of statistical control, the following threshold values were used

ρ	0	.1	.2	.3	.4	.5	.6	.7	.8
h	4.584	1.023	1.043	1.077	1.139	1.225	1.331	1.469	1.668

Some of the results obtained are shown in Figures 1, 2, 3 and 4. The ARLs obtained with the standard CuSum procedure are also plotted to allow a quick comparison of the performance of the two schemes. Setting $\mu_0 = 0$ and $\mu_1 = 1$ in Q_n' results in a scheme that is designed to detect a shift of size 1. We see that the GCuSum scheme detects a shift of this size more quickly on average than the standard CuSum scheme, regardless of the value of ρ. (The standard CuSum scheme considered here is the optimal scheme for detecting a shift of size 1 when $\rho = 0$.) Figures 1-4 also show that the GCuSum performs uniformly better than the standard CuSum for shifts of size < 1. It also performs uniformly better than the standard CuSum for shifts of size > 1 provided p is not to large.

4. CONCLUSIONS

The scheme proposed here for detecting a change in the mean of an autocorrelated sequence is easy to implement on-line but it is not the exact likelihood ratio procedure. Nevertheless it performs better than the standard CuSum procedure unless the degree of autocorrelation is high and the shift in the mean is large. It remains to be seen whether its performance can be enhanced by, for example, adjusting the value of the smoothing constant p.

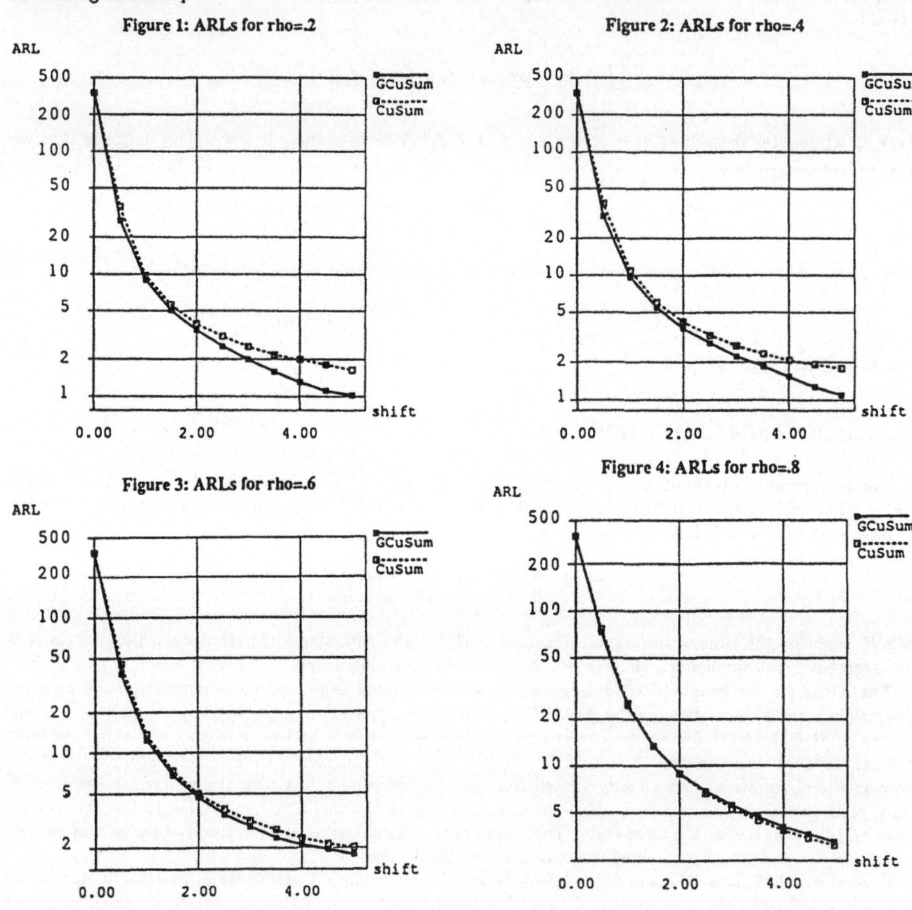

REFERENCES

[1] Shewhart, W.A., (1931). *Economic control for manufactored product*, Van Nostrand, New York. (Republished in 1981 by the American Society for Quality Control, Milwaukee, W.I.)

[2] Box, G.E.P. and G.B. Jenkins,(1976). *Time series analysis: forecasting and control*, Holden-Day, USA.

[3] Page, E.S., (1954). *Continuous inspection schemes* , Biometrika, 41, 100-14.

[4] Moustakides, G.V., (1986). *Optimal stopping rules for detecting changes in distributions*, Annals of Statistics, 14, 1379-87.

ROBOT CONTROL BASED ON NEURAL NETWORKS

G.M. Pfaffenzeller, P. Rentrop, G. Schmidt

Abstract

The multibody system approach leads to a well approved model for a robot. Its numerical simulation, however, cannot always meet real-time demands. Therefore a control based on feed-forward neural nets is studied. Numerical simulations with a 2-D robot in domains with obstacles are discussed.

1 Introduction

The multibody system approach, see [6] describes a robot in terms of **bodies** and **connections** . The bodies are coupled by massless connections, which cause forces acting on the bodies or constrain their relative motion. The motion of the bodies is described by position coordinates ϕ. Due to famous mechanical principles the equations of motion in state space form can be written as

$$M(\phi, t)\, \ddot{\phi} = F(\dot{\phi}, \phi, t) \tag{1}$$

Their treatment can suffer under severe shortcomings, e.g.

- in practice it is very difficult to measure all technical data like forces and moments,

- real time computations of (1) are not always possible, since the numerical solution of the initial value problem is too time consuming [5].

Therefore an alternative concept is studied. In a first step a **neural net** should control a robot to a target point with a rough precision. Then the precise adjustment should be done by sensor techniques.

2 Mathematical Properties of Neural Nets

Among the large number of models for neural nets we have concentrated on feed-forward nets, see Hertz et al. [1], or Ritter et al. [3]. They consist of input neurons, output neurons and neurons in hidden layers. Connections of neurons are only allowed between different layers, but not inside a layer. Figure 1 shows a simple net with input-output neurons, see sketch (A); and with one hidden layer, see sketch (B).
Each neuron except the input cells performs the computations

(A)   (B)

Figure 1: (A) single layer net of type 3-2,   (B) two layer net of type 2-3-1

- neuron i sums up the weighted inputs e_j of the m neurons of a preceeding layer

$$net_i = \sum_{j=1}^{m} w_{ij} e_j \tag{2}$$

- the input sum is transformed by an activity function f(x) of sigmoidal type [1],
  e.g. $f(x) = \frac{1}{1+e^{-x}}$ leading to an output $f(net_i)$. A threshold can be introduced
  as a neuron with a fixed weight $w_{ij} = -1$.

From the point of computer science we can consider a neural net as a distributed set of specialized processors. The degrees of freedom to adjust this potentially powerful, parallel machine to a special task are the weights w_{ij}. They are determined by learning strategies. Our tests were based on a supervised learning, i.e. given output values are compared with the net output. Then the famous **backpropagation rule** with learning parameter ϵ was applied for an improvement of the weights w_{ij}, see Rumelhart et al. [4]. From the point of mathematics we can classify a feed-forward net with regard to its approximation and its numerical properties.

- Approximation Properties
  A single layer net (with no hidden layer) essentially generates a **linear combination** of the input data. If we choose the input data as an appropriate base, we may expect a satisfactory approximation.
  Nets with hidden layers generate a **nonlinear combination** of the input data.

  $$(f \circ f \circ \ldots \circ f)(net_i) \qquad \text{f is applied k-times for a k-layer net}$$

  For two-layer nets exists a Weierstrass-type Approximation Theorem for continuous functions.

- Numerical Properties
  The backpropagation rule can be interpreted as a **gradient method**, especially as a steepest descent method for the minimization of a nonlinear objective function. The objective function consists of the sum of the square deviations of exact output and net output. The learning parameter of the backpropagation rule corresponds to an incomplete line search. The typical properties of gradient methods are documented in Luenberger [2].

---

[1] A real differentiable function f(x) is of sigmoidal type, if f(x) is limited and has a positive first derivative. f(x) takes its limits and behaves like a linear function between them.

3 Two-Dimensional Robot Model

The technical base for our tests is a robot with two degrees of freedom. The robot can rotate around the z-axis and his arm can be extended in a horizontal direction. The kinematic behaviour of the robot should be controlled. In a simple case, the kinetic requires the conversing of a ring domain in the first quadrant from cartesian into polar coordinates. This is a task, where the performance of the neural net can be easily supervised.

Figure 2A shows the 16 training values - 4 values on each line in 30 degree distance. The convergence process after 1000 iterations is documented. As expected, the largest deviations between exact and simulated values occur in the corners, where the training values are spreaded out. The trained net reproduces quite satisfactorily a straight line, see Figure 2B.

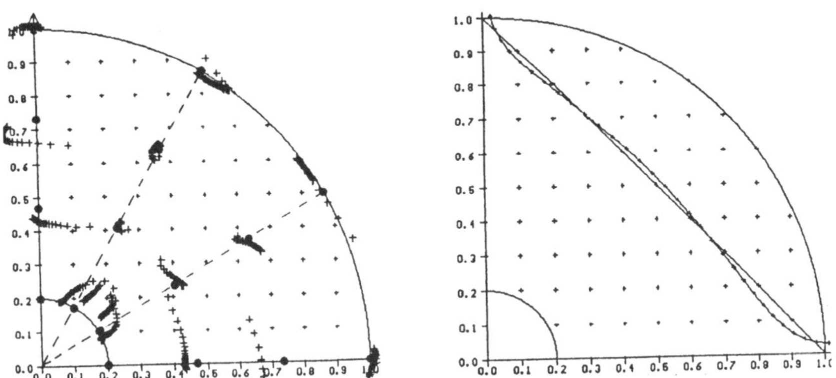

Figure 2: (A) Single layer net of type 10-2 after 1000 iterations
training values and convergence for a learning parameter $\epsilon = 0.15$,
input values (x^k, y^j), $j, k = 1, 2, 3$ and output values ϕ, r.
(B) The trained net reproduces a straight line.

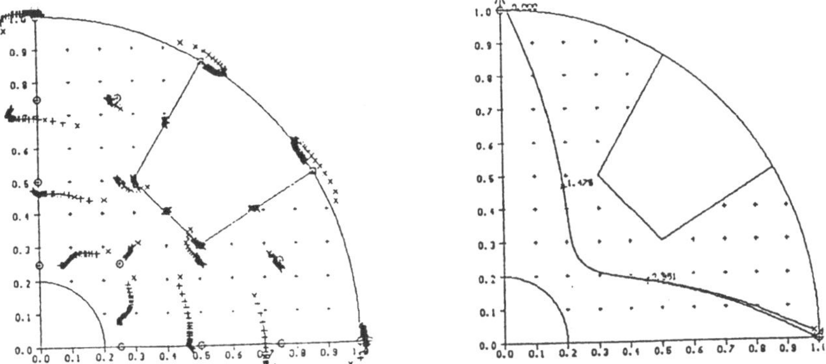

Figure 3: (A) Training history of a 3-10-3 net for a domain with obstacles.
(B) The trained net – 2000 iterations – reproduces a curve.

In the next step the net was tested for a more difficult task. At first a 3-10-3 net with input values: 1, x, y and output values: ϕ, r, and 0 for forbidden region or 1 for allowed region was trained with the marked values in Figure 3A. Also the learning history for 2000 iterations is shown. Then the trained net was used to reproduce a typical curve, see Figure 3B. Similiar satisfactory results were obtained for acceleration-limited trajectories.

Conclusion

- The use of more training values and less iterations is favorable. The number of training values must be large enough in order to prevent the net to be overparametrized. If there are less training values, the nonlinear least square problem, which is actually solved by a feed–forward net with supervised learning strategy, is ill–conditioned. Ill–conditioning can cause a zig–zag pattern in the learning progress.

- The initialization of the weights w_{ij} usually costs a large number of learning cycles. Since many neural net simulations are performed on a serial digital computer, one can think of an initialization by well-approved numerical techniques for nonlinear least square problems.

4 References

1. Hertz, J., Krogh, A., Palmer, R.G.: Introduction to the theory of neural computation. Santa Fe Series by Addison Wesley 1991

2. Luenberger, D.G.: Linear and nonlinear programming. Addison Wesley 1984

3. Ritter, H., Martinetz, T., Schulten, K.: Neuronale Netze – Eine Einführung in die Neuroinformatik. Addison Wesley 1990

4. Rumelhart, D.E., McCelland, J.L., PDP Research Group: Parallel Distributed Programming. Vol. 1 Foundations. MIT Press 1987 (fifth ed.)

5. Schmidt, G., Xu, X.: A comparison of model based path algorithms for direct-drive SACRA robots. J. Intelligent and Robotic Systems, to appear 1992

6. Simeon, B., Führer, C., Rentrop, P.: Differential-algebraic equations in vehicle system dynamics. Surveys Math. Industry 1 (1991) 1-37

Address of the authors:

Prof. Dr. P. Rentrop
Dipl.-math. G.M. Pfaffenzeller
Mathematisches Institut
TU München
Arcisstr. 21
D–8000 München 2

Prof. Dr.-Ing. G. Schmidt
Lehrstuhl für Steuerungs–
und Regelungstechnik
TU München
Theresienstr. N5
D–8000 München 2

COMPUTER-AIDED LAYOUT DESIGN

Z. POGÁNY, I. ZSIGMOND
Eötvös Loránd University, Computer Centre
Budapest
Bogdánfy u. 10/b
H-1117 Hungary

ABSTRACT. The location of the facilities in a plant may heavily influence production — first of all material handling — costs. In this paper concepts and methods developed and implemented in a program system of optimal layout design for a Hungarian bus factory are presented. The basic ideas of two new approximate solution methods for the Quadratic Assignment Problem are outlined: a construction algorithm based on cluster-analysis and an iterative improvement algorithm based on feasible directions. These algorithms were incorporated in an interactive program system developed for IBM PC/AT compatible computers and tested on several problems known from the literature. The computational results are very good, e.g., for the Steinberg test problem solutions were obtained which seem to be the best ever found. Our experiments suggest that in many real-life situations computer program systems can provide effective tools to support layout design.

1. Introduction

A Hungarian bus factory contracted with our university to develop a computer program system (for IBM/PC AT or compatible) to support layout design in order to reduce the material-handling costs. About 20 machines were to be placed in a 10 x 30 meter workshop-hall where raw-materials and products could be transported by overhead cranes and by trucks or trolleys. Graphic input of the workshop-hall and the machines, man-machine interactivity, display of the intermediate results and graphic output of the selected layouts were requirements. Problems concerning databases, graphic information management, determination of the material-handling costs etc. cannot be discussed here (see [6]).

The layout problem was formulated mathematically as a Quadratic Assignment Problem (QAP); in this paper attention is focused on the solution of QAP.

2. The Plant Layout Problem — Mathematical Model

The following problem is considered: Given a plant and n machines (facilities), locate the machines in the plant so that the technological conditions of the production are satisfied (feasible layout) and the material-handling costs are minimal (optimal layout). Raw-materials and products can be transported by different transportation facilities along routes (Euclidean or rectilinear distances are also allowed). To transform the original practical problem into a mathematically tractable problem, almost inevitable simplifications are necessary. A possible approach can be the introduction of "reference points", which means that reference points of the facilities are to be assigned to reference points in the plant. Feasibility conditions which cannot be formulated in terms of reference points can be treated only in an interactive way.

The objective function is supposed to depend on the amounts of material-flow between the facilities (called "connections") and on the distances of their locations.
This is the usual Quadratic Assignment formulation of the plant layout problem.

3. The Quadratic Assignment Problem (QAP)

3.1. PROBLEM FORMULATION

The QAP can be formulated as follows: Given $N = 1, 2, .., n$ and real numbers a_{ik}, c_{kl} and d_{ij} $(i, j, k, l = 1, 2, .., n)$, find the permutation $p \in P_n$ for which

$$f(p) = \sum_{i=1}^{n} \sum_{j=1}^{n} d_{ij} c_{p(i)p(j)} + \sum_{i=1}^{n} a_{i,p(i)} \tag{1}$$

becomes minimal, where $P_n = (p : N \mapsto N)$ is the group of n-order permutations, a_{ik} is the cost of assigning facility k to location i, d_{ij} denotes the distance of locations i and j and c_{kl} denotes the connection of facilities k and l. Suppose that the problem is symmetric, that is, $c_{kl} = c_{lk}$ and $d_{ij} = d_{ji}$.

3.2. SOLUTION METHODS

As the QAP belongs to the "hard core" of NP-complete problems, only approximate methods can be taken into consideration even for medium size problems, say, for $n \geq 20$.
The most frequently used methods can be classified as
- construction methods: usually greedy-type algorithms which build up the approximate solution by assigning a facility to a location (and thereby excluding a subset of P_n from further consideration) at each step so as to minimize the immediate cost increase. However, the quality of solutions obtained by this "short-sighted" approach leaves much to be desired in most cases ([1]).
- iterative improvement methods: local search methods which begin with a given feasible solution (permutation) and attempt to improve it sequentially by interchanges of single assignments. The known improvement methods are essentially variants of the pair-wise exchange algorithm. 3-way or higher level exchanges are reported to be computationally infeasible for larger problems ([1]).

In this section we describe two approximate algorithms (a construction and an improvement method) which were incorporated into our program system. The algorithms have been implemented in several variants and tested on Nugent's test problems ([5]) and on the Steinberg problem (7]).

Our algorithms produced for *all* tested problems better solutions than other algorithms in their categories ([1],[4],[5]). For the Steinberg problem even nine better solutions were obtained than the best solution known so far ([4]).

3.2.1. *A Construction Method Based on Cluster-Analysis.* As the distances in (1) depend on the relative positions of the facilities, we are looking for permutations for which *all* facilities having large "relative connections" are relatively close to each other. Relative connection means a comparison, e.g.

$$r_{ij} = c_{ij} - \frac{\sum\limits_{k \neq i, k \neq j} (c_{ik} + c_{jk})}{2(n-2)} \tag{2}$$

is one of the possible definitions of the relative connection of facilities i and j. Using relative connections as a measure of similarity, cluster analysis can be used to determine hierarchically organized groups of facilities. At start, each facility is regarded as a cluster of one element. Then, at each step, the two clusters having the greatest similarity are contracted until a single cluster containing all the facilities is obtained. (Relative connection can be generalized for clusters containing more than one elements.)

Then, in a reversed order of their formation, clusters of facilities are assigned to appropriately selected subsets of locations; at each step a subset of locations is partitioned into two subsets so as to minimize the expectation value of the objective function. Assuming that each permutation can be chosen with equal probability, we can determine the expectation value of the objective function, see [3]. The crucial point of the algorithm is the definition of similarity. We used definition (2) in the program system. Recent experiments show that in some cases better solutions are obtained using results of multidimensional scaling ([2]).

3.2.2. *An Improvement Method Based on Feasible Directions.* QAP is considered as a quadratic optimization problem on binary variables, and the information available for the corresponding continuous problem is utilized in finding improving directions.

The quadratic programming formulation of QAP can be written as

$$z^* = \min_{\mathbf{x} \in L} f(\mathbf{x}), \tag{3}$$

where

$$f(\mathbf{x}) = \sum_{i=1}^{n} \sum_{j=1}^{n} \sum_{k=1}^{n} \sum_{l=1}^{n} d_{ij} c_{kl} x_{ik} x_{jl} + \sum_{i=1}^{n} \sum_{k=1}^{n} a_{ik} x_{ik} = \mathbf{x}^T \mathbf{Q} \mathbf{x} + \mathbf{a}^T \mathbf{x}, \tag{4}$$

$$L = \{\mathbf{x} \in B^{n*n} : \sum_{k=1}^{n} x_{ik} = 1, i = 1, ..., n; \sum_{i=1}^{n} x_{ik} = 1, k = 1, ..., n\}, \tag{5}$$

D, **C** and **A** are given as in (1), **Q** is the matrix of the coefficients $d_{ij} c_{kr}$, **a** and **x** are n^2 vectors of the elements of **A**, resp. **X**, and the interpretation of the variables x_{ik} in terms of the layout planning problem: $x_{ik} = 1$ if facility k is assigned to location i, and 0 otherwise.

The change of the objective function along a direction **r** can be expressed as

$$h(\mathbf{x}, \mathbf{r}, \lambda) = f(\mathbf{x} + \lambda \mathbf{r}) - f(\mathbf{x}) = \lambda \mathbf{g}(\mathbf{x})^T \mathbf{r} + \lambda^2 \mathbf{r}^T \mathbf{Q} \mathbf{r}, \tag{6}$$

where $\mathbf{g}(\mathbf{x})$ is the gradient of $f(\mathbf{x})$. The t-th step of the algorithm searches for directions **r** for which

$$\mathbf{x} + \mathbf{r} \in L \quad and \quad h(\mathbf{x}^t, \mathbf{r}, 1) < 0, \tag{7}$$

and finds the improved solution $\mathbf{x}^{t+1} = \mathbf{x}^t + \mathbf{r}$.

Although nothing can be stated in general about the sign of $\mathbf{r}^T \mathbf{Q} \mathbf{r}$ in (6), (the matrix $\mathbf{Q}$ is indefinite) it seems to be reasonable (and it is also justified by the computational results) to limit the scope of candidate directions for feasible directions with negative reduced gradient, i. e. for directions $\mathbf{r}$ satisfying $g(\mathbf{x})^T \mathbf{r} < 0$. This limitation accelerates considerably the evaluation of possible directions. Furthermore, it can be proved that if a direction $\mathbf{r}$ corresponds to a pair-wise exchange, the second term in (6) is always nonnegative (assuming that $\mathbf{C} \geq 0$, $\mathbf{D} \geq 0$, and $c_{ii} = d_{ii} = 0$ for $i = 1, ..., n$).

The actual realization of the algorithm depends on the kind of directions allowed to be taken into consideration in the search, i. e. on the definition of the neighbours of a given solution. If only directions corresponding to pair-wise exchanges are allowed, the algorithm corresponds to the known pair-wise exchange algorithms, but it is computationally much more efficient. A more general approach is to define the neighbours as the neighbouring extremal points of the feasible set L. Then the resulting algorithm is an adaptation of the simplex method of quadratic programming (with restricted pivoting). In this case a considerable part of the computations is analoguous to that of the solution of the classical transportation problem.

4. Conclusions

Our experience suggests that interactive computer program systems can provide effective tools to support layout design. The general concepts of our solution methods allow considerable freedom in the actual realization of the algorithms; they can be modified to treat also special or more general problems. The good computational results suggest that perspectives of further research are promising.

References

[1] Burkard, R. E. (1984) 'Quadratic assignment problems', European J. of Operational Research 15, 283–289.

[2] Goetze, B. and Nehrlich, W. (1986) 'Scaling of random variables and arrangement problems in layout design', Akademie der Wissenschaften der DDR, Berlin (preprint).

[3] Graves, G. W. and Whinston, A. B. (1970) 'An algorithm for the quadratic assignment problem', in Abadie (ed.), Integer and Nonlinear Programming, North-Holland Publishing Company, Amsterdam, 473–498.

[4] Liggett, R. S. (1981) 'The quadratic assignment problem: an experimental evaluation of solution strategies', Management Science 27, 442–458.

[5] Nugent, C. E., Vollmann, T. E. and Ruml, J. (1968) 'An exprimental comparison of techniques for the assignment of facilities to locations', J.Operations Res. Soc. Amer. 16, 150–173.

[6] Pogány, Z. et al. (1987) 'Computer-aided layout design', in Kovács, M. (ed.), Optimal Planning of Production Systems – Research Report (in Hungarian), ELTE Computer Centre, Budapest, 1–90.

[7] Steinberg, L. (1961) 'The backboard wiring problem: a placement algorithm', SIAM Review 3, 37–50.

OPTIMAL MINIMAX CONTROLLERS WITH STRUCTURAL CONSTRAINTS

SEPPO POHJOLAINEN, MERJA LAAKSONEN
Tampere University of Technology
Department of Mathematics
P.O.Box 692, SF-33101 Tampere
Finland

ABSTRACT. The problem of optimal tuning a stabilizing controller under a minimax criterion in the frequency domain will be discussed. The solution will be found with the aid of modified Youla parametrization, perturbation results for both the system and the cost functional show how the controller should be modified, and finally a complex Remez method is used to compute the optimal controllers.

1. Introduction

A common problem in control engineering is that of tuning a feedback controller. The purpose of a controller C (Figure 1) often is 1) to stabilize the feedback system 2) to regulate the measured output y to follow a given reference signal y_{ref} 3) to decouple the perturbation signal w and the output. The structure of the controller should be as simple as possible; a typical practical selection is a P, PI, or PID controller. In this paper the problem of tuning a given stabilizing controller under a minimax criterion will be discussed. The basic idea is to use modified Youla parametrization [1], to show how much a given controller may be varied without losing stability. Then a cost function is set up to measure the quality of the control. A formula is then derived to show how controller perturbations will effect the quality of the control. Tuning then turns out to be a complex nonlinear minimax problem in which complex Remez method [3] will be used. An example is given to demonstrate the applicability of the method.

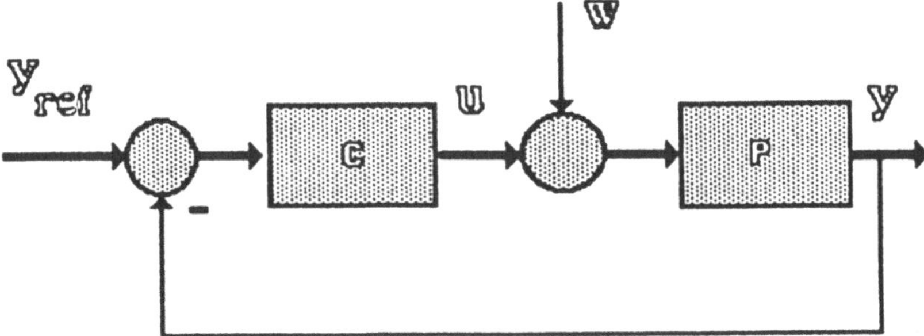

Figure 1. The basic feedback loop.

2. Materials and Methods

2.1 NOTATIONS AND PRELIMINARIES

The **system** $P(s)$ is assumed to be a strictly proper real rational function, i.e. $P(s) \to 0$, as $s \to \infty$ and the coefficients of $P(s)$ are real. The **controller** $C(s)$ is a proper real rational function, i.e $C(s)$ remains bounded as $s \to \infty$. A real rational transfer function is **stable** if all its poles are in the open left half plane. For (stable) transfer functions the (H_∞) L_∞ -**norm** is defined as

$$\| P \| = \sup_{\omega \in \mathbb{R}} | P(i\omega) | . \tag{1}$$

Q-parameter is the transfer function from the perturbation signal w to the output y (Figure 1)

$$Q(s) = \frac{P(s)}{1 + C(s)P(s)} . \tag{2}$$

Q is strictly proper, because $P(s)$ is strictly proper. The **overall transfer function** is

$$\begin{pmatrix} y \\ u \end{pmatrix} = \begin{pmatrix} QC & Q \\ C(1-QC) & -CQ \end{pmatrix} \begin{pmatrix} y_{ref} \\ w \end{pmatrix} . \tag{3}$$

The above closed loop system is **stable**, if all its component transfer functions are stable. In this case C **stabilizes** P.

3. Results

3.1 ADMISSIBLE CONTROLLER VARIATIONS

Assuming that a stabilizing controller C has been found and changed to $C + \Delta C$, then the feedback system remains stable, if C and $C + \Delta C$ have the same number (counted according to multiplicity) of unstable poles, and $\|Q\Delta C\| < 1$ [4].

3.2 COST FUNCTION

In order to evaluate the quality of the control the following cost function is set up

$$J(\omega, C) = f(\omega)|1 - Q(i\omega)C(i\omega)| + g(\omega)|Q(i\omega)|. \tag{4}$$

The functions f and g are nonnegative, bounded, continuous, and even in ω, and $\lim_{\omega \to \infty} f(\omega) = 0$. The first component measures the deviation between the given reference signal y_{ref} and the measurement signal y and the second component the effect of the perturbation signal w on the output y. Let $J(C) = \|J(\omega, C)\|$, and $Max = \{\omega | J(\omega, C) = J(C)\}$.

3.3 LINEARIZATION OF THE COST FUNCTION

Let the controller C to be changed by the amount of ΔC and assume that $\|Q\Delta C\| < 1$. The criterion may be written as

$$J(\omega, C+\Delta C) = \frac{J(\omega, C)}{|1+Q(i\omega)\Delta C(i\omega)|} = J(\omega, C)|1-Q(i\omega)\Delta C(i\omega)| + O(\|Q\Delta C\|^2) \tag{5}$$

Under these assumptions it may be proved [2]:

Theorem. Select ΔC so that

$$|1-Q(i\omega)\Delta C(i\omega)| < 1 \ , \quad \text{for all} \ \ \omega \in \text{Max} \ . \tag{6}$$

Then for sufficiently small positive values of ε,

$$J(C+\varepsilon \Delta C) < J(C) \ , \tag{7}$$

i.e the quality of the control has improved.

3.4 COMPLEX REMEZ METHOD

A complex Remez method is used to solve the above linear problem (6). For simplicity we make the assumption $\Delta C(i\omega) = I(i\omega) \, \mathbf{k}$,where $I(i\omega) \in \mathbb{C}^{1 \times n}$ and $\mathbf{k} \in \mathbb{R}^n$. The problem is to find the parameters $\mathbf{k}$, such that

$$\min_{\mathbf{k}} \ \max_{\omega} \ | \ J(\omega, C)[1-Q(i\omega)I(i\omega) \, \mathbf{k}] \ | \ . \tag{8}$$

Finding a nontrivial solution to the above problem certainly the implies (6), which in turn suggests to modify the controller by the amount of $\varepsilon\Delta C$.

The problem (8) will be solved with the algorithm proposed by Tang [3]. Actually it solves the dual of (8) as follows : maximize the inner product $\mathbf{c}^T\mathbf{r} = h$, where $\mathbf{r}$ is a non-negative vector satisfying $A \, \mathbf{r} = [1,0,...,0]^T$, and $\mathbf{c}$ is the cost vector with respect to points $\{(\omega_j, \alpha_j)|j=1,...,n+1\}$, where $\alpha_j = \text{Arg}(\, J(\omega_j, C)[1-Q(i\omega_j)I(i\omega_j) \, \mathbf{k}] \,)$. The vector $[h,\mathbf{k}] = \mathbf{c}^T A^{-1}$. The matrix A is determined by the problem data, the error and a convexity condition.

4. An Example for tuning of a P-controller

Let the process be

$$P(s) = \frac{1}{s^3+6s^2+11s+6} \ ,$$

and the cost function

$$J(\omega,C) = \frac{1}{1 + |\omega|} \, (\, |1-Q(i\omega)C(i\omega)| + |Q(i\omega)| \,).$$

A P-controller, that is $C(s)=k_p$, was tuned for the process. The algorithm converged to the value $k_p=8.15$. The cost function has two global maximums at the points $\omega=0$ and $\omega=1.63$ (Figure 2).

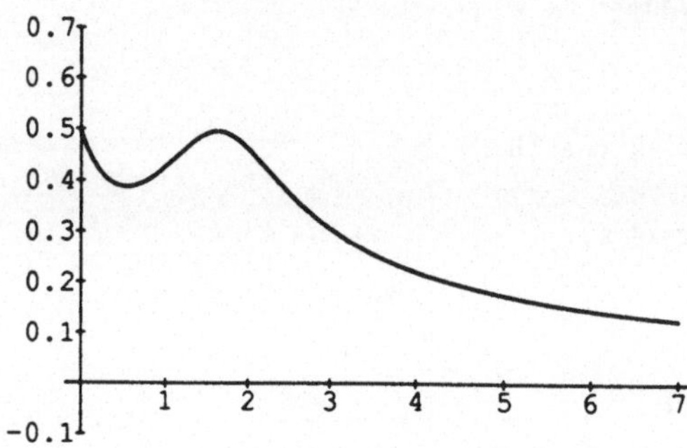

Figure 2. The criterion $J(\omega,8.15)$.

5. Conclusions

A modified Q-parameter method was used to improve the quality of the control under an minimax criterion. Well-known robustness results were used to estimate how much a controller could be varied without losing closed loop stability. A linearized version of the criterion was derived and complex Remez method was used to solve the tuning problem.Finally we note that controllers are locally optimal, global optimality is not guaranteed.

6. References

[1] Maciejowski J. M.'Multivariable Feedback Design', Addison-Wesley, Reading, Massachusetts, 1989.

[2] Pohjolainen S.'Optimal H_∞-SISO-controllers with structural constraints', Proc. MTNS-89, Volume II, Birkhäuser, Boston 1990, pp. 329-338.

[3] Tang, Pingtak Peter 'Chebyshev Approximation on the Complex Plane', Ph.D.-thesis, University of California, Berkeley 1987.

[4] Vidyasagar M. 'Control System Synthesis, A Factorization Approach', The MIT Press, Cambridge, Massachusetts, 1985.

SYSTEM FOR STATISTICAL SIMULATION, ANALYSIS AND OPTIMIZATION OF LINEAR HYBRID CIRCUITS

Leonidas L. SAKALAUSKAS
Institute of Mathematics and Informatics
2600 Vilnius, Akademijos st.4, LITHUANIA

ABSTRACT. Computer simulation has become an important tool in the creating of new electronic devices. In the paper we consider a system devoted to optimal statistical design of linear hybrid circuits of bypolar transistors. The Monte–Carlo method for simulation of circuits is applied. When computing the influence of temperature, radiation and working regime under current is taken into account. In statistical processing of simulation results the coefficients of sensibility of FP (functional parameters: gain, phase, input and output resistances and capasities) and of the yield (the percent of circuits, sattisfying restrictions of FP) to the dispersion of the component parameters (CP) are evaluated. Then the maximization of the yield by correction of resistances of substrate resistors for their fixed tolerancies is carried out.

1. Simulation of components

It is assumed, that the passive elements (resistors, capacitors, inductivities), full-filled on the same substrate, are determined by one parameter (resistance, capacity, inductivity). These parameters are correlated by the same correlation, distributed normally and are statistically independent of the parameters other components, suspended on the substrate (Sakalauskas (1989), Logan (1970)). The parameters of different suspended components are statistically independent. It is experimentally established, that the parameters $\beta_0, \beta_F, H_{11}, C_k, T_k, R_b$ of the transistor model have the lognormal distribution, where only the β_0 and β_F are correlated with R_b, other correlations being equal to zero (Kriukov et.al. (1984)). In such a way the correlation matrix of CP is decomposed into small submatrices. That enables us to generate large samples of CP efficiently, using the MCB (Matrices with Constant Blocks) techniques (Sakalauskas (1989)). Further these samples are applied to statistical simulation of the yield and FP of circuits.

2. Statistical simulation and evaluation of the influence of CP to FP and the yield

Let a circuit be tested l times. Then, the means and variances of FP, the yield and the sensibility of FP and of yield to the parameters of substrate and suspended components is evaluated. The sensibility of the j-th FP to the i-th parameter of the k-th component is determined by the expression:

$$S_{ik}^{j} = \frac{\Delta x_i \cdot \sum_{m=1}^{n^k} \sum_{p=1}^{l} y_j^p (x_{mk}^p - \bar{x}_{mk}) \cdot R_{mi}}{\sum_{j=1}^{l} y_j^p}, \qquad (1)$$

where $(y_j^p,\ 1 \leqslant j \leqslant 6,\ 1 \leqslant p \leqslant l)$ is the sample of FP, $(x_{mk}^p,\ 1 \leqslant p \leqslant l,\ 1 \leqslant m \leqslant n^k)$ is the sample of parameters of the k-th component (or substrate), $(\bar{x}_{mk},\ 1 \leqslant m \leqslant n^k)$ are their nominal values, n^k is the number of parameters of one component, Δx_{ik} is the tolerance, R_{mi} are the elements of the inversed correlation matrix of CP. It is of importance to evaluate the significance of S_{ik}^j. It is considered, that a parameter has a considerable influence on the FP, if

$$S_{ik}^j \sqrt{l/R_{ii}}/\Delta x_{ik} \geqslant \eta(\alpha), \qquad (2)$$

where $\eta(\alpha)$ is the quantile of the Gaussian distribution. The whole component essentially influences the FP if

$$l \cdot \sum_{i=1}^{n^k} \sum_{m=1}^{n^k} S_{ik}^j S_{mk}^j R_{im}/\Delta x_{ik} \Delta x_{mk} \geqslant \chi_\alpha^2(n^k). \qquad (3)$$

The sensibility and significance of the influence to the yield may be evaluated replacing y_j^p by ψ^p in (1), (2), (3), where $\psi^p = 1$, if the circuit is good, and $\psi^p = 0$, if the circuit is defective.

3. The optimization of resistors of the substrate

If during the analysis it appears that the resistors of the substrate have a significant influence on FP and the yield, then their correction is carried out by Monte–Carlo sets (Sakalauskas (1990)):

$$\bar{r}_s^{m+1} = u_s^m - \Delta R \cdot \rho_R \cdot a^m /(3(1 + (n_R - t - 1)\rho_R)),\ 1 \leqslant s \leqslant t,$$

where $a^m = \sum_{k=t+1}^{n_R} 3(u_j^m - \bar{r}_k^m)/(\Delta R \cdot \bar{r}_k^m)$, u_k^m are mean of resistances of good circuits, n_R is the number of resistors, t is the number of the resistors optimized, s is the number of the resistor corrected, ΔR is the tolerance, ρ_R is the correlation among resistors. The number of tests is choossen according to the rule:

$$l_{m+1} \geqslant \frac{\chi_\beta^2(t) \cdot (1 - \rho_R)}{P_m \big(h^m - \rho_R \cdot (a^m)^2 \big) /\big(1 + (n_R - 1)\rho_R \big)},$$

where P_m is the estimation of the yield, $h^m = \sum_{i=1}^{t}(u_i^m - a^m)^2$. The nominals of resistances are iterated while the influence of resistors to the yield is siqnificant:

$$l_m \cdot P_m \cdot \left(h^m - \rho_R \cdot (a^m)^2\right)/\left(1 + (n_R - 1)\rho_R\right))/(1 - \rho_R) \geqslant \chi_\alpha^2(t)$$

and the tolerance interval of the yield exceeds the accuracy ε. In (Sakalauskas (1992)) it is proved, that such a procedure guarantees the convergence to the maximum of the yield.

4. The simulation and optimization of an amplifier

Let us consider the results of statical design of an amplifier from 7 resistors, 3 capacitors and 2 transistors (Figure 1). The tolerancies of CP are the following: $\Delta R = \Delta C = 10\%$, $\Delta B_0 = \Delta B_F = \Delta H_{11} = 20\%$, $\Delta T_K = \Delta C_K = 2\%, \Delta R_B = 5\%$, the resistances are correlated by the correlation $\rho_R = 0.8$. After 100 tests was established, that the influence of resistors to FP and the yield is significant, and that the influence of transistors and capacitors is insignificant. The nominals of resistances were optimized. In TABLE 1 the dependencies of the yield and of the number of l_m on the iteration number m are illustrated. After 5 iteration and 416 statistical tests the yield increased from $20.0 \pm 11.1\%$ to $86.7 \pm 4.98\%$.

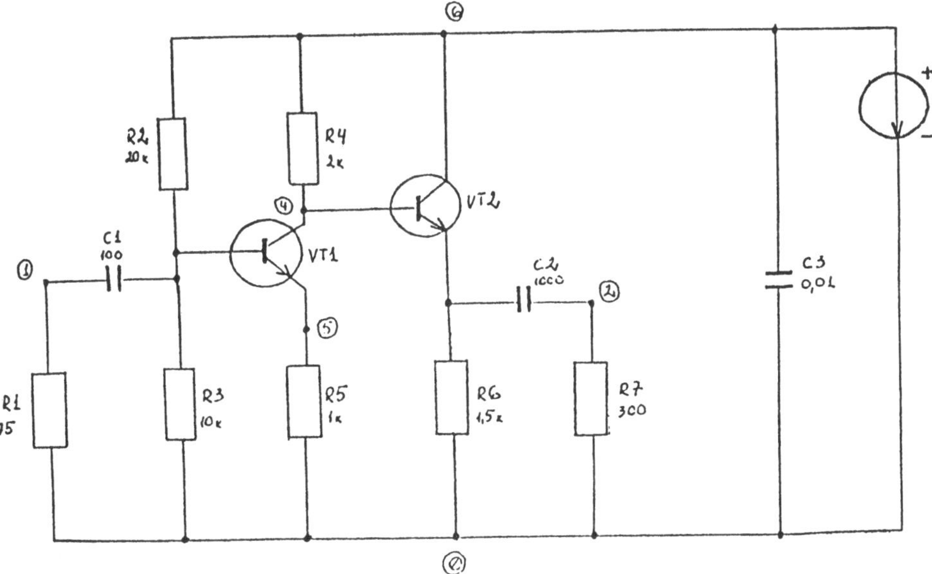

Figure 1. Linear amplifier.

TABLE 1. Dependency of yield on the iteration number m

m	l_m	$P_m \pm \Delta P_m \%$
1	50	20. ± 11.1
2	37	44.4 ± 16.24
3	52	82.7 ± 10.3
4	102	82.35 ± 7.4
5	176	86.7 ± 4.98

References

Kriukov J., Shishkin V., Gorlova Z., Pitalin V. (1984) Statistical Analysis, Prognosis and Optimization of Parameters of Analogues Hybrid Integrated Circuit // Inf. bull. "Algorithms and Programes", No. 6/63. Moscow: VNTT. CENTR. (in Russian).

Logan J. (1970) Characterization and Modeling for Statistical Design // The Bell System Technical Journal, V. 50, No. 4, 1105–1147.

Sakalauskas L.L. (1989) Application of Matrices of Special Structure to Statistical Analysis of Integrated Circuits // Elektonnaya tekhnika, ser. 3 (Mikroelektronika), No. 3(132), 29–33 (in Russian).

Sakalauskas L.L. (1990) Adaptive Methods of Statistical Optimization of Integrated Circuits. // Avtomatizaciya projektirovaniya v elektronike, No. 41, Kiev, 118–124 (in Russian).

Sakalauskas L.L. (1992) On the convergence of one centering procedure // Lithuanian Mathematical Journal, V. 31, No. 4, 670–677 (in Russian).

NON-REGULAR ADAPTIVE GRIDS FOR 2-D GAS DYNAMICS

A.A. SAMARSKY, A.P. FAVORSKY, V.F. TISHKIN,
V.F. VASILEVSKY, K.V. VYAZNIKOV.
Keldysh Institute of Applied Mathematics,
USSR, 125047, Moscow, Miusskaya sq. 4.

ABSTRACT. Non-regular grid adaptation technique is presented. Dirichlet tesselation is used as non-regular computational grid because of its well known metric properties. This method proved to give high accuracy in 1-D case and was applied to 2-D gas dynamics problems.

1. Introduction

Technique for dynamic adaptation of Dirichlet tesselation to shape of computational domain and singularities of solution is presented [1]. Technique is based on using special weight function controlling distribution of cells [2]. Several examples of weight functions for different applications are described.

The technique was used for computation of gas flows contaning shock waves. We used monotonous difference schemes of 2-nd and 3-rd order. Technique of generation these schemes is based on special modification of 1-st order monotonous schemes and was carefully tested.

Results of computations have good agreement with theory and demonstrate advantages of dynamic adaptation of Dirichlet tesselation.

2. Adaptation Technique

2.1. ADAPTATION TECHNIQUE IN 1-D CASE

In 1-D case equidistribution consept requires

$$(\Delta x w)_{i+1/2} = const \tag{1}$$

all over the grid, where $\Delta x_{i+1/2} = x_{i+1} - x_i$, $w_{i+1/2}$ – value of weight function. We may rewrite (1)

$$(\Delta x w)_{i+1/2} - (\Delta x w)_{i-1/2} = 0. \tag{2}$$

For solving (2) we use simple iteration procedure

$$x_i^{S+1} = \left(\frac{w_{i+1/2} x_{i+1/2} + w_{i-1/2} x_{i-1/2}}{w_{i+1/2} + w_{i-1/2}} \right)^S. \tag{3}$$

Exact form of weight function depends on singularities of solution. As it follows from (1) values of weight function shall be greater in those regions where it is nessesary to concentrate grid's cells. For example, to obtain adaptation to region of high variation of function f we may use weight function in the following form

$$w_{i+1/2} = |f_i - f_{i-1}|(x_i - x_{i-1})^p, \tag{4}$$

where $p = const$ – parameter may be customised.

264

We tested (4) on following ODE problem

$$\varepsilon u'' - u = 0, \qquad u(0) = 1, \quad u(1) = 0, \tag{5}$$

where ε – small parameter. Highest precision was reached when $p = 2$ in (4). For this particular problem such method of grid adaptation leads to wanishing leading term of approximation error. Numerical results for different values of ε show uniform convergence to exact solution. In tab.1 you can see comparison of results obtaned using described technique (left side) and technique based on reducing approximation error [3] (right side) when $\varepsilon = 1.e - 4$, $N = 21$, $i = 1 \ldots N$.

2.2. ADAPTATION TECHNIQUE IN 2-D CASE

Direct generalisation of (2) for Dirichlet cells is

$$\sum_j w_{ij}(x_i - x_{ij})S_{ij} = 0, \qquad \sum_j w_{ij}(y_i - y_{ij})S_{ij} = 0. \tag{6}$$

where summation performs through all neighbours of node P_i – center of Dirichlet cell with number i; x_{ij}, y_{ij} – coordinates of center of edge between i and j cells; S_{ij} – area of triangle, formad by the edge and node P_i. Corresponding iteration formulae

$$x_i^{S+1} = \left(\frac{\sum x_{ij} w_{ij} S_{ij}}{\sum w_{ij} S_{ij}} \right)^S, \qquad y_i^{S+1} = \left(\frac{\sum y_{ij} w_{ij} S_{ij}}{\sum w_{ij} S_{ij}} \right)^S. \tag{7}$$

To obtain nearly equal sizes of cells we use

$$w_{ij} = const. \tag{8}$$

Such function can be used for construction of equidistributed grids in domains of complicated shape. We can take arbitrary initial distribution of grid nodes (fig.1) and move them, using adaptation technique. Finally we obtain equidistributed grid (fig.2).

To obtain adaptation to region of high variation of function f we use

$$w_{ij} = |f_i - f_{ij}|l_{ij}, \tag{9}$$

where $l_{ij} = ((x_i - x_{ij})^2 + (y_i - y_{ij})^2)^{1/2}$. For practical usage we take combination of (8) and (9):

$$w_{ij} = \theta_0 + \theta_1|f_i - f_{ij}|l_{ij}^p. \tag{10}$$

On fig.3,4 you can see result of adaptation and 3-D view for function $f(x,y) = 1/(r+\varepsilon)$; r – distance between point (x,y) and (x_0, y_0); ε – small parameter.

3. 2-D Gas Dynamics

To solve numerically system

$$\frac{\partial U}{\partial t} + \frac{\partial F(U)}{\partial x} + \frac{\partial G(U)}{\partial y} = 0, \qquad U = \begin{pmatrix} \rho \\ u \\ v \\ e \end{pmatrix} \tag{11}$$

we use conservative difference scheme

$$\frac{\partial(US)_i}{\partial t} + \sum_j (Ql)_{ij} = 0, \tag{12}$$

where S_i – area of i cell, l_{ij} – length of ij edge, Q_{ij} – flux. For this equation we choose basic 2-points monotonous flux such as Godunov, Lax-Friederix or kinetic-consistant flux. These schemes have first order approximation. To increase order of approximation without violating monotonous property of the scheme we use interpolation procedure like in TVD-schemes.

Finally we present results of solving oblique shock problem using grid adaptation technique. We use velocity v (y-component of velocity) as function for adaptation. Grid captures moving fronts of shocks and following them. Contours of specific internal energy and grid configuration are presented on the fig.5,6.

4. Conclusion

We had created and tested original adaptation technique using Dirichlet tesselation. Presented technique provides effective computation of boundary and internal layers and may be generalized to 3-D. Technique also may be used for generation of initial computational grids in domains of complicated shape.

5. References

[1] В.Ф. Василевский, К.В. Вязников, В.Ф. Тишкин, *Квазимонотонные разностные схемы повышенного порядка точности на адаптивных сетках нерегулярной структуры*, препринт ИПМ им.М.В. Келдыша АН СССР, N124, 1990.

[2] J.F. Thompson, A.U.A. Warsi, C.W. Mastin, *Numerical Grid Generation, Foundations and Applications*, North-Holland, New-York, 1985.

[3] Л.М. Дегтярев, В.В. Дроздов, Т.С. Иванова, *Метод адаптивных к решению сеток в одномерных краевых задачах с пограничным слоем*, препринт ИПМ им.М.В. Келдыша АН СССР, N164, 1986.

Tab.1

	this article			article [3]		
i	$x_i - x_{i-1}$	u_i	$\|u_i - u_*\|$	$x_i - x_{i-1}$	u_i	$\|u_i - u_*\|$
1	$.000e + 0$	$.100e + 1$	$.000e + 0$	$.000e + 0$	$.100e + 1$	$.000e + 0$
2	$.222e - 2$	$.801e + 0$	$.226e - 6$	$.300e - 2$	$.777e + 0$	$.421e - 6$
3	$.235e - 2$	$.633e + 0$	$.417e - 6$	$.200e - 2$	$.593e + 0$	$.765e - 6$
4	$.250e - 2$	$.493e + 0$	$.575e - 6$	$.300e - 2$	$.444e + 0$	$.104e - 5$
5	$.266e - 2$	$.378e + 0$	$.706e - 6$	$.300e - 2$	$.325e + 0$	$.127e - 5$
6	$.285e - 2$	$.284e + 0$	$.812e - 6$	$.400e - 2$	$.232e + 0$	$.144e - 5$
7	$.307e - 2$	$.209e + 0$	$.897e - 6$	$.300e - 2$	$.160e + 0$	$.157e - 5$
8	$.333e - 2$	$.150e + 0$	$.964e - 6$	$.400e - 2$	$.107e + 0$	$.167e - 5$
9	$.363e - 2$	$.104e + 0$	$.102e - 5$	$.500e - 2$	$.679e - 1$	$.174e - 5$
10	$.399e - 2$	$.698e - 1$	$.105e - 5$	$.500e - 2$	$.408e - 1$	$.179e - 5$
11	$.444e - 2$	$.448e - 1$	$.108e - 5$	$.600e - 2$	$.227e - 1$	$.182e - 5$
12	$.499e - 2$	$.272e - 1$	$.110e - 5$	$.700e - 2$	$.115e - 1$	$.183e - 5$
13	$.570e - 2$	$.154e - 1$	$.111e - 5$	the rest data is not avalable		
14	$.664e - 2$	$.793e - 2$	$.112e - 5$	—		
15	$.795e - 2$	$.358e - 2$	$.111e - 5$	—		
16	$.990e - 2$	$.133e - 2$	$.111e - 5$	—		
17	$.131e - 1$	$.360e - 3$	$.108e - 5$	—		
18	$.192e - 1$	$.534e - 4$	$.101e - 5$	—		
19	$.349e - 1$	$.213e - 5$	$.536e - 6$	—		
20	$.101e + 0$	$.486e - 8$	$.480e - 8$	—		
21	$.766e + 0$	$.000e + 0$	$.000e + 0$	—		

u_* – exact solution.

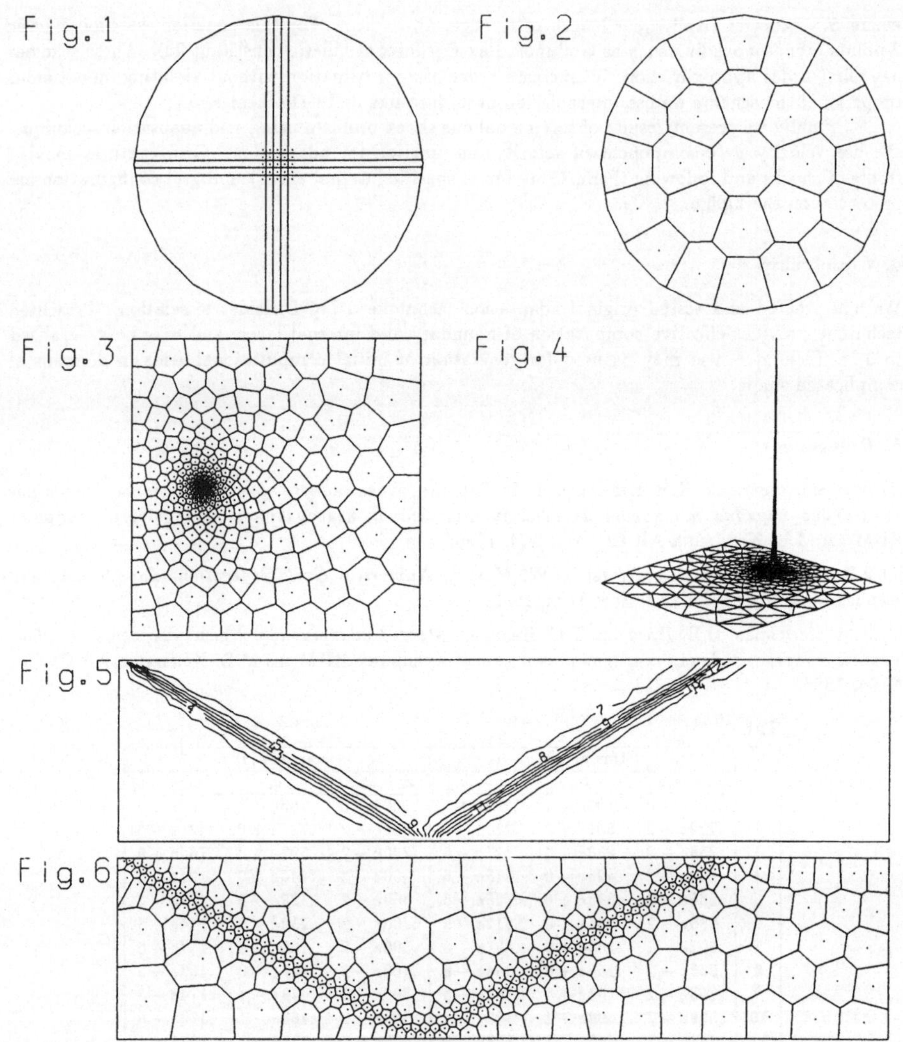
Fig.1
Fig.2
Fig.3
Fig.4
Fig.5
Fig.6

STEADY SOLUTIONS FOR THE FLOW OF AN ELECTRO-RHEOLOGICAL FLUID BETWEEN CYLINDRICAL ELECTRODES

XIAO SHI[**], R. J. ATKIN[*] and W. A. BULLOUGH[**]
[*]Department of Applied and Computational Mathematics,
PO Box 597, The University, Sheffield S10 2UN, England

[**]Department of Mechanical and Process Engineering

ABSTRACT. Results are presented for steady flow of an ideal ER fluid in a rotating clutch with a potential difference between the cylindrical electrodes. The continuum model is based on a Bingham fluid in which the viscosity is constant and the yield stress depends upon the magnitude of the electric field, E, and is proportional to either E^2 or $E - E_o$. The electric field is assumed to vary radially and to be unaffected by the flow. Flow patterns are discussed for different ranges of applied voltage and the differences between flat plate and radial devices are examined.

1.   Introduction

Electro-rheological fluids are mixtures of micron sized particles carried in a dielectric liquid. Their fluid characteristics change, even to the point of becoming solid, when they are exposed to strong electric fields. When the field is imposed transverse to the direction of motion of the fluid the solid particles form chains or fibres when viewed under the microscope and the resistance to flow increases. With zero volts applied most fluids exhibit near Newtonian properties except possibly at very low shear rates. An important part of electro-rheological fluid design is the matching of the suspension and the base liquid. Like much of innovation today research in this area is a multi-disciplinary activity involving people with expertise in different areas of engineering and science. It therefore provides a case study for such activity and the problems associated with it. This paper describes some of the results of one such collaboration between an engineer and a mathematician. To be successful it seems essential to us that the collaborators combine their expertise at the outset of the task and work on a combined and rational basis.

Here we concentrate on the clutch which is one of the tests for the characterisation of electro-rheological fluids and has possible application in high speed (quick time) voltage-controlled torque transmission. Application dictates using excited concentric cylindrical electrodes the flow being caused by the cylinders rotating relative to each other (Couette flow). In this geometry the velocity and electric fields must vary with radius. The practical question arising is what geometry and speed of rotation produce relatively uniform conditions for both velocity and field so that the characteristics of the fluid can be determined at set conditions.

The fluid is modelled as a Bingham plastic in which the yield stress, τ_e, depends upon the magnitude of the electric field E. Two forms which appear in the literature are $\tau_e = \alpha E^2$ and $\tau_e = \alpha^*(E - E_o)$ for $E > E_o$ with little yield stress being generated below the threshold value E_o. Typical values of the constants are $\alpha = 2.5 \times 10^{-10}$, $\alpha^* = 0.843 \times 10^{-3}$, $E_o = 0.6 \times 10^6$ all in basic SI units. It is through this relationship, which is found to be a significant factor in our analysis, that the field affects the flow since the viscosity, η is taken to be independent of E. The fluid is assumed to be a perfect dielectric in which current flow has no effect on the voltage distribution. All fluid flows are fully developed and laminar and the dimensions of the device are such that end effects can be neglected. Isothermal conditions are assumed.

2.   Couette Shear Mode

2.1   FLOW BETWEEN TWO INFINITE PARALLEL FLAT PLATES $y = 0$, $y = h$

If the fixed plate $y = 0$ is earthed and the plate $y = h$ which is moving in the x-direction with constant speed U is raised to a potential V the electric field has magnitude $E = V/h = $ constant and either $\tau_e = \alpha E^2 = \alpha V^2/h^2$ or $\tau_e = \alpha^*(E - E_o) = \alpha^*(V/h - E_o)$. In the latter case for $\tau_e > 0$ the voltage must satisfy $V > h E_o$. Assuming that flow is in the x-direction and varies with y the equations of motion are satisfied if the pressure and the only non-zero shear stress τ_{xy} are constant. When $|\tau_{xy}| > \tau_e$ there is flow across the whole gap with Newtonian profile $v_x = Uy/h$ and $\tau_{xy} = \tau_e + \eta U/h$ so that the graphs of shear stress v shear rate are parallel straight lines with the same gradient as in the Newtonian case and intercepts depending upon V. There is no flow if $|\tau_{xy}| < \tau_e$.

2.2   FLOW BETWEEN TWO CONCENTRIC CYLINDERS OF RADII R_1, R_2 $(R_1 < R_2)$

If the inner cylinder is earthed and the outer one raised to a potential V for a perfect dielectric the electric field is radial with magnitude $E = V/r\ell n(R_2/R_1)$. Here the flow is caused by the outer cylinder rotating with constant angular speed Ω under the action of a torque T whilst the inner cylinder is fixed. If $\omega(r)$ is the angular speed at a point in the fluid distant r from the axis so that $v_r = 0$, $v_\theta = r\omega(r)$, $v_z = 0$ then

$$\tau_{r\theta} = \tau_e + \eta \, rd\omega/dr = T/2\pi Lr^2 ,$$

where L is the length of the cylinder so that the only non-zero shear stress $\tau_{r\theta}$ is a decreasing function of r.

Case (a) $\tau_e = \alpha E^2$· Here $\tau_e = C/r^2$ where $C = \alpha \{V/\ell n(R_2/R_1)\}^2$ and the critical value of torque for flow to occur is the torque due to the yield stress $T_e = 2\pi Lr^2\tau_e = 2\pi LC$ which is independent of r but varies with applied voltage and the geometry of the device. If $T < T_e$ there is no flow whereas if $T > T_e$ there is flow across the whole gap with Newtonian profile

$$\omega = \frac{\Omega \, R_1^2 \, R_2^2}{(R_2^2 - R_1^2)} \left\{ \frac{1}{R_1^2} - \frac{1}{r^2} \right\} \quad \text{and} \quad T = T_e + \frac{4\pi \, \eta \, \Omega \, L \, R_1^2 R_2^2}{(R_2^2 - R_1^2)}$$

so that the graphs of T v Ω are parallel straight lines with the same gradient as in the Newtonian case and intercepts which depend upon V and R_2/R_1 through C. Also for small gaps $T \simeq T_{//}$ where

$$T_{//} = (\tau_{xy})2\pi \, L \, R_2^2 = \{(\alpha V^2/(R_2 - R_1)^2 + \eta \, R_2 \, \Omega/(R_2 - R_1))\} \, 2\pi \, L \, R_2^2$$

is the corresponding expression for the torque obtained from the parallel plate solution. The electric field and shear rate in the annular case can also be approximated by the corresponding constant cartesian expressions (see [1]).

Case (b) $\tau_e = \alpha^*(E - E_o)$. The yield stress - electric field relationship now takes the form $\tau_e = \alpha^* E_o \, (\bar{r}/r - 1)$ for $r < \bar{r}$, $\tau_e = 0$ for $\bar{r} \le r$, where $\bar{r} = V/E_o \ell n(R_2/R_1)$. If $\bar{r} \ge R_2$ the voltage is sufficiently high for $V > E_o R_2 \ell n(R_2/R_1)$ and $\tau_e > 0$ for $r \le R_2$ so that plastic flow is possible across the whole gap. Attention is restricted to this range and to the case when the gap width is less than the inner radius so the $R_2 - R_1 < R_1$. The torque due to the yield stress is now given by $T_e(r) = 2\pi \, L \, r^2 \tau_e = 2\pi \, L \, \alpha^* E_o r \, (\bar{r} - r)$ and so depends upon r. This quadratic vanishes at $r = 0$ or $\bar{r}$ and at $r = \bar{r}/2$ has a maximum value $T_m = \pi \, L \, \alpha^* \, E_o \, \bar{r}^2/2 = \pi \, L \, \alpha^* \, V^2/2E_o[\ell n(R_2/R_1)]^2$. At all points where $T > T_e(r)$ there is flow and yield surfaces occur at $r = r_1$ and $r = r_2$ where

$$r_1 = \bar{r} \, (1 - \sqrt{1 - T/T_m})/2, \quad r_2 = \bar{r} \, (1 + \sqrt{1 - T/T_m})/2$$

are the roots of the quadratic equation $T = T_e(r)$ and depend upon V, R_2/R_1, α^*, L and E_o. For a given material when the voltage and geometry are fixed as T increases r_1 increases and r_2 decreases, $r_1 < \bar{r}/2$ and $\bar{r}/2 < r_2 < \bar{r}$.

For torques $T > T_m$ no yield surfaces occur and

$$\eta \, \omega = \frac{R_1^2 \, R_2^2}{R_2^2 - R_1^2} \left\{ \frac{1}{R_1^2} - \frac{1}{r^2} \right\} \left[\eta\Omega + \frac{\alpha^* \, V}{\ell n(R_2/R_1)} \left\{ \frac{1}{R_1} - \frac{1}{R_2} \right\} - \alpha^* E_o \ell n \left\{ \frac{R_2}{R_1} \right\} \right]$$

$$\frac{- \alpha^* \, V}{\ell n(R_2/R_1)} \left\{ \frac{1}{R_1} - \frac{1}{r} \right\} + \alpha^* \, E_o \, \ell n \left\{ \frac{r}{R_1} \right\},$$

$$T = \frac{4\pi \, L \, R_1^2 \, R_2^2}{R_2^2 - R_1^2} \left[\eta \, \Omega + \frac{\alpha^* \, V}{\ell n(R_2/R_1)} \left\{ \frac{1}{R_1} - \frac{1}{R_2} \right\} - \alpha^* E_o \ell n \left\{ \frac{R_2}{R_1} \right\} \right]$$

so that the graphs of T v Ω are parallel straight lines with the same gradient as in the Newtonian case but with intercepts which depend upon α^*, V, E_o, R_1 and R_2. In some cases ((i) and (iv) below) this field can arise for torques less than T_m.

For torques $T < T_m$ one or two yield surfaces are possible and the flow patterns which occur depend upon the relative sizes of $\bar{r}$, R_1 and R_2. Four possibilities arise:

(i) $R_2 < \bar{r} < 2R_1$. The fluid is stationary until $T > T_2$ $(= T_e(R_2))$. For $T_2 < T < T_1$ $(= T_e(R_1))$ there is a stationary plug attached to the inner cylinder with plastic flow in the rest of the gap. As T approaches T_1 the plug disappears giving plastic flow across the whole gap.

(ii) $2R_1 \le \bar{r} < R_1 + R_2$. T_2 is again the critical value of torque for the onset of flow which first occurs near the outer cylinder and for $T_2 < T \le T_1$ the flow pattern is as in (i) but in this case critical values of T are T_1, T_2 and T_m. For $T_1 < T < T_m$ $R_1 < r_1 < r_2 < R_2$ and there is a central plug with plastic flow near both cylinders. As T approaches T_m the plug disappears and finally for $T \ge T_m$ there is plastic flow across the whole gap.

(iii) $R_1 + R_2 \le \bar{r} < 2R_1$. For this range $T_1 < T_2$ and T_1, T_2, T_m are again critical values. A similar order of flow patterns emerge as under (ii) except now the flow occurs first near the inner cylinder then near the outer and for higher torques gives a central plug which finally disappears resulting in plastic flow across the whole gap.

(iv) $\bar{r} \ge 2R_2$. Here $T_e(r)$ is an increasing function of r with minimum T_1 and maximum T_2. T_1 is the critical value for the onset of flow. For $T_1 < T < T_2$ there is a plug attached to the outer cylinder rotating with angular speed Ω with plastic flow in the rest of the gap. When $T \ge T_2$ all the fluid flows plastically. The qualitative behaviour is similar to the case when E is assumed to be constant between the cylindrical electrodes (see [1]) but the details differ.

3.   Discussion and Conclusions

Comparison of the solutions for the parallel plate and concentric cylinder cases for $\tau_e = \alpha E^2$ shows that the simpler approach may be used without error. This situation will always produce relatively constant field gradients in velocity and electric potential. However, if $\tau_e = \alpha^*(E - E_o)$ similar desirable constancy can only be achieved on a restricted speed of test basis; this indicates a significant development from established non Newtonian rheometry. In both cases there is little to be achieved in torque generation by changing the radius ratio. The form of the yield stress - electric field relationship can have a dramatic effect on the problem, not least by the assumption of constant yield stress in the concentric cylinder case. This is detailed in [1] which also gives more general information on the research field.

Reference

1.   Atkin, R.J., Xiao Shi and Bullough, W.A. (1991) "Solutions of the constitutive equations for the flow of an electro-rheological fluid in radial configurations", J. Rheology 35, 1441-1461.

Application of Interval Arithmetic to Solve a Problem in Liquid-Cristal Modelling

S. Siebert

Inst. für Theor. Elektrotechnik und Meßtechnik
Universität Karlsruhe
Kaiserstr. 12, D–7500 Karlsruhe 1
Germany

G. Schumacher

Institut für Angewandte Mathematik
Universität Karlsruhe
Kaiserstr. 12, D–7500 Karlsruhe 1
Germany

ABSTRACT: Newly developed numerical methods with guaranteed results have been successfully applied to a simulation problem in liquid-cristal research.

1. Introduction

Computer simulations have become a widely used tool in science and engineering. Moreover, simulations are no longer just useful additions to experiments. Instead, they are used to decrease the costs. Further advantages are: parameter variations can be done easily and the areas of reliability of a model can be evaluated.

Unfortunately, many people are not aware of the fact that calculations on a computer are distorted by several kinds of errors and that results might be severely wrong. Very often a rigorous error analysis is replaced by assumptions about the solution derived from additional test measurements. But this is heuristic and contradicts the pretension of reducing costs.

Beyond that, newly developed numerical methods are available which give always reliable results. If they couldn't be successful, they will end with an error message rather than an uncommented approximation. All error analysis is done by the computer itself.

This paper will give a short overview about that. An example from electrical engineering will demonstrate how these methods can be applied to deal with problems in simulation processes and how to improve todays scientists work.

2. Reliable Numerics

The error of a simulation process on a computer arises from the following sources: the model error (approximation of the real world), the algorithmic error (only finite calculation steps), and the rounding error (only finite number of digits). A reliable numerical program takes all these errors into account and computes bounds for the solution. This guarantees also the existence of the solution which is not always obvious. If the program fails (which may happen), then the problem is either extremely ill-conditioned (in which case a refined version may apply), or there is no solution at all. No uncommented approximation is produced.

3. Mathematical Tools

One of the basic "tricks" of reliable numerics is the use of interval arithmetic [1]. It is something like an arithmetic for numbers with tolerances $x \pm tol$, storing them as upper and lower bounds of an interval $[a, b] := [x - tol, x + tol]$. This allows us to store errors within such intervals and to work with them.

An arithmetic for two intervals $[a, b]$ and $[c, d]$ consists of rules like

$$
\begin{array}{rcl}
[a, b] \; + \; [c, d] & := & [a + c, b + d] \\
[a, b] \; - \; [c, d] & := & [a - d, b - c] \\
[a, b] \; \cdot \; [c, d] & := & [\min\{ac, ad, bc, bd\}, \max\{ac, ad, bc, bd\}]
\end{array}
$$

Mathematically spoken, this means: for any two operands taken from each of the operand intervals on the left side, the result of the corresponding real operation is contained in the interval on the right side. Similar definitions apply for functions on intervals, e.g.

$$
[a, b]^2 := \left\{ \begin{array}{ll}
[0, \max\{a^2, b^2\}] & \text{if } 0 \in [a, b] \\
[\min\{a^2, b^2\}, \max\{a^2, b^2\}] & \text{else}
\end{array} \right.
$$

Suppose now we have a function $f(x) := 3 + 2x - x^2$, where the variable x is only known to be in $X := [0, 3]$. What is the range of f over X? We replace the real operations by corresponding interval operations and find either $F_1(X) := 3 + 2 \cdot X - X^2 = [-6, 6]$ or $F_2(X) := 3 + X \cdot (2 - X) = [0, 9]$, both satisfying $f(x) \in F_i(X)$, $i = 1, 2$, for all $x \in X$. The real range of f is $[0, 4]$. Thus, both interval evaluations are over-estimations.

Note that for real numbers the formulas for F_1 and F_2 are equivalent – but not for intervals! An immediate improvement of our result could be to intersect $F_1(X)$ and $F_2(X)$ leading to the statement that $f(x) \in [0, 6]$. But this is not a general strategy.

What we learn from this example is that the simple rules above do not guarantee a simple theory. It is still a matter of research and development to deal with interval arithmetic and to apply it to certain problems. A naive use is not recommended. Nevertheless, up to now many numerical techniques have been developed based on interval arithmetic which are superior to traditional ones [4], [5].

It is a remarkable fact that interval arithmetic is not necessary throughout the whole computation to guarantee the final result. Vice versa, a given algorithm can be improved by detecting its critical parts by interval arithmetic. At that very parts the effect of over-estimation becomes significant. Then, these critical program pieces may be treated with more effort (e.g., multiple-precision arithmetic), or by a totally new method (e.g., switch from a direct method to an iterative one). The following example is characteristic of this idea.

4. An Application in Electrical Engineering

During the design process of liquid-crystal-displays (LCD) the reflectance and transmittance of the display has to be computed. A widely used method in science and industry for this task is the characteristic matrix method, [2]. Figure 1 shows a typical display. In general this is a system of anisotropic, inhomogeneous layered media.

To calculate reflectance and transmittance we have to describe how light will propagate through the system. It can be shown, [2] that this leads to a differential equation for each layer of the form:

$$
\frac{\partial}{\partial z}\vec{\psi} = -jk_0\Delta\vec{\psi} \quad \text{with} \quad \vec{\psi} = \left(E_x, \sqrt{\frac{\mu_0}{\epsilon_0}}H_y, E_y, -\sqrt{\frac{\mu_0}{\epsilon_0}}H_x \right)^T,
$$

where $\vec{\psi}$ describes the light and Δ is a matrix containing the material parameters of each layer. The solution of the differential equation is given by the *characteristic matrix P*:

$$
\vec{\psi}(z_0 + h) = \underbrace{e^{-jk_0\Delta h}}_{P}\vec{\psi}(z_0) .
$$

A common way to determine the matrix P was to evaluate an infinite serie. Recently, a more elegant way was found to determine $\mathbf{P}$. By the theorem by Cayley-Hamilton, $\mathbf{P}$ has the form [6]:

$$
P(h) = \beta_0 I + \beta_1\Delta + \beta_2\Delta^2 + \beta_3\Delta^3 ,
$$

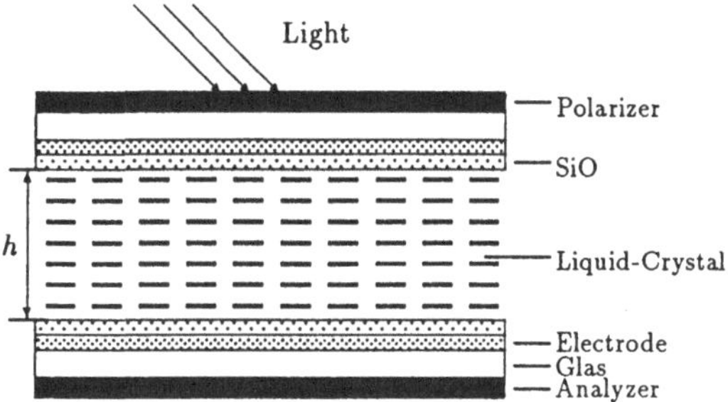

Figure 1: Liquid-Crystal-Display

where the coefficients are determined by solving the set of linear equations

$$e^{-jk_0\lambda_i h} = \beta_0 + \beta_1\lambda_i + \beta_2\lambda_i^2 + \beta_3\lambda_i^3, \quad i = 1, ..., 4 \, .$$

Finally, the reflectance and transmittance are determined by simply solving $\vec{\psi}_t = P(\vec{\psi}_i + \vec{\psi}_r)$.
The key point of this method is that the solution is known explicitely. A corresponding algorithm
consists of a straight-forward computation of some (complicated) formulas. *But:* it has turned out
that such a program fails for certain sets of input parameters.

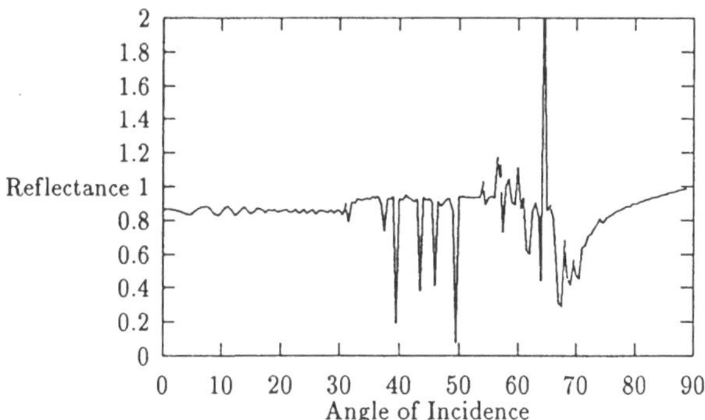

Figure 2: Reflectance using normal arithmetic

Numerical problems related to the characteristic matrix method were first reported after using
the method for the calculation of material parameters at liquid-crystal-metal interfaces. These
parameters can be determined using surface-plasmons. One would expect a curve of reflectance
with a characteristic minimum indicating the surface-plasmon. Furthermore, reflectance between
0 and 1 is expected because of conservation of energy. Figure 2 shows a typical result using normal
arithmetic. The characteristic minimum can not be recognized and there are values larger than 1.

Figure 3 displays the results for the same problem, this time using a refined algorithm with interval techniques and multiple-precision arithmetic. The characteristic minimum can be quantified at about 68 degrees.

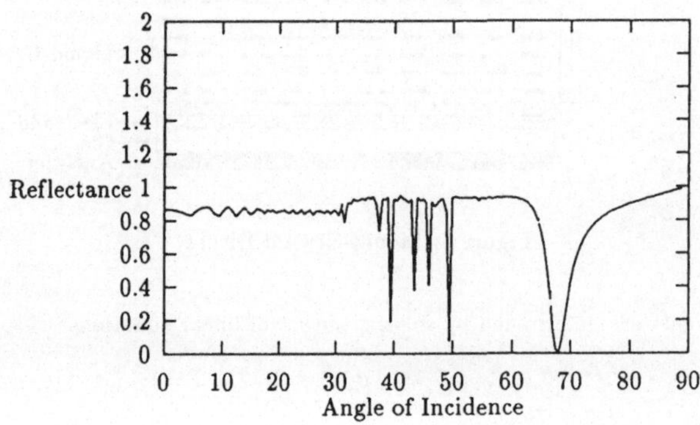

Figure 3: Reflectance using interval arithmetic

5. References

[1] Alefeld, G.; Herzberger, J.: Introduction to Interval Computations. Academic Press, New York, 1983.

[2] Berreman, D. W.: Optics in stratified and anisotropic media. J. Opt. Soc. Am. , 62: 502-510, 1972.

[3] Kretschmann, E.: Die Bestimmung optischer Konstanten von Metallen durch Anregung von Oberflächenplasmaschwingungen. Z. Phys. , 241: 313-324, 1971.

[4] Kulisch, U. (ed.): Wissenschaftliches Rechnen mit Ergebnisverifikation – Eine Einführung. Vieweg, Wiesbaden, 1989.

[5] Kulisch, U.; Miranker, W. L.: A New Approach to Scientific Computation. Academic Press, New York, 1983.

[6] Wöhler, H.: Analytische und numerische Methoden zur Berechnung der elektrooptischen Eigenschaften von Flüssigkristall-Zellen. PhD thesis, University of Karlsruhe, 1990.

HYDRODYNAMIC PROBLEMS IN FIBRE SPINNING

Louise Terrill

Introduction

In the industrial process of wet spinning, a collection or 'tow' consisting of a large number of fibres (typically 100,000) is drawn by means of rotating rollers through a succession of baths of fluid. In the bath considered in this paper, which is shown in Figures 1 and 2, the downstream roller rotates about 50% faster than the upstream roller so that the fibres are stretched causing fluid to be drawn into the moving tow and the fibres to be displaced. This process may be unstable; the tow can split across its width into bands of fibres separated by thin regions free of fibres. Mathematically, we must solve a coupled problem for the fluid flow and fibre displacements and examine its stability.

The Model

Details of the derivation of the model are given in [1]. The model equations are

$$\nabla p \;=\; \frac{4\mu\sigma}{a^2}M(\boldsymbol{u} - \boldsymbol{U}), \tag{1}$$

$$\nabla \cdot \boldsymbol{u} \;=\; 0, \tag{2}$$

$$\frac{d}{dl}(T\boldsymbol{t}) + M(\boldsymbol{u} - \boldsymbol{U}) \;=\; 0, \tag{3}$$

$$\frac{\partial \sigma}{\partial t} + \nabla \cdot (\sigma \boldsymbol{U}) \;=\; 0, \tag{4}$$

where p, μ and $\boldsymbol{u}$ are, respectively, the fluid pressure, viscosity and velocity; a is the fibre radius, σ is the volume fraction of fibres, $\boldsymbol{U}$ is the fibre velocity, l is the fibre arc length, T the fibre tension, $\boldsymbol{t}$ the tangent to a fibre and, with respect to axes tangential and normal to the fibre axes,

$$M = \begin{pmatrix} 1 & 0 \\ 0 & 2 \end{pmatrix}.$$

The fluid momentum equation, (1), is an anisotropic form of Darcy's Law so that the tow acts as an anisotropic porous medium. The fibre momentum equation, (3), is given by balancing the tension force acting on a fibre with the fluid drag acting on it. Equations

(2) and (4) are statements of conservation of fluid mass and number of fibres respectively.

Two-Dimensional Solution

We seek a steady solution of equations (1)–(4) which is independent of distance across the width of the bath. We choose co-ordinate axes (x, y) with origin at the centre of the upstream roller such that x is distance along the bath length and y is vertical distance. We non-dimensionalize the variables as follows;

$$x = L\hat{x}, \quad y = \delta L\hat{y}, \quad u = U_0\hat{u}, \qquad v = \delta U_0\hat{v}, \quad U = U_0\hat{u},$$

$$\sigma = \sigma_0\hat{\sigma}, \quad a = a_0\hat{a}, \quad p = \frac{8\mu\sigma_0 UL}{a_0^2}\hat{p}, \quad l = L\hat{l}, \qquad T = T_0\hat{T},$$

where $u = (u, v)$, L is the bath length, $2\delta L$ is the tow height at $x = 0$; U_0, σ_0, a_0 and T_0 are the fibre speed, volume fraction, radius and tension at $x = 0$ which are assumed to be known constants. We will henceforth drop the hats so all variables are non-dimensional.

Since the problem is steady, the fibres travel along their tangents so that $U = U(x)t$ where U is a prescribed increasing function of x. We define $h(x, \tilde{y})$ to be the displacement in the y direction of the fibre with endpoint $(0, \tilde{y})$. Then the interfaces between the bath and the tow, which are unknowns of the problem, are given by $y = h(x, \pm 1)$.

The boundary conditions for this problem are

$$\sigma = 1, \qquad u = 1 \quad T = 1 \qquad \text{at} \quad x = 0, \tag{5}$$

$$\sigma = 1, \qquad \text{at} \quad x = 1, \tag{6}$$

$$v = h = 0 \qquad \text{at} \quad y = 0, \tag{7}$$

$$p = 0 \qquad \text{at} \quad y = h(x, 1). \tag{8}$$

$$\tag{9}$$

In (4), (5) and (6) we have assumed that the fluid travels with the fibres at $x = 0$, that the rollers give equal uniform packing to the fibres and that the tow is symmetrical about its centreline $y = 0$. Equation (7) follows from the requirement that the pressure be continuous across the tow/bath interface and the fact that pressure variations in the tow are much higher than those in the surrounding bath.

We now exploit the fact that $\delta \ll 1$ to find solutions of the equations and boundary conditions which are the leading terms in asymptotic expansions in powers of δ. It can then be shown that

$$u = U(x) + O(\delta^2), \tag{10}$$

$$v = -U'(x)y + O(\delta^2), \tag{11}$$

$$T = 1 + O(\delta^2), \tag{12}$$

$$h = F(x)\tilde{y} + O(\delta^2), \tag{13}$$

$$\sigma = \frac{1}{Uh_{\tilde{y}}} + O(\delta^2), \tag{14}$$

$$p = \frac{\delta^2 F''(x)}{\alpha F^2(x)}\{y^2 - F^2(x)\} + O(\delta^4), \tag{15}$$

where F satisfies

$$F'' = \alpha[UF]' \tag{16}$$

with

$$F(0) = 1, \qquad F(1) = 1/U(1),$$

and

$$\alpha = \mu U L / T_0.$$

Equation (10) has an exact solution which we will not detail here. We note that for any function $U(x)$, there is a region in which $F'' > 0$, $p_y > 0$, and fluid flows *into* the tow from the bath and a region in which $F'' < 0$, $p_y < 0$, and fluid flows *out of* the tow to the bath.

Stability to two-dimensional disturbances

When the flow is two-dimensional and unsteady, the only change to the previous analysis is that the fibre velocity is now given by $U = Ut + U(0, h_t)$. The analogous equation to equation (10) is

$$h_t + (Uh)_x = \frac{1}{\alpha} h_{xx}, \tag{17}$$

whose solutions are stable. Thus, the flow is stable to two-dimensional disturbances.

Stability to three-dimensional disturbances

There is a mechanism for instability to three-dimensional disturbances. To show this, consider a region in which bath fluid is being entrained into the tow so that the unperturbed pressure p_I is positive. As shown in Figure 3, we consider small perturbations $\hat{H}(x, z)$ to the tow/bath interface where z is the *widthways* direction. We have the boundary condition

$$p_I + \hat{p} = 0 \qquad \text{on} \qquad y = F(x) + \hat{H}(x, z),$$

where $\hat{p}$ is the perturbed pressure. Linearizing this condition for small disturbances gives

$$\hat{p} = -p_{Iy}\hat{H} \qquad \text{on} \qquad y = F(x).$$

Thus, if $\hat{H} > 0$, then $\hat{p} < 0$, whilst if $\hat{H} < 0$, then $\hat{p} > 0$. Hence, in the case shown in Figure 2 a sideways pressure gradient will be produced which will drive a flow from the trough to the crest, clearly an unstable process. In regions where fluid is being expelled from the tow a similar argument implies that perturbations to the tow height produce stabilizing pressure gradients. The question of whether the destabilizing effects of inflow regions can exceed the stabilizing effect of outflow regions has been addressed in [1], where cases are detailed in which the existence of an inflow region does give rise to instability.

References

[1] TERRILL, E.L. 1990 *Mathematical Modelling of Some Spinning Processes* D. Phil. Thesis, Oxford University.

278

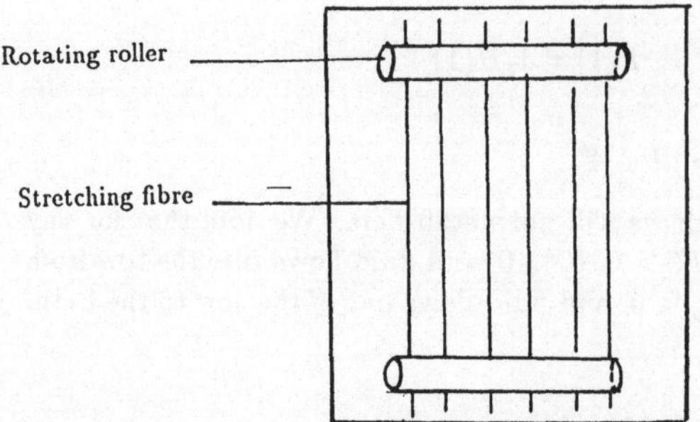

Figure 1: Plan view of the bath

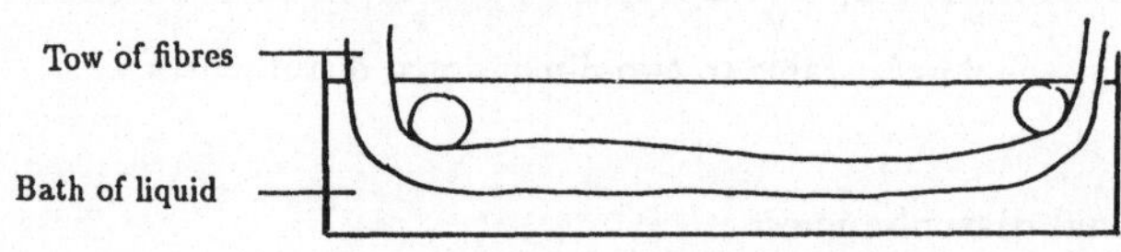

Figure 2: Side view of the bath

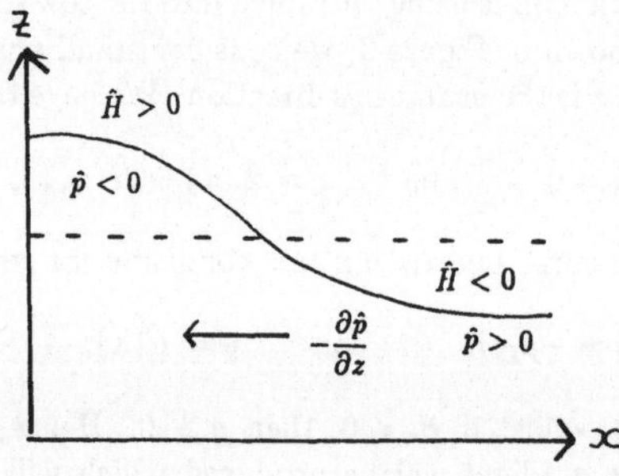

Figure 3: Cross-section of the tow

E.L.TERRILL
University of Oxford
Mathematical Institute
24-29 St. Giles'
Oxford OX1 3LB
United Kingdom

NUMERICAL ANALYSIS OF HIGH-VISCOSITY JET BUCKLING

B. M. TCHAVDAROV   S. P. RADEV[1]
Institute of Mechanics and Biomechanics
1113 Sofia, Bulgaria

A. L. YARIN
Israel Institute of Technology
Haifa 32000, Israel

ABSTRACT. The onset of a lateral instability (buckling) observed in a high-viscosity jet flowing onto a plate is studied numerically by a quasi-one-dimensional model accounting for the viscous and gravitational forces. The critical buckling height and the frequency of the corresponding oscillations are compared with experimental data.

1. Introduction

The oscillations of the jet axis observed when a high-viscosity fluid jet flows slowly vertically against a plate is referred as jet buckling. First qualitative description of this type of instability is given in the work of Taylor [1]. Recently, the phenomenon of jet buckling has been investigated both experimentally and theoretically by Cruickshank and Munson in a number of papers. Detailed experimental analysis during and subsequent to the onset of buckling is provided in [2]. The frequency of the coiling oscillations for a buckled axisymmetric viscous-gravity jet is predicted in [3] by a a semi-empirical model. A perturbation one-dimensional analysis of a cylindrical infinite weightless jet is adopted in [4] to predict the critical plate-orifice distance at which buckling is first initiated.

In the present work a linear analysis is applied to study the onset of buckling in viscous-gravity jet using a quasi-one-dimensional model of the actual jet flow.

2. Basic Equations and Method of Solution

The onset of buckling in Newtonian liquid jets flowing vertically against a plate will be considered accounting for the viscous and gravitational forces while neglecting the surface tension and inertial effects. The jet axis is regarded as a planar curve. The general equations of the quasi-one-dimensional model of jet flows are derived in [5]. Under the assumption made above, the quasi-one-dimensional balance equations of momentum and momentum of momentum together with the continuity equation acquire the form:

$$\frac{1}{\rho}\frac{\partial}{\partial x}\left(P\vec{\tau}+\vec{Q}\right)+f\vec{F}=0, \quad \frac{1}{\rho}\left(\frac{\partial \vec{M}}{\partial x}+\vec{\tau}\times\vec{Q}\right)-kI\vec{n}\times\vec{F}=0, \quad \frac{\partial f}{\partial t}+\frac{\partial(fv_\tau)}{\partial x}=0$$

[1]Supported by a National grant No. MM-34/91

Here x is the longitudinal coordinate; t is the time: k is the curvature of the jet axis; ρ is density; $f = \pi r^2$ is the jet cross section area; $I = \pi r^4/4$ is the moment of inertia of the jet cross section; $\vec{\tau}$ and $\vec{n}$ are respectively the tangential and normal to the jet axis unit vectors; $P\vec{\tau}$ and $\vec{Q}$ is the longitudinal and shearing forces respectively; $\vec{M}$ is the momentum of internal viscous stresses; $\vec{F} = g\left(\vec{\tau} - \vec{n}\,\partial h/\partial x\right)$ is the outer force per unit mass where g is the acceleration due to gravity and $h(x,t)$ is the amplitude of the planar deflection of the jet axis. $\vec{v} = (v_\tau, v_n)$ is the jet velocity (the subscript τ, n and b denote the tangential, normal and binormal component, respectively);

From the n-component of momentum equation and the b-component of the momentum of momentum equation it is easy to obtain

$$\frac{1}{\rho}\left(Pk - \frac{\partial^2 M_b}{\partial x^2}\right) - g\left[\frac{\partial(kI)}{\partial x} + f\frac{\partial h}{\partial x}\right] = 0 \tag{1}$$

The connection between dynamic and kinematic parameters of the flow in case of Newtonian liquid has been derived in [5]. In the present case we have:

$$P = 3\mu f\frac{\partial v_\tau}{\partial x}, \qquad M_b = 3\mu I\left(\frac{\partial \Omega_b}{\partial x} - \frac{3}{2}k\frac{\partial v_\tau}{\partial x}\right), \qquad \Omega_b = \frac{\partial v_n}{\partial x} + kv_\tau \tag{2}$$

where μ is viscosity, $\vec{\Omega}$ is the angular velocity of the material jet cross section.

In case of small disturbances $k = \partial^2 h/\partial x^2$ and $v_n = \partial h/\partial t$. Assuming $h(x,t) = F(x)exp(\lambda t)$ equations (1) with account for the continuity equation and (2) lead to

$$R^{-1}R'F'' + \frac{1}{8L^2}\frac{d^2}{dx^2}\left(\lambda R^4 F'' + R^2 F''' + RR'F''\right) + GL^2\left[R^2 F' + \frac{1}{4L^2}(R^4 F'')'\right] = 0 \tag{3}$$

were $R(x)$ is the jet radius of the steady-state flow (prime denotes the derivative with respect to x). Eq. (3) is written in a non-dimensional form. The exit velocity $\bar{U}_0$ and the plate-orifice distance $\bar{L}$ are chosen as a scale for velocity and length ($t = \bar{t}\,\bar{U}_0/\bar{L}_0$) while $R(x)$ and $h(x,t)$ are scaled by the orifice radius $\bar{R}_0$. The parameter $G = \left(\rho g\bar{R}_0^2\right)/\left(6\mu\bar{U}_0\right)$ represents the ratio of gravitational to viscous forces and $L = \bar{L}/\bar{R}_0$ is the non-dimentional plate-orifice distance.

Eq. 3 is considered under the following boundary conditions: smooth junction of the jet axis to the nozzle, zero angular velocity of the material cross section at the nozzle exit and "clamped" jet axis at the plate. These boundary conditions give respectively

$$F = F' = F'' = 0, \quad x = 0; \qquad F = F' = 0, \quad x = 1 \tag{4}$$

The steady-state radius $R(x)$ is obtained from the τ-component of momentum equation (accounting for the continuity equation) as a solution of the boundary value problem

$$RR'' - (R')^2 = \beta^2 R^4, \ \left(\beta = LG^{1/2}\right), \ x \in (0,1); \quad R = 1, \ x = 0; \ R(x) \to \infty, \ x \to 1 \tag{5}$$

The problem (3)-(4) is the eigenvalue problem about the parameter λ when L , G and $R(x)$ are given. The critical plate-orifice distance L_c above which buckling occurs will be those value of the parameter L at which the eigenvalue with the greatest real part λ_1 crosses the imaginary axis of the complex plane. Hence, at the critical buckling height L_c we have $Real(\lambda_1) = 0$ whereas $Im(\lambda_1)$ corresponds to the buckling frequency.

Using the substitutions $y_k = (1-x)^{k-1}F^{(k)}(x)$, $(k = 0 \div 4)$, $\mathbf{Y} = (y_0, y_1, y_2, y_3, y_4)^{\mathsf{T}}$ the problem (3)-(4) is transformed into an eigenvalue problem for a system of first order linear differential equations

$$(1-x)^2\mathbf{Y}'(x) + \mathbf{A}(x;\lambda)\mathbf{Y}(x) = 0, \qquad x \in [0,1] \tag{6}$$

$$\mathbf{\Psi}_0^{\mathsf{T}}\mathbf{Y} = 0, \quad x = 0; \qquad \mathbf{\Psi}_1^{\mathsf{T}}\mathbf{Y} = 0, \quad x = 1 \tag{7}$$

where $\mathbf{A}(x;\lambda)$ is a 5×5-matrix; $\mathbf{\Psi}_0^{\mathsf{T}}\,and\,\mathbf{\Psi}_1^{\mathsf{T}}$ are scalar matrixes of order 5×3 and 5×2, respectively ($\mathbf{\Psi}^{\mathsf{T}}$ denotes a transpose matrix of $\mathbf{\Psi}$).

To solve the singular eigenvalue problem (6)-(7) we transfer the boundary condition at a singular point $x = 1$ to the other end of the interval by the initial value problem [6]:

$$\mathbf{\Psi}' - \left(\mathbf{A}^{\mathsf{T}} + \mathbf{\Psi}(\mathbf{\Psi}^{\mathsf{T}}\mathbf{\Psi})^{-1}\mathbf{\Psi}^{\mathsf{T}}\mathbf{A}^{\mathsf{T}}\right)\mathbf{\Psi} = 0, \qquad x \in [0,1] \tag{8}$$

$$\mathbf{\Psi} = \mathbf{\Psi}_1, \qquad x = 1 - \varepsilon \qquad (\varepsilon \text{ is small}) \tag{9}$$

Hence, the required eigenvalue relation is obtained in the form $\quad det\left(\begin{array}{c}\mathbf{\Psi}_0^{\mathsf{T}} \\ \mathbf{\Psi}_{1,0}^{\mathsf{T}}(\lambda)\end{array}\right) = 0$ where $\mathbf{\Psi}_{1,0}(\lambda)$ denotes the solution of (8)-(9) at $x = 0$.

3. Theory Versus Experiments

The numerical results are compared with the experimental data given in [2] and [4]. Data with negligible surface tension effect have been chosen (small values of the parameter $\sigma/\gamma d^2$ representing the ratio of surface tension forces to gravitational forces; σ - surface tension, $\gamma = \rho g$, $d = 2\bar{R}_0$). The second parameter gd^3/ν^2 ($\nu = \mu/\rho$ is kinematic viscosity) represents the ratio of inertial forces times gravity forces to viscous forces.

Figure 1 shows the buckling height L_c as a function of the gravitational forces to viscous forces ratio G. The theory is in a good agreement with the experiments in a large region of variation of G. A possible reason for a greater buckling height predicted by the theory could be the boundary conditions of 'clamped' jet axis at the plate which seems to be rather 'stiff' condition, producing more stable jet than, in fact, it is.

The frequency of the oscillations at the buckling height L_c is shown in Figure 2. In order to uniform the frequency scales, the comparison with experimental data is made on the basis of the non-dimensional frequency $\omega_L = 2Im(\lambda_1)/L_c$. The numerical results are in a very good agreement with the experimental data in case of negligible small inertial forces. As it has been shown experimentally in [2] the inertial forces are insignificant as far as the buckling height is considered but even small the inertial forces influence the frequency of the oscillations. The present theory does not account for the inertial forces which explains the divergence between theory and the experimental data (□) with a relatively large value of the parameter gd^3/ν^2.

References

[1] G. I. Taylor (1968) "Instability of jets, threads and sheets of viscous fluid", in *Proc. Intl. Congr. Appl. Mech.*, Springer-Verlag.

[2] J. O. Cruickshank and B. R. Munson (1981) "Viscous fluid buckling of plane and axisymmetric jets", *J. Fluid Mech.*, vol. 113, pp. 221–239.

[3] J. O. Cruickshank and B. R. Munson (1983) "A theoretical prediction of the fluid buckling frequency", *Phys. Fluids*, vol. 26, no. 4, pp. 928–930.

[4] J. O. Cruickshank (1988) "Low-Reynolds-number instabilities in stagnating jet flows", *J. Fluid Mech.*, vol. 193, pp. 111–127.

[5] V. M. Entov and A. L. Yarin (1984) "The dynamics of thin liquid jets in air", *J. Fluid Mech.*, vol. 140, pp. 91–111.

[6] A. A. Abramov (1961) "On the transfer of the boundary conditions for systems ODEs", *Journal of Computational Mathematics and Mathematical Physics*, vol. 1, no. 3, pp. 542–545. Academy of Sciences of the USSR (in Russian).

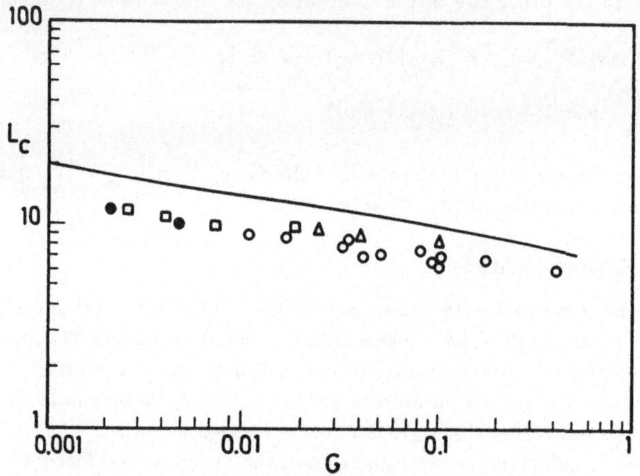

Figure 1: Critical buckling height L_c as a function of the parameter $G = \left(\rho g \bar{R}_0^2\right) / \left(6\mu \bar{U}_0\right)$ — theory; Cruickshank and Munson [2] data: $\bigcirc - \sigma/\gamma d^2 = 1.15 \times 10^{-2}$; Cruickshank [4] data: Values of gd^3/ν^2: $\bullet$, 2.85×10^{-3}; $\square$, 2.57×10^{-2}; $\triangle$, 3.33×10^{-4}.

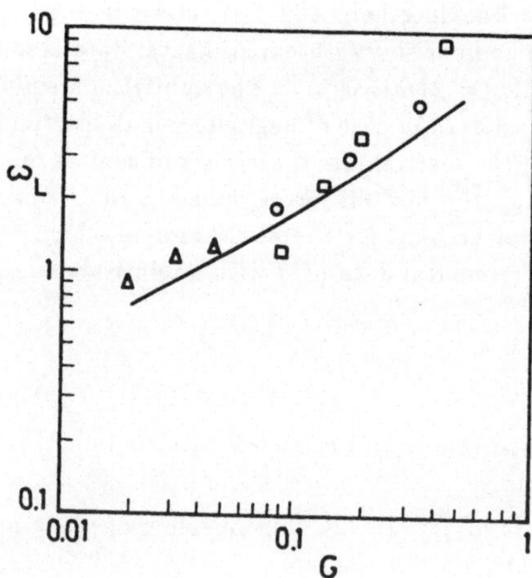

Figure 2: Buckling frequency $\omega_L = 2Im(\lambda_1)/L_c$ as a function of the parameter G.
— theory;
Cruickshank and Munson data:

$\sigma/\gamma d^2$	gd^3/ν^2
$\bigcirc - 6.33 \times 10^{-3}$	$7.59 \times 10-2,$
$\triangle - 2.54 \times 10^{-2}$	$8.48 \times 10-2,$
$\square - 1.43 \times 10^{-2}$	$0.66.$

STOCHASTIC SCHEDULING FOR RISK-LEVELLING MAINTENANCE SCHEDULING IN ELECTRIC POWER SYSTEMS

L. VARGA
Hungarian Electricity Board
H-1011 Budapest, Vám u. 5-7.
Hungary

ABSTRACT. A heuristic algorithm is presented for risk-levelling maintenance scheduling in electric power systems. We point out the connection
between the stochastic scheduling on parallel machines and the risk-
levelling approach.

1. Introduction

Several methods are usually applied for annual maintenance scheduling
[1]. The commonly applied methods are so called reserve-levelling ones
[2]. They aim at keeping the amount of reserve almost level through the
year by withdrawing units in the low load periods. The other group of
automated maintenance scheduling techniques are the risk-levelling
methods. These methods seek to spread risk evenly as measured by some
kind of reliability index.

2. Probabilistic Approach to Maintenance Scheduling

The reserve-levelling technique does not consider the uncertainty inherent in power systems. The risk-levelling methods attempt to enhance
the system reliability taking into account the unit availabilities and
the uncertainty in the forecast peak load.

Consider a power system consisting of n generating units, where
the capacity of unit i is c_i with forced outage rate f_i. Define X_i as
the unavailability capacity for unit i. X_i is a random variable with a
distribution of $X_i = c_i$ with probability f_i

$$= 0 \quad \text{with probability } 1-f_i$$

It is customary to assume that $X_1, \ldots, X_n$ is a sequence of independent
random variables.

Let P_j denote the system load in the j-th time period. In power
generation planning the forecast loads will be uncertain and regarded
as random variables. Then the loss of load probability (LOLP) index in

the period j is defined by

$$LOLP_j = Pr\{ \sum_{i=1}^{n} X_i > \sum_{i=1}^{n} c_i - P_j \} \qquad j=1,2,\ldots,m$$

where m is the number of the time periods in one year.

The LOLP index measures the probability that a given system's available capacity is insufficient to meet the system peak load in a given time period.

Let n_j units be scheduled for maintenance in period j, consequently, $n-n_j$ units are available. The LOLP index for this period is:

$$LOLP_j = Pr\{ \sum_{k=1}^{n-n_j} X_{j_k} > \sum_{k=1}^{n-n_j} c_{j_k} - P_j \} \qquad j=1,2,\ldots,m$$

where j_k is the index of the unit in the k-th position of the sequence operated in the j-th period.

Another commonly used reliability index which measures the expectation of the energy not served(ENS) is:

$$ENS_j = T \cdot E\{ (\sum_{i=1}^{n} c_i - \sum_{i=1}^{n} X_i - P_j)^- \} \qquad j=1,2,\ldots,m$$

where T is the load duration in hours and $(.)^-$ denotes the negative part of a random variable describing the power deficit in the system.

The main objective of the maintenance scheduling is to keep the electricity supply as reliable as possible. An appropriate scheduling ensures a sufficient reliability level over the whole year. Consequently, we have to minimize the largest LOLP index over the time periods of the scheduling horizon:

$$(1) \qquad LOLP_{opt} = min(\max_{1 \le j \le m} LOLP_j)$$

For the reliability index ENS the objective function is:

$$(2) \qquad ENS_{opt} = min(\max_{1 \le j \le m} ENS_j)$$

3. A Heuristic Algorithm for Risk-levelling Maintenance Scheduling

The maintenance scheduling specifies the periods of the year when the generating units are to be shut down for preventive maintenance, taking into account the forecast yearly load profile, the time demand of the preventive maintenance and the forced outage rates of the units. Based on the previous section the objective of maintenance scheduling is to solve an optimization task (1) or (2).

In this section a simple heuristic algorithm is suggested to levelize the risk over all the time periods of the year. The algorithm requires calculation of the reliability indices for each time period of the year which is performed by utilizing a convolution formula of the independent random variables. Assume that the maintenance is performed in only one time period for any generating unit and any generating unit can be scheduled for maintenance in any time period. The algorithm for risk-levelling is as follows:

Step 0: For each time period, the $LOLP_j$ with all units available is computed from the predicted peak load (P_j), the capacities (c_i) and forced outage rates (f_i) of the generating units.

Step 1: Remove each unit i in turn from the system $(i=1,\ldots,n)$ and compute the new reliability indices $LOLP_j^i$ for each time period j $(j=1,\ldots,m)$. Store the minimum values of $LOLP_{j(i)}^i$ $(i=1,\ldots,n)$ with the unit indices i and the time period number $j(i)$.

Step 2: Select the maximum value $LOLP_{j(s)}^s$ from the previously stored LOLP indices. Schedule the unit s for maintenance in the period $j(s)$.

Step 3: Remove the scheduled unit s from the time interval $j(s)$.

Steps 1, 2 and **3** are repeated until all generating units are scheduled for maintenance.

In the power system there are usually units of different capacity which require maintenance periods of different duration. We can easily modify the proposed algorithm to solve this more typical case (see [2]).

4. The Probabilistic Approach and Stochastic Scheduling

We point out that under some simplifying assumptions, the risk-levelling maintenance scheduling can be considered as a stochastic scheduling problem for parallel machines. For the reader's convenience we define the stochastic scheduling task [3]: Suppose that m machines are available to perform n tasks with random processing times $Y_1,\ldots,Y_n$. Each machines can process one task, each task can be processed by one machine and any machine can process any task. Performing task i requires one of the machines for a duration Y_i. As a task is complete another task is put on the machine that is freed. One of the most classic objective functions is the one that minimizes the expected makespan that is the time from zero until all tasks have been completed.

Let the simplifying assumptions be for the maintenance scheduling task is as follows: the time demand for maintenance is one period for any generating unit and any generating unit can be scheduled for maintenance in any time period. In addition assume that the system load $P_j=p$ $j=1,\ldots,m$ is constant over the whole year.

Observe the following equivalences:

- the j-th time period in the maintenance scheduling task corresponds to the j-th machine in the parallel machine problem,

- the available capacities of the generating units (c_i-X_i) correspond to the random processing times of the tasks.

Since the total installed capacity of the power system and the peak load are assumed to be constant over all scheduling periods, the task (2) is equivalent to minimizing the expected value of the maximum planned outages over the time periods:

$$\max_{1\leq j\leq m} E\{ \sum_{k=1}^{n_j} (c_{j_k} - X_{j_k}) \}$$

where j_k is the index of the unit in the k-th position of the sequence scheduled for maintenance in the j-th time period.

It can be shown [3] that if we have m<n machines with exponentially distributed processing times $Y_1,\ldots,Y_n$, the so called LEPT (Largest Expected Processing Time) rule which uses the scheduling order $i_1,\ldots,i_n$ where $E(Y_{i_1})\geq\ldots \geq E(Y_{i_n})$ provides an optimal non-preemptive schedule. It should be noted that the LEPT policy does not remain optimal when one allows general random processing times [3], and we are forced to use heuristic algorithms like described in Section 3.

Acknowledgement

The author is grateful to Gábor Tusnády for valuable discussions and to Zsolt Soós for assistance in the reliability calculations.

References

[1] El-Sheiki, F. A., Billinton, R. (1985) 'Maintenance considerations in Generating capacity reliability assessment' Maintenance Management International 5, 63-76.

[2] Soós, Zs., Varga, L. (1991) 'Greedy Algorithm for Maintenance Scheduling in Electric Power System' in M. Heiliö (ed), Proc. of the Fifth ECMI Conf. B. G. Teubner and Kluwer, 115-118.

[3] Weiss, G. (1982) 'Multiserver Stochastic Scheduling' in M. A. H. Dempster et al. (eds) Deterministic and Stochastic Scheduling D. Reidel, 157-179.

TWO CRITERIA FOR TIME SERIES MODELS COMPARISON

B. VOVK, B. MOTNIKAR, D. ČEPAR
'Jožef Stefan' Institute
Jamova 39
61111 Ljubljana
Slovenia

ABSTRACT. Many methods for time series modelling have been proposed in literature. A general measure describing how a given model is suitable for representation of a given time series may be of interest in a broad field of applications. We suggest two criteria for time series models comparison. The first criterion estimates the degree to which a time series is explained by a given model. The criterion may be used for determining the optimal model obtained by the given method and for comparison of models obtained by different methods. We define dynamics of time series as mean square difference between neighbouring observations and a *model activity* criterion as the proportion of the explained time series dynamics. It will be shown that models with low value of model activity and high value of the explanatory criterion do not capture neither the long term components of the process nor the short term ones. Suggested criteria are compared with some other well–known criteria. Use of criteria is presented on the ARIMA models by some examples.

1. Introduction and motivation

Let us consider time series (abbr. TS) observations z_t, $t = 1, \ldots, n$ and a model describing their structure. Based on the model, one–step–ahead forecast of z_t denoted by $\hat{z}_{t-1}(1)$ can be calculated. Residuals a_t represent model deviations from the observed data: $a_t = z_t - \hat{z}_{t-1}(1)$. By Box–Jenkins approach [1], TS should first be transformed to the stationary TS denoted by w_t and corresponding residuals by $b_t = w_t - \hat{w}_{t-1}(1)$. A seasonal ARIMA $(p, d, q) \times (P, D, Q)$ model for stationary TS is defined as

$$(1 - \Phi_1 B - \Phi_2 B^2 - \ldots - \Phi_p B^p)(1 - \varphi_1 B^s - \varphi_2 B^{2s} - \ldots - \varphi_P B^{Ps}) w_t =$$
$$= (1 - \Theta_1 B - \Theta_2 B^2 - \ldots - \Theta_q B^q)(1 - \vartheta_1 B^s - \vartheta_2 B^{2s} - \ldots - \vartheta_Q B^{Qs}) b_t + \vartheta_0 ,$$

where B is a backward shift operator: $Bz_t = z_{t-1}$; s is a season length; p, q, P, Q are model orders; $\varphi, \Phi, \vartheta, \Theta$ are model parameters and d, D are orders of differencing and seasonal differencing. The criterion for models comparison is the residual variance estimator

$$\hat{\sigma}_b^2 = SS/(n - k) , \quad \text{where} \quad SS = \sum_{t=Q}^{n} b_t^2 \tag{1}$$

and k is the number of model parameters. Residuals b_t converge to zero as t goes to $-\infty$. Residuals with index less than Q ($Q < 0$) are small enough to neglect them.

Transformations of a particular nonstationary TS (bellow) may cause some difficulties in model comparison. At various transformations, residuals b_t become 'small enough' at different time index Q, therefore different numbers of residuals are captured in the SS.

Logarithmic and power transformations change TS unit. At (seasonal) differencing, the number of observations is decreased, therefore a loss of information is unavoidable.

120 observations of an almost straight line were generated by the model described with the difference equation $z_t = 0.98\,z_{t-1} + a_t + 1$, $z_0 = 1$, where the distribution function of a_t is $N(0, 0.2)$. We will forecast the last 20 observations using the first 100. Box–Jenkins approach suggests differencing to eliminate trend. Model ARIMA(0,1,0) minimizes (1). Figure shows that forecasts generated by this model are higher than original data.

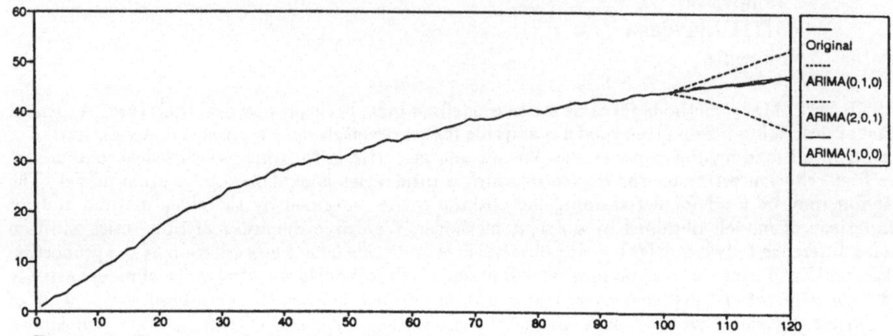

Figure 1: Forecasts obtained by all proposed models compared to the original data.

The model obtained without differentiation is nonstationary (negative residuals do not converge). We tried this approach with $Q = -49$. Minimum of (1) is reached with the model ARIMA(2,0,1), which also generates inappropriate forecasts (figure).

2. A new criterion for models comparison

To avoid these troubles we propose a more general criterion which allows comparison of models obtained by various transformations or even various methods. We define the criterion $FQ1$ as the proportion of explained TS standard deviation:

$$FQ1 = 1 - \sqrt{\sum_{t=1}^{n} a_t^2 \Big/ \sum_{t=1}^{n} (z_t - \bar{z})^2}\,, \tag{2}$$

where $\bar{z}$ is mean value of $z_1, \ldots, z_n$. In case that we are unable to calculate the first few residuals, we omit them from the summation. The value of the criterion $FQ1$ equals 1 if and only if the model exactly fits the data, and equals 0 if mean of squared residuals and TS variance are the same (e.g. model contains only a constant term $\bar{z}$). For the model selection problem, the criterion should be adjusted to balance between the goodness of fit and the complexity of the model: $FQ1^* = 1 - (1 - FQ1)n/(n - k)$. Criterion allows comparison of models with or without transformations because changes in TS unit measure do not influence the residuals unit measure (we consider a_t instead of b_t). On the other hand, negative residuals are not considered, so transformations to obtain their convergence are unnecessary. In the above example ARIMA(1,0,0) model is chosen by maximizing the adjusted $FQ1$ criterion. Corresponding forecasts seem to be adequate (figure).

When TS has a strong trend (its variance is high, relative to the variances of TS subintervals) we can easily obtain a high value of $FQ1$ criterion solely by differencing (model $\hat{z}_t(1) = z_t$). For TS B considered in [1] for example, $FQ1 = 0.91$. It is easy to check that such a model does not cover local trends and is also inappropriate for long term forecasts. We want to test the model ability to determine TS local trend.

3. Model activity criterion

We therefore define dynamics of TS as a mean square difference of neighbouring observations and model activity criterion $FQ2$ as the proportion of the explained TS dynamics:

$$FQ2 = 1 - \sqrt{\sum\nolimits_{t=1}^{n} a_t^2 \Big/ \sum\nolimits_{t=1}^{n} d_t^2}\,, \quad \text{and} \quad FQ2^* = 1 - \frac{(1 - FQ2)n}{n - k} \tag{3}$$

where $d_t = z_t - z_{t-1}$. Criterion $FQ2$ equals 1 if and only if the model exactly fits the data. If the model follows data with a delay of a time unit: $\hat{z}_t(1) = z_t$, criterion $FQ2$ equals 0 as in the above example for TS B.

4. Some other aspects of suggested criteria

In the following example TS G described in [1] is used. Various models are compared by $\hat{\sigma}_b$, $FQ1$ and $FQ2$ criteria, illustrated in Table 1. Contribution of the particular transformation or parameter to the goodness of fit may be recognized. TS has strong trend and seasonal component so the model can be built in the following steps: Using only differencing, $FQ1$ criterion is improved since global trend of TS is explained. By seasonal differencing local behaviour of TS is mainly explained, this results in improving the $FQ2$ criterion and then by improving the $FQ1$ criterion. Addition of ARMA model parameters almost does not have any effect to the goodness of fit.

$(p,d,q)\times(P,D,Q)$	$\hat{\sigma}_b$	$FQ1^*\%$	$FQ2^*\%$
$(0,0,0)\times(0,0,0)$	120.0	0.0	-253.4
$(1,0,0)\times(0,0,0)$	33.7	72.0	0.3
$\log(1,0,0)\times(0,0,0)$	0.1	71.9	-0.1
$(0,1,0)\times(0,0,0)$	33.8	71.0	0.2
$(0,0,0)\times(0,1,0)$	17.7	85.9	49.9
$(0,1,0)\times(0,1,0)$	12.4	90.1	65.1
$(1,0,0)\times(0,1,0)$	11.6	90.8	66.8
$(0,1,1)\times(1,0,1)$	10.5	91.3	68.5
$(0,1,1)\times(1,1,1)$	10.5	91.4	69.6
$(2,1,0)\times(1,1,1)$	10.4	91.4	69.7

Table 1: Values of criteria for various ARIMA models of TS G

TS	(p,d,q)	$FQ1^*\%$	$FQ2^*\%$	$\frac{1-FQ1}{1-FQ2}$
A	$(1,0,1)$	20.8	14.4	0.93
A	$(0,1,1)$	20.2	13.8	0.93
B	$(0,1,1)$	91.4	0.3	0.09
C	$(1,1,0)$	93.4	42.4	0.11
C	$(0,2,2)$	93.2	39.8	0.11
D	$(1,0,0)$	49.7	3.4	0.52
D	$(0,2,1)$	48.1	0.2	0.52
E	$(2,0,0)$	59.0	31.7	0.60
E	$(3,0,0)$	59.5	32.6	0.60
F	$(2,0,0)$	7.7	43.8	1.64

Table 2: Values of criteria for ARIMA models of TS A to F

Criteria $FQ1$ and $FQ2$ are indiscriminate to the TS unit measure. This allows comparison of more TS as is shown on the data and models for TS from 'A' to 'F', described in [1]. Models are compared to test whether the Box–Jenkins approach is suitable for a particular TS or not and which of the proposed models should be selected. We suggest another measure $(1 - FQ1)/(1 - FQ2)$ which may reveal some characteristics of a given TS. The value of the measure depends only on the TS not on the model or method.

In Table 2 it can be seen for each TS whether the deterministic component prevails (low values of the above measure). Examples of such TS are B and C, the difference between the two is the higher value of the $FQ2$ criterion for the latter, which suggests that short term forecasts could be beneficial. In the case of B all forecasts are equal. In conjunction with the zero value of the $FQ2$ criterion it can be concluded that they are useless. At TS F, the high value of the fraction indicates a process without significant deterministic components. Short term forecasts generated by the given model could be useful.

5. Comparison with some standard criteria

There are no assumptions about density functions of data or residuals, so in general nothing can be said about density function or statistical properties of $FQ1$ or $FQ2$. However, we compare them with some standard criteria for model comparison.

$FQ1$ criterion in TS analysis corresponds to *determination coefficient* in regression analysis. The essential difference is in the domain of models.

Akaike criterion is commonly used for the comparison of models. In [2] it is defined as $AIC = 2n \log \hat{\sigma}_b + 2k/n$. The first part of AIC measures goodness of fit, while the second part is a penalty term. The last one is independent of TS unit measure, so its relative contribution to the AIC depends on high or low absolute values of observations. In the adjusted $FQ1$ and $FQ2$ criteria, the penalty term is included in the relative sense.

A very common criterion is *relative error*. The only problem is presented by TS with observations very close to zero.

Residual variance estimator may cause some difficulties with a particular method as was shown in the Box–Jenkins approach.

Conclusions

Standard *fit quality* estimators for TS based on one–step–ahead forecast errors are useful for comparison of models within one method and are limited to the same TS transformations. An example of nonstationary TS shows that the residual variance estimator in the Box–Jenkins approach does not always lead to sensible forecasts. To overcome this trouble, we suggested two criteria for model comparison: $FQ1$ and $FQ2$. They can be used also for testing whether a particular forecasting method is suitable for a given TS.

It should be emphasized that values of criteria $FQ1$ and $FQ2$ avoid influence of the TS unit measure. They may be understood as per cent of the process explanation. Models including various TS transformations and/or models obtained by different methods are also directly comparable. In this sense a comparison of the methods and not only models is possible.

References

[1] G. E. P. Box, G. M. Jenkins (1970), Time Series Analysis: Forecasting and Control, Holden–Day, San Francisco.
[2] J. G. Gooijer, B. Abraham, A. Gould, L. Robinson (1985), 'Methods for Determining the Order of an ARMA Process: A Survey', International Statistical Review.

Coupled Heat and Mass Transfer in a Weld Pool

M. VYNNYCKY
School of Mathematics
University of East Anglia
Norwich
NR4 7TJ
England

July 9, 1992

ABSTRACT. A mathematical model is proposed for coupled laminar high Reynolds (Re) number shear-driven flow and high Péclet (Pe) number heat flow in a semi-circular tungsten inert gas (TIG) weld pool. The fluid flow within the right-hand side of pool is assumed to be driven in a clockwise direction by a surface-tension gradient on the pool surface, as a result of a high temperature gradient induced by an arc source distributed about the centre of the pool. Numerical solutions of the full heat and momentum equations indicate that a consistent description for the flow structure can only be obtained if the arc source is concentrated at the pool centre; in addition, this structure can be recovered using asymptotic analysis for $Re, Pe \gg 1$. Further numerical computations indicate that high Re and Pe anti-clockwise flow, caused by the presence of surface impurities, is possible regardless of the type of heat source used.

1   Introduction

In the TIG (tungsten inert gas) welding process, an arc is struck between an inert electrode and the metal substrate, the thermal energy thus generated causing the base metal, usually steel or aluminium, to melt (fig. 1). The solidification of the molten metal in the weld pool forms the bond between the pieces of metal that are joined in the process. The heat input for the process is typically $O(10^3)$ W and the current is $O(10^2)$ A.

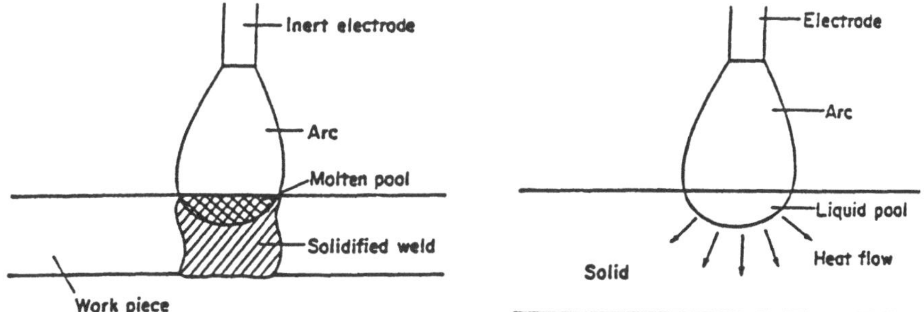

Fig. 1 Schematic sketch of the TIG welding process

One of the most dominant factors which affect weld-pool shape is the surface-tension (Marangoni) force which is generated at the pool surface as a result of the large temperature gradients induced by the arc source; in addition, the nature of these flows can be dramatically affected [1] by the presence of impurities, such as sulphur. In this note, we shall make use of earlier work [3] by the author to provide a mathematical description of laminar Marangoni flow under the assumption of high Reynolds and Péclet $(Re, Pe \gg 1)$ number, with the Prandtl number (Pr) $O(1)$ so that $Re \sim Pe$.

2 Formulation

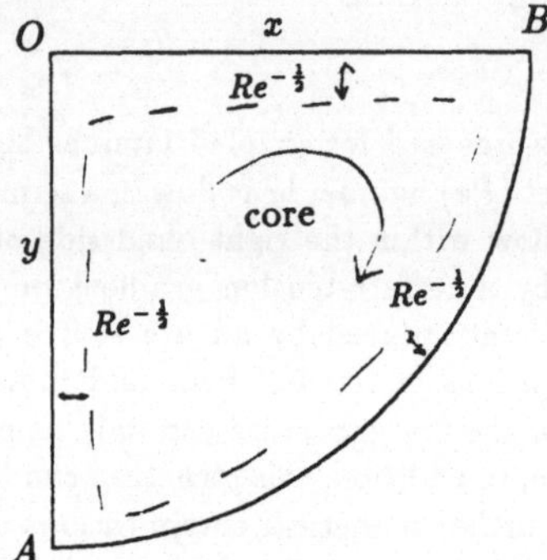

Fig. 2 Sketch of fluid-flow structure

For simplicity, we consider the right-hand side of an idealised two-dimensional semi-circular weld pool (fig. 2), in which fluid flow is driven by a shear stress proportional to the horizontal temperature gradient, applied at OB; a shear applied in the direction from O to B, as arises in an impurity-free pool, results in clockwise flow in the right-hand side of the pool, whereas a shear applied in the opposite direction gives rise to anti-clockwise flow. In addition, AB is a no-slip boundary, OA is a symmetry plane at which there is no shear stress, and there is no fluid flow out of the region OAB. With regard to the heat flow, AB is an isothermal boundary, OA is a symmetry plane across which there is no heat flux, and there is a prescribed heat flux, $q(x)$, at OB. Since $Re, Pe \gg 1$, we expect a core region where the vorticity and the temperature are constant (by the Prandtl-Batchelor theorem), thermal and viscous boundary layers of thickness $O(Re^{-\frac{1}{2}})$ at OA, OB and AB, and corner layers at O, A and B. A momentum balance for the driving force in the Marangoni shear layer OB and the frictional force at AB, and a heat balance for the input at OB and output at AB, determines the appropriate velocity and temperature scales, U and ΔT, respectively:

$$U = \left(\frac{Q\left|\frac{\partial \gamma}{\partial T}\right|}{\rho a k} \right)^{\frac{1}{2}} , \qquad \Delta T = \left(\frac{Q^3 a^3 \mu^2 c_p}{k^3 \rho \left|\frac{\partial \gamma}{\partial T}\right|} \right)^{\frac{1}{4}} ,$$

where Q is the characteristic heat input, ρ the metal density, a the pool radius, k the thermal conductivity, μ the viscosity, c_p the specific heat capacity and $\dfrac{\partial \gamma}{\partial T}$ the surface-tension coefficient, with γ the surface tension and T the temperature.

3 Numerical solution

The governing heat fluid flow equations, subject to these scalings, were solved numerically using the Galerkin formulation of the finite-element method for values of Re of order 10^3, corresponding to $\Delta T \sim 10^3\mathrm{K}$, $U \sim 1\mathrm{ms}^{-1}$ and $\dfrac{\partial \gamma}{\partial T} \approx 5 \times 10^{-4}\mathrm{Jm}^{-2}\mathrm{K}^{-1}$. The expression used for the prescribed heat flux was

$$q(x) = Q\mathrm{e}^{-\chi x^2}, \chi = \frac{1}{100}, 1, 100;$$

this is a form that is commonly used in the welding literature [2].

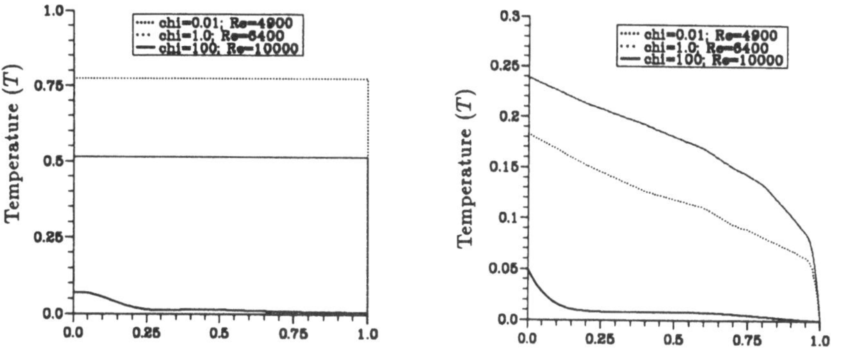

Fig. 3 Surface temperature vs. distance (x) for $\chi = \dfrac{1}{100}, 1, 100$ for clockwise flow

Fig. 4 Surface temperature vs. distance (x) for $\chi = \dfrac{1}{100}, 1, 100$ for anti-clockwise flow

Solutions were obtained for fluid flow in both clockwise and anti-clockwise directions, and may be categorised by considering the temperature profile at the pool surface OB (figs. 3 and 4 respectively). Fig. 3 indicates that for clockwise flow there is an appreciable horizontal temperature gradient only for the most concentrated source ($\chi = 100$); for the two more diffuse sources, the velocity field that results is inconsistent with the initial assumption of a high Reynolds number flow. Fig. 4, on the other hand, indicates that the horizontal surface temperature gradient is appreciable for all of the sources considered. Figs. 5 and 6 show the streamlines and isotherms, respectively, for clockwise flow with

$\chi = 100$. The above results are discussed in greater detail in [4], wherein are also presented further plots of the pool temperature and velocity fields.

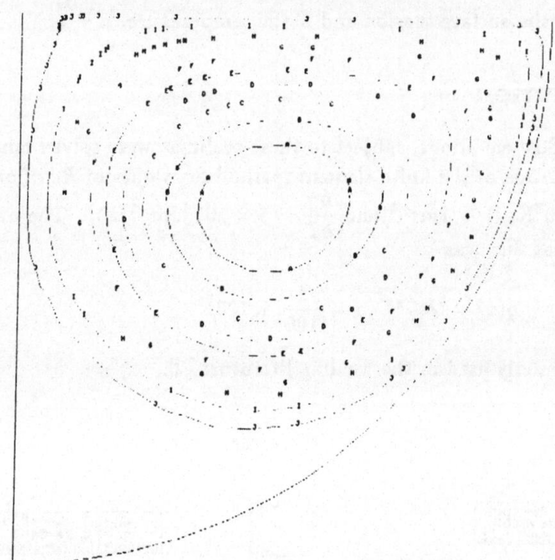

Fig. 5 Plot of streamlines for $\chi = 100$ (anti-clockwise flow)

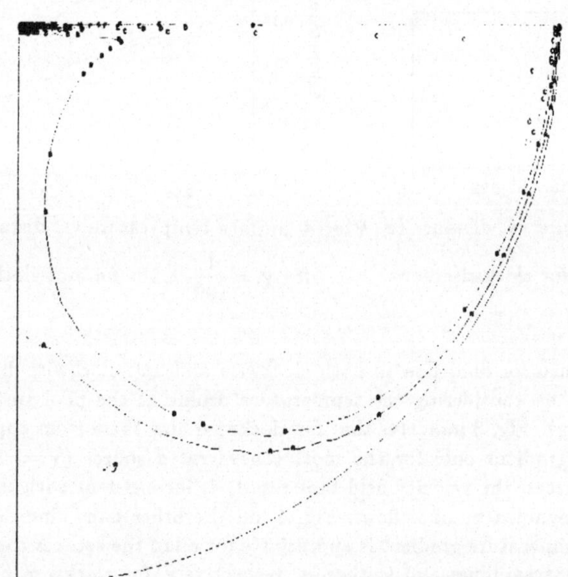

Fig. 6 Plot of isotherms for $\chi = 100$ (anti-clockwise flow)

4   Asymptotic solution

A parallel approach to that in section 3 was adopted by considering an asymptotic solution for $Re, Pe \gg 1$; this entailed appending a temperature field to the velocity field already determined in [3]. There it was demonstrated, for the clockwise case, that the viscous boundary layer at the no-slip boundary detaches into the body of the pool, and that the flow above the layer is far more vigorous than the flow beneath. In the present coupled problem, this still remains the case, but with the additional feature that the thermal boundary layer at the isothermal boundary detaches with the viscous boundary layer into the core of the pool. This proposed asymptotic structure is indeed recovered in the numerical solutions for high values of Re and Pe [4].

5   Conclusion

In this account, we have used two different methods to determine the structure of laminar Marangoni flow at high Re and Pe in a weld pool. By considering different surface heat-flux profiles, we have demonstrated using a finite-element solution to the governing flow equations that, whereas for anti-clockwise flow in the right-hand side of the pool, a high Re and Pe fluid- and heat-flow structure is possible for all types of physically-realistic heat flux profile, for clockwise flow this is only the case when the heat flux is concentrated at the pool centre. The qualitative features of the numerical solution may be recovered by recourse to asymptotic analysis for $Re, Pe \gg 1$.

In reality, however, the shape of a weld pool is not prescribed, but must determined by coupling the problem of heat transfer in the pool with that of heat conduction in the surrounding metal substrate. Consequently, the shape may differ significantly from semi-circular, thereby influencing the nature of the fluid flow that occurs; furthermore, clockwise/anti-clockwise flows will result in wider/deeper pools respectively. Finally, we observe that in practice $\chi \sim O(1)$, in which case a high Reynolds number regime is plausible for anti-clockwise flow only.

A fuller account of the above may be found in [4].

References

[1] BURGARDT, P. & HEIPLE, C. R. 1986 Interaction between impurities and GTA weld shape. *Weld. J.* **65** *Res. Suppl.* 150-s to 155-s.

[2] OREPER, G. M. & SZEKELY, J. 1984 Heat- and fluid-flow phenomena in weld pools. *J. Fluid Mech.* **147**,53-79.

[3] VYNNYCKY, M. 1991 Marangoni Convection in a weld pool. *Proceedings of the Fifth European Conference on Mathematics in Industry*, M. Heiliö (ed.), 381-385.

[4] VYNNYCKY, M. 1991 D. Phil. Thesis, University of Oxford.

THE MODELLING OF NAPKINS

JOACHIM WEICKERT
University of Kaiserslautern
Laboratory of Technomathematics
P.O. Box 3049
D-W-6750 Kaiserslautern
Germany

ABSTRACT. The behaviour of Ultra napkins is investigated by a mathematical model, focussing on liquid transport within the cellulose and liquid exchange between cellulose and granules of a superabsorbent. The model ends up in a nonlinear parabolic PDE strongly coupled with an ODE. Numerically this system is treated by means of a symmetrical time splitting, leading to pure diffusion and pure exchange processes. While the exchange equation can be solved analytically, an ADI-predictor-corrector method is applied to the diffusion equation. The resultant computer programme can be used to propose improved napkins with inhomogeneous granule distributions.

1. Introduction

Ultra napkins contain cellulose fibres with small, homogeneously distributed granules of a superabsorbent (Fig. 1a) which is capable of absorbing up to 65 times its own weight in fluid. Since the baby is not sensible of any liquid inside the granules this technique helps to keep it dry. However, a big disadvantage arises from the fact that due to the liquid absorbtion, the granules swell up and hinder further fluid transport in the intermediate spaces of the cellulose (Fig. 1b). Therefore the time for an Ultra napkin to completely absorb the liquid is larger than the time for a napkin without granules. This phenomenon is called *gluing effect*.

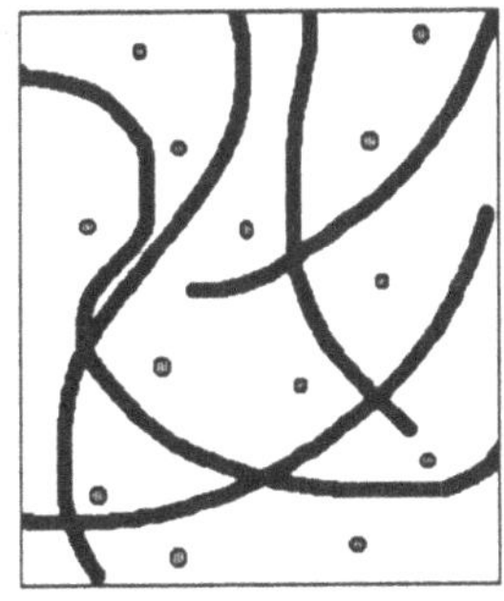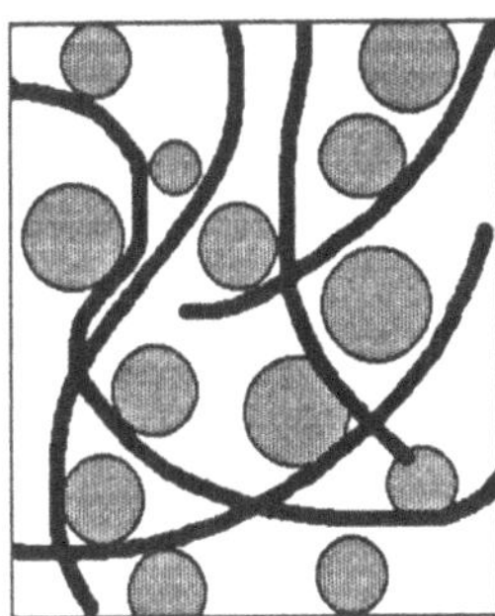

Fig. 1a (left).
Structure of an Ultra napkin
before liquid absorbtion

Fig. 1b (right).
Structure of an Ultra napkin
after liquid absorbtion

Our goal now is to find a mathematical model which handles these effects and allows us to improve napkins. A proposal for such a model is derived in the sequel section. Section 3 deals with a numerical approximation of the napkin equations and in section 4 the computer simulation is applied to real-life problems.

2. The Mathematical Model

In order to arrive at a mathematical model which is not too complicated we have to make some *basic assumptions*. In our case we neglect influences arising from temperature or gravitation and we assume the napkin to possess a constant outer shape. This means e.g. that we do not take into account outer deformations caused by the baby's own weight. Our model napkin is a block with a liquid cylinder of constant radius (Fig. 2). We assume that the granules can only store liquid, while the cellulose exhibits transport and storage properties. So we have to consider a transport process inside the cellulose and, because of their competitive storage properties, a process of fluid exchange between cellulose and granules.

The fluid transport within the cellulose is described as a *nonlinear diffusion phenomenon* which takes place in the intermediate spaces. Fick's law related to the intermediate spaces reads as

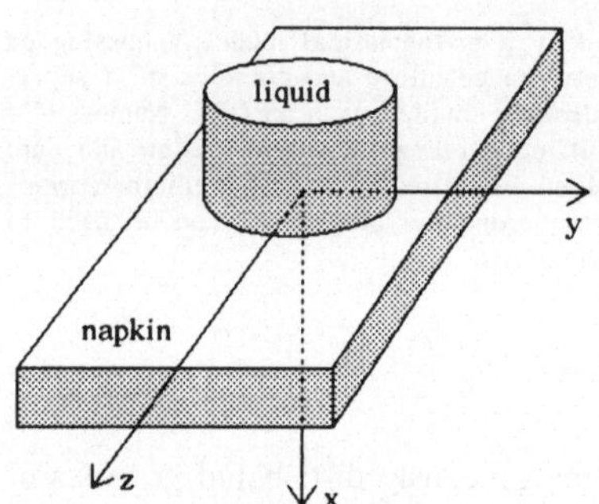

Fig. 2. Model geometry

$$(1) \qquad \widetilde{\mathbf{j}}(\mathbf{x},t) = - D\left(\widetilde{c}_C(\mathbf{x},t)\right) \cdot \nabla\left(\widetilde{c}_C(\mathbf{x},t)\right),$$

where $\widetilde{\mathbf{j}}$ is the flux in the intermediate spaces, u_C the liquid mass in the cellulose, V_I the intermediate space volume, $\widetilde{c}_C := du_C/dV_I$ the intermediate space saturation, t the time and $\mathbf{x}$ the place. For the diffusivity D it is possible to motivate an exponential ansatz of the form [3]

$$(2) \qquad D(\widetilde{c}_C) = D_0 \cdot \exp\left(\beta \cdot \widetilde{c}_C\right) \qquad\qquad (D_0, \beta > 0).$$

Denoting the liquid mass in the granules by u_G, the dependance of V_I on the complete volume V and the liquid concentration in the granules $c_G := du_G/dV$ can be expressed by [3]

$$(3) \qquad V_I = V \cdot \left(A - B\, c_G\right) \qquad\qquad (A, B > 0).$$

Using (1), (3) and the continuity equation one ends up with the diffusion equation

$$(4) \qquad \frac{\partial c_C(\mathbf{x},t)}{\partial t} = \nabla \cdot \left(\left(A - B\, c_G(\mathbf{x},t)\right) \cdot D\left(\widetilde{c}_C(\mathbf{x},t)\right) \cdot \nabla\left(\widetilde{c}_C(\mathbf{x},t)\right)\right).$$

Hereby, $c_C := du_C/dV$ denotes the liquid concentration in the cellulose.
The exchange of fluid between granules and cellulose is a competitive process, which can be assumed to be reversible [3]. Our model uses the *exchange equation*

$$(5) \qquad \frac{\partial c_G(\mathbf{x},t)}{\partial t} \;=\; K_G(\mathbf{x}) \cdot c_C(\mathbf{x},t) \;-\; K_C \cdot c_G(\mathbf{x},t).$$

K_G and K_C (>0) represent "affinities" of the granules and the cellulose, respectively. In a usual Ultra napkin with homogeneous granule distribution, K_G and K_C exhibit the same order of magnitude [3]. If the granule distribution is inhomogeneous, then K_G is a function of $\mathbf{x}$.

Changes in the fluid levels within the cellulose are either caused by diffusion in the intermediate spaces or by liquid exchange between cellulose and granules. Changes in the fluid levels in the granules are described according to the exchange equation (5). Hence we obtain the *napkin equations*

$$(6) \qquad \frac{\partial c_C(\mathbf{x},t)}{\partial t} \;=\; \nabla \cdot \left(\left(A - B\, c_G(\mathbf{x},t) \right) \cdot D\left(\tilde{c}_C(\mathbf{x},t) \right) \cdot \nabla \left(\tilde{c}_C(\mathbf{x},t) \right) \right) \;-\; \frac{\partial c_G(\mathbf{x},t)}{\partial t}$$

$$(7) \qquad \frac{\partial c_G(\mathbf{x},t)}{\partial t} \;=\; K_G(\mathbf{x}) \cdot c_C(\mathbf{x},t) \;-\; K_C \cdot c_G(\mathbf{x},t)$$

supplemented with suitable initial conditions. The boundary conditions below the liquid cylinder are given by intermediate space saturation (as long as the cylinder exists). Otherwise homogeneous Neumann boundary conditions are assumed.

The napkin equations (6), (7) form a system of a PDE strongly coupled with an ODE. The strong coupling is due to the intermediate space fraction $(A - B c_G)$ and the intermediate space saturation $\tilde{c}_C$, both dependent on c_G. This system is too complicated for an exact solution, so we have to look for numerical approximations.

3. Numerical Approximation

In order to avoid numerical problems arising from strong coupling, a *symmetrical time splitting* into pure exchange and pure diffusion steps was applied [3].

For the *pure exchange process* the sum of liquid concentration in the cellulose and in the granules is constant for each $\mathbf{x}$. Using this information the exchange equation (5) becomes a simple ODE in c_G, which is easily solved analytically.

For the *nonlinear diffusion equation* (4) several finite difference solvers, explicit and implicit, with 2 or 3 time levels [2], were implemented and compared. Finally, an ADI-predictor-corrector method [1, 3] turned out to be superior. The predictor-corrector technique avoids nonlinear systems (caused by the diffusivity) and the ADI technique leads to tridiagonal systems, which are solved in linear effort by a modified Gaussian algorithm. So the whole computational and storage effort depends linearly on the number of grid points, hence establishing a fast three dimensional method. Furthermore, the complete scheme is consistent up to second order.

An *automatic step size control* ensures that at the beginning (when dynamic changes occur) very small time steps are used. Later on the time step size increases drastically. For our test cases, this resulted in a speed up of two orders of magnitude [3].

4. Application to Real-Life Problems

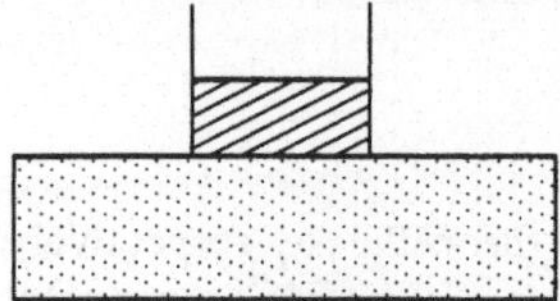

Fig. 3 a. Conventional napkin

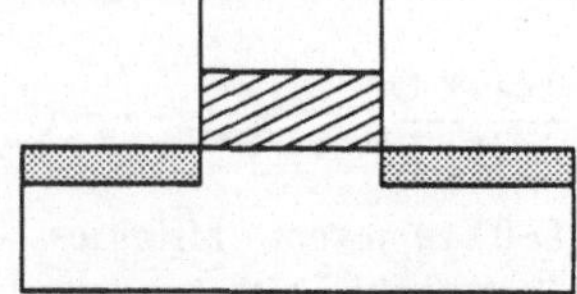

Fig. 3 b. New proposal

Fig. 3a shows a conventional Ultra napkin. It depicts a homogeneous mixture of granules and cellulose with a liquid cylinder at the top. A new proposal for an improved napkin is described in Fig. 3b. While using the same amount of granules and cellulose, it differs by an inhomogeneous granule distribution. In order to achieve a short time for complete fluid absorbtion, no granules are placed below the liquid cylinder. Indeed, numerical results show [3] that this proposal reduces the influence of gluing by approximately 60 %. Since it is aimed to keep the baby dry, the granules are concentrated inside a layer at the top of the napkin (excluding only the granule-free zone below the cylinder). In this layer, the large concentration of granules absorbs the major part of the fluid in the intermediate spaces. Hence, by calculating the average humidity of the upper napkin surface, it is found [3] that after the second wetting the new proposal is nearly as dry as a conventional napkin after the first wetting.

5. Conclusions

In this paper a mathematical description for diffusion and exchange phenomena in Ultra napkins was developed. The use of this model was demonstated by investigating a simple proposal for a new napkin, which turned out to be a remarkable improvement. We are confident that further progress can be obtained by more sophisticated studies of suitable granule distributions. Applications of our model are widespread, since it can be used everywhere similiar Ultra techniques with a cellulose-granule composite are involved, e.g. in sanitary towels.

ACKNOWLEDGEMENT. The author thanks Prof. H. Neunzert for initiating this interesting subject and Markus Tücks and John Brudowsky for valuable hints.

References

[1]   Douglas, J. jr., Gunn, J. E. (1964) 'A general formulation of alternating direction methods. Part I. Parabolic and hyperbolic problems', Numer. Math. 6, 428-453.

[2]   Mitchell, A. R., Griffiths, D. F. (1980) The finite difference method in partial differential equations, Wiley, Chichester.

[3]   Weickert, J. (1992) 'A mathematical model for diffusion and exchange phenomena in Ultra napkins', report no. 72, University of Kaiserslautern, Laboratory of Technomathematics, P.O. Box 3049, D-W-6750 Kaiserslautern, Germany. Submitted to Math. Meth. in the Appl. Sci.

APPLICATIONS OF THE GEOMETRICAL SHOCK CORRECTION METHOD FOR SOLVING NONLINEAR HYPERBOLIC CONSERVATION LAWS

K.-D. WERNER
GKSS Forschungszentrum
Max-Planck-Str.
2054 Geesthacht
Germany

ABSTRACT. In this paper we present an overview of applications of the Geometrical Shock Correction (GSC) Method, introduced in [1]. These include scalar hyperbolic conservation laws with source term in 1- and 2-space dimensions and systems with a suitable structure. We start with two-phase flow problems in porous media and consider a system appearing in the field of oil recovery. Furthermore, for some hyperbolic systems of chromatography and distillation we show that GSC is the optimal solution method. In the final category we provide examples which can be reduced to a nonlinear kinematic wave problem, easily solvable with GSC for generic initial data. Numerical results are given in [1], showing that the exact solutions at an arbitrary fixed time are computed in much less time than with finite-difference methods.

1. Introduction

The GSC method is designed for constructing the entropy solution of the Cauchy problem of general scalar 1-d and special 2-d conservation laws at an arbitrary time $T > 0$ in just <u>one</u> time step, and is based on the method of characteristics and a new geometrical averaging principle for determing shocks and resolving the various interactions of the waves from the multivalued relation, see [1]. Since this method is not based on a time discretisation, questions (and problems) concerning stability and convergence do not arise. The accuracy of the calculated shock positions depend on the used quadrature rule and the high accuracy of the computed solutions results from the sufficiently fine discretisation of the x-axis. This has only marginal effects on the CPU time, since it affects only the transport along characteristics, in contrast to any finite difference scheme.

In the field of two-phase flow problems in porous media the Buckley-Leverett equation can be solved by GSC in much less CPU time than any other method. This is demonstrated by examples in [1]. However, in the 2-d case the fluxes are restricted to permit a reduction to a parameter dependent 1-d problem. For practical applications this is not a serious limitation. Then, exploiding the triangular structure of the Jacobian of the flux, some problems in the field of oil recovery are addressed and for some hyperbolic systems of chromatography and distillation we explore the special structure inherent in the systems, suited for computing solutions by GSC. If a problem permits the use of Riemann invariants, a simple wave argument reduces the system to a nonlinear kinematic wave problem. This occurs e.g. for the dam break and piston problem in the shallow water theory. This is also true for the gas dynamics equations if in addition the assumption of small entropy variation is justified, or if only weak shocks occur.

2. Two-phase flow in porous media

The equations of incompressible and immiscible two-phase flow, excluding capillary effects, read

$$\partial_t(\Phi s) + \nabla \cdot \underline{v} f(s) = 0 \tag{2.1}$$

$$\underline{v} = -K\lambda \cdot \nabla p, \quad \nabla \cdot \underline{v} = 0 \tag{2.2}$$

Here, $s(\underline{x}, t)$ is the saturation, $\Phi(\underline{x})$ the porosity, $\underline{v}(\underline{x}, t)$ denotes the total fluid velocity, p the pressure, $K(\underline{x})$ the rock permeability tensor, $\lambda(s)$ the total fluid transmissibility function and $f(s)$ is the fractional flow function. Equation (2.1) and $\nabla \cdot \underline{v} = 0$ represent conservation of fluids and the other equation in (2.2) is Darcy's law. This system represents an hyperbolic elliptic problem. In one space dimension, $v \equiv \text{const}$ which is set to 1, choosing $\Phi = 1$, then (2.1) is the Buckley-Leverett equation, written as

$$\partial_t s + \partial_x [1 + (\alpha K_o(s))/K_w(s)]^{-1} = 0$$

K_w, K_o denote the permeability of sand to water resp. to oil and determine the flux f, α is a constant to be chosen. This is a conservation law of the form

$$\partial_t u + \partial_x f(u) = g, \tag{2.3}$$

where the source term g is zero. We mention that a variable porosity Φ affects the curvature of the characteristics in space-time diagramm and the 'time of shock formation' when starting with smooth data. In [1] some numerical results for problems with different f and one involving a source term are given.

In the 2-d case of (2.3), with equal fluxes, the transformation $\zeta = 0.5(x + y)$, $\eta = 0.5(x - y)$ leads to

$$\partial_t s + \partial_\zeta f(s; \eta) = g(s, \zeta, \eta)$$

if $g = g(s, x, y)$. This is a 1-d problem for every fixed parameter η. The fractional flow function s depends, of course, on this parameter. In [4] it is shown that some problems of two-phase flow in 2-d can be formulated in this way and that (2.3) need not have a global weak solution, due to properties of the source function g.

Using Stone's assumption that the fractional flow function of gas depends only on the gas saturation, three phase flow can be modelled by

$$\partial_t v + \partial_x G(v) = 0, \quad \partial_t u + \partial_x H(u, v) = 0, \quad w = 1 - u - v,$$

see [3]. Such systems are hyperbolic since the Jacobian is lower triangular. Then, v is determined by the first equation and if v is smooth, $H_v v_x$ acts as a source term for computing u. If a jump in v occurs, use the limit values v_l, v_r as parameters. A jump in u (at the same position) is then given by the Rankine-Hugoniot condition and the transport in u is only used in regions where v does not jump. The solution of the Riemann problem is unique in the class elementary waves, provided the two constant initial states are nearby.

3. Chemical Engineering

In chromatography and distillation we have m+1 species with concentration $c_i^j(z,t)$ of component i in phase j=1,2; $c_0^1 = const.$ Assuming the Langmuir isotherm

$$c_i^2 = N K_i c_i^1 / (1 + \sum_{l=1}^m K_l c_l^1), \quad 0 < K_1 < ... < K_m,$$

where N is a normalisation constant, and introducing $w_i = K_i c_i^1$, $D(w) = 1 + \sum_{l=1}^m w_l$, $h_l(w) = K_l w_l / D(w)$, the system in chromatography reads

$$\partial_z \underline{w} + \partial_t \underline{f}(\underline{w}) = \underline{0},$$

with $\underline{f}(\underline{w}) = \frac{1}{u}(\underline{w} + N \frac{(1-\epsilon)}{\epsilon} \underline{h}(\underline{w}))$, $\epsilon \in (0,1)$ fractional void space of column and u the constant vector fluid velocity. In [2] the corresponding system for distillation is also derived:

$$\partial_z(-F_1 \underline{w} + F_2 \underline{h}(\underline{w})) + \partial_t(f_1 \underline{w} + f_2 \underline{h}(\underline{w})) = 0$$

Here, F_j is the molar flow in phase j, $x_i^j = c_i^j / \sum_{l=0}^m c_l^j$, $i = 0, ..., m$, $j = 1, 2$ are the molar fractions. Again, the Langmuir isotherm relates x_i^2 to x_i^1 and is given by

$$x_i^2 = \beta_i x_i^1 / (1 + \sum_{l=1}^m (\beta_l - 1) x_l^1), \quad \beta_l = const., \ l = 1, ..., m$$

The vector $\underline{h}$ is defined as above and $w_i \equiv (\beta_i - 1) x_i^1$, $\beta_i = const.$
Remark. If the equilibrium matrix $D(\underline{w}) \underline{h}'(\underline{w})$ is diagonalizable, both systems are hyperbolic and the following equivalent features result from properties of the Langmuir isotherm:
Rarefaction and shock curves coincide $\Longleftrightarrow$ the integral curves of the eigenvectors are straight lines in $\underline{w}$ space $\Longleftrightarrow$ the system reduces to a __single__ conservation law on such manifolds. This last statement means that weak solutions are given by $\underline{w} = \Phi(u(x,t))$, where u is a weak solution of the scalar conservation law

$$\partial_z u + \lambda(\Phi(u)) \partial_t u = 0,$$

and $\Phi : I\!R \to I\!R^m$, $u \mapsto \underline{w}$ is a parametrisation of the manifold w.r.t. some component of $\underline{w}$, see [4].

4. Simple waves as kinematic waves

The idea is using n-1 integrals as simple waves, which results in a reduction to one equation. As an example consider the 1-d shallow water equations

$$\partial_t s^2 + \partial_x(us^2) = 0 \tag{4.1}$$

$$\partial_t u + \partial_x(0.5u^2 + s^2) = 0, \tag{4.2}$$

where $s^2 \equiv g\eta$, $\eta =$ equation of the surface, g is the constant gravitational acceleration and u the velocity. The Riemann invariants are $u \pm 2s$, which are const. along $dx/dt = u \pm s$.

Assuming $u \pm 2s = 2s_0 \equiv const.$, then (4.1) or (4.2) provides the other relation for determing u or s, for example (4.1) gives

$$\partial_t s^2 + \partial_x (2s_0 s^2 - 2s^3) = 0$$

We remark that using one of the Riemann invariants in the conservation equation of mass resp. momentum equation leads to the correct shock relations, which would be not true by using the characteristic equations!

In case of isentropic flow, a simple wave argument reduces the Euler equations of gasdynamics to a scalar problem in either the density ρ or the velocity u. Let $c(\rho)$ denote the velocity of sound. The Riemann invariant is

$$\int \frac{c(\rho)}{\rho} \, d\rho \pm u = const. \ \ along \ \frac{dx}{dt} = u \pm c$$

Hence, for a polytropic gas we have $u = 2(c - c_0)/(\gamma - 1)$, which provides the relation between ρ and u

$$u = V(\rho) = \frac{2c_0}{\gamma - 1}\{(\rho/\rho_0)^{\frac{\gamma-1}{2}} - 1\}$$

ρ is now determined from conservation of mass, involving c, the velocity of sound:

$$\partial_t \rho + h(\rho)\partial_x \rho = 0, \ \ h(\rho) = V(\rho) + c(\rho)$$

In terms of u we get

$$\partial_t u + (c_0 + \frac{\gamma + 1}{2}u)\partial_x u = 0 \tag{4.3}$$

For weak shocks changes in the entropy and Riemann invariants can be neglected because they are of third order w.r.t. the shock strength. Shock relations involving (4.3) agree to first order compared to the original ones.

5. REFERENCES

1. Böing, H., Werner, K.-D., Jackisch, H. (1991) 'Construction of the Entropy Solution of Hyperbolic Conservation Laws by a Geometrical Interpretation of the Conservation Principle', J. Comp. Phys., Vol.95, 40-58.
2. Canon, E., James, F. (1990) 'Resolution of the Cauchy problem for several hyperbolic systems arising in chemical engineering', in B.Engquist, B.Gustafson (eds.), Proc. 3. Int. Conf. on Hyperbolic Problems, Vol.1, Studentlitteratur, Lund, Sweden, pp. 212-225.
3. Holden, L., Hoegh-Krohn, R., 'A class of N nonlinear hyperbolic conservation laws' (1987), Preprint Series, No. 6, Dept. of Maths., Univ. of Oslo, Norway.
4. Temple, B. (1983) 'Systems of conservation laws with invariant submanifolds', Trans. A.M.S., Vol.280, 781-795.
5. Werner, K.-D., (1988) 'The Evolution of Discontinuities in Solutions of Inhomogeneous Scalar Hyperbolic Conservation Laws in Several Space Dimensions', J. Appl. Analysis, Vol.31, 11-33.

SELECTION, PURCHASE, CAPITAL RECOVERY, MULTI - LEVEL MAINTENANCE & REPLACEMENT MODEL FOR A STOCHASTI-CALLY DETERIORATING PRODUCTION SYSTEM

G. B. WILLIAMS, AND R. S. HIRANI
School of Manufacturing & Mechanical Engineering
University of Birmingham
B15 2TT
U.K.

ABSTRACT. The model described in this paper considers all the relevant independent cost parameters, in conjunction with the process capability of the production
system, market demand, technological development and the required capital recovery factor. System deterioration is modeled as a non decreasing semi-Markov process
with the state space being the deterioration level of the process capability of the production system. Whenever, there is a change of state, the decision maker has several
options from which to choose; (i) the system may be left as it is, (ii) the system may
be replaced by a new and identical system or its modern alternative, (iii) the system
may be repaired to some better state. The model facilitates these decisions and is
suitable for a variety of industrial problems.

Introduction

Production systems stochastically deteriorate with age, system failure risk and deterioration rates can be substantially reduced by a properly planned maintenance policy.
The parts of a system are categorised into two categories from the maintenance point
of view:(i) those which have a significant effect on the process capability of the system
and (ii) those which have no or negligible effect on the process capability. This is further classified into four basic groups (a) non-repairable parts which fail suddenly [4],
(b) repairable parts with self announcing states [1] (c) repairable parts whose states
are determined by inspection [2] and (d) repairable parts which when inspected give
prior indication of failure [3]. Each subsystem has different characteristics, therefore
separate models have been developed. This model deals with category (i) along with,
the overall optimum long run benefit rate, selection, purchase and capital recovery
factor of the production system. Benefit related costs and optimum long run repair
maintenance cost rates are taken from other models as inputs [1,2,3,4].

Design and Derivation of Mathematical Model

Here the states of the subsystem are determined by the process capability of the system, with the state space $\Im = \{0, 1, 2, ..., L\}$, where a state can only change to higher states. Let us assume that set $\aleph^c = \{\beta, \beta + 1, ..., L\}$ is composed of states which call for intervention, *set* $\aleph = \{0, 1, 2, ..., \beta - 1\}$ is the set of the states which call for no action and set $\wp \subseteq \aleph$ is the set of states, where the system is placed in repair or replacement epochs. Mathematically such a pseudo-control limit policy $\delta(i)$ can be expressed as:

$$\delta(i) = \begin{cases} i & \text{for } i < \beta \in \aleph \\ j < i & \text{for } i \geq \beta \in \aleph^c \end{cases} \tag{0.1}$$

Normal production capacity P_{nc} will always be less than the installed production capacity P_{ic} (usually $\phi = 0.8$), due to uncertainty in production processes and inputs. There will always be tool setting error T_{se} – usually 25% of job tolerance T_{tol}, even if tool setting is approved scientifically. If the expected demand rate for the product is $\bar{F}_{E(T_{i,\star})}$ then the optimum relationships are[2]:

$$P_{nc} = \phi P_{ic}, \ P_{nc} = \bar{F}_{E(T_{i,\star})} \ and \ P_{c,i} = 6\sigma = (T - T_e) \tag{0.2}$$

When such a production system is not available, then the optimum option is the one which has the optimum long run benefit rate. The value of percentage defectives $D_{p,i}$, rejection percentage $D_{r,i,\star}$ for allowable percentage defectives D_a can easily be determined by using statistical tables Fig.1.

For Controlled Production system
with No or Negligible Tool Wear
$T_{se} > 0.00$ *and* $D_a > 0.00$
If $T_{se} \leq (T_{tol} - P_{c,i})$ *Then*
$D_{p,i} = 0.00$ *and* $D_{r,i} = 0.00$
In Other Cases:
$D_{r,i} = (D_{p,i} - D_a)$ *for* $D_{p,i} > D_a$
& $D_{r,i} = 0.00$ *for* $D_{p,i} \leq D_a$

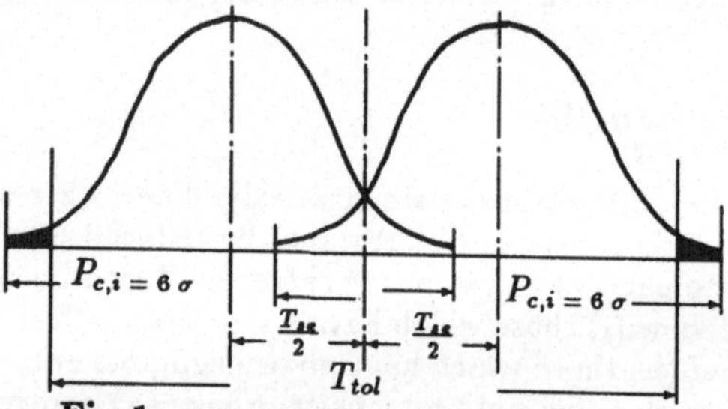

Fig.1

Whereas the logical options for optimum production rate $P_{p,i,\star}$, sale rate $P_{s,i,\star}$ and inspection rate $P_{i,i,\star}$, under various conditions are[2]:

(i) if $D_{p,i} \leq D_a$ and $\bar{F}_{E(T_{i,\star})} \geq P_{nc}$ then; $P_{p,i,\star} = P_{nc}$, $P_{s,i,\star} = P_{p,i,\star}$, and $P_{i,i,\star} = 0.00$,

(ii) if $D_{p,i} \leq D_a$ and $\bar{F}_{E(T_{i,\star})} < P_{nc}$ then; $P_{p,i,\star} = \bar{F}_{E(T_{i,\star})}$, $P_{s,i,\star} = P_{p,i,\star}$, and $P_{i,i,\star} = 0.00$

(iii) if $D_{p,i} > D_a$, $\bar{F}_{E(T_{i,\star})} < P_{nc}$ and $P_{nc} \frac{(100 - D_{p,i})}{(100 - D_a)} < \bar{F}_{E(T_{i,\star})}$ then;
$P_{p,i,\star} = P_{nc}$, $P_{s,i,\star} = P_{p,i,\star} \frac{(100 - D_{p,i})}{(100 - D_a)}$, and $P_{i,i,\star} = 100 \frac{(P_{p,i,\star} - P_{s,i,\star})}{D_{p,i}}$,

(iv) if $D_{p,i} > D_a$, $\bar{F}_{E(T_{i,\star})} < P_{nc}$ and $P_{nc} \frac{(100 - D_{p,i})}{(100 - D_a)} > \bar{F}_{E(T_{i,\star})}$ then;
$P_{s,i,\star} = \bar{F}_{E(T_{i,\star})}$, $P_{p,i,\star} = P_{s,i,\star} \frac{(100 - D_a)}{(100 - D_{p,i})}$, and $P_{i,i,\star} = 100 \frac{(P_{p,i,\star} - P_{s,i,\star})}{D_{p,i}}$,

(v) and if $D_{p,i} > D_a$ and $\bar{F}_{E(T_{i,\star})} \geq P_{nc}$ then; $P_{p,i,\star} = P_{nc}$,
$P_{s,i,\star} = P_{p,i,\star} \frac{(100 - D_{p,i})}{(100 - D_a)}$, and $P_{i,i,\star} = 100 \frac{(P_{p,i,\star} - P_{s,i,\star})}{D_{p,i}}$.

Given the following notations; the one step transition probability from state i to state j $P(i, j)$, expected time of this transition $E(T_{i, j})$, present value of a machine in State i $E(C_{p, i, 0})$, expected purchase plus installation time $E(\mu_{*,i})$, present expected unit costs for tools $E(C_{t, i, 0})$, energy $E(C_{e, i, 0})$, material $E(C_{m, 0})$, direct labour $E(C_{l, 0})$, shortage $E(C_{sh, 0})$, and overheads $E(C_{o, 0})$, expected long run cost rates for non-repairable parts $E(P_{nr}^*)$ [4], parts with self announcing states $E(P_{cofr,*,*}^*)$ [1], parts with states being determined by inspection $E(P_{cof\&ir,*,*}^*)$ [2], and parts which give prior indication of pending failure $E(P_{ocpmr,*,*}^*)$ [3], the expected selling price of the product $E(S_{p, 0})$, time value factor V_1 [2], and net inflation factor V_2 [2]. The net present value (NPV) of one step expected benefit irrespective of the destination state $E(P_{osb, i, *})$ during waiting time $E(T_{i, *})$[2] in state i from state i enterance time $E(\mu_{*, i})$ is:

$$
\begin{aligned}
E(P_{osb, i, *}) \;=\; & \sum_{j=i+1}^{L} P(i, j)\,[\,E(T_{i, j})\{\,E(S_{p, 0})\,P_{s, i, *} - \\
& \{\,P_{p,i,*}\,\{E(C_{t, i, 0}) + E(C_{e, i, 0}) + E(C_{m, 0}) + \\
& E(C_{l, 0})\,\} + E(C_{sh, 0})\,\{\bar{F}_{T_{i,*}} - P_{s, , i, *}\,\}\} + \\
& E(C_{o, 0}) + E(P_{nr}^*) + E(P_{cofr,*,*}^*) + \\
& E(P_{cof\&ir,*,*}^*) + E(P_{ocpmr,*,*}^*)\,\}\} + \{E(C_{p, i, 0}) - \\
& E(C_{p, j, 0})\}\,]\,(V_1\,V_2)^{[E(\mu_{*,i}) + E(T_{i, j})]} \qquad (0.3)
\end{aligned}
$$

The NPV of cumulative benefit $E(P_{cosb, i, \aleph^c})$ at the time of its first transition from state $i \in \aleph$ to any state $m \in \aleph^c$ independent of its destination state would be:

$$
\begin{aligned}
E(P_{cosb, i, \aleph^c}) \;=\; & E(P_{osb, i, *}) + \sum_{j=i+1}^{\beta-1} P(i, j)E(P_{osb, j, *})(V_1\,V_2)^{[E(T_{i, j})]} + \\
& \sum_{j=i+1}^{\beta-2}\sum_{k=j+1}^{\beta-1} P(i, j)\,P(j, k)\,E(P_{osb, k, *}) \\
& (V_1\,V_2)^{[E(T_{i, j}) + E(T_{J, k})]} \; + \;................\; + \\
& P(i, j)\,P(j, k)................P(\beta - 2, \beta - 1) \\
& E(P_{osb, \beta-1, *})\,(V_1\,V_2)^{[E(T_{i, j}) + E(T_{J, k}) + ...+ E(T_{\beta-2, \beta-1})]} \\
& for\; i, \in \wp \subseteq \aleph\; and \; \leq\, j\, \leq\, k\,.\leq\, \beta - 1 \in \aleph \quad (0.4)
\end{aligned}
$$

Since the NPV of benefit rate $E(P_{br, i, \aleph^c})$ is a concave function of control state β, the optimum β state for $E(\tau_{i, \aleph^c})$ represents the expected first passage time to any state $\in \aleph^c$ [2] and would be one which results in maximum NPV of benefit rate:

$$
E(P_{br, i, \aleph^c}^*) \;=\; \max_{\beta > i \leq L}\left[\frac{E(P_{cosb, i, \aleph^c})}{E(\mu_{*,i}) + E(\tau_{i, \aleph^c})}\right]\; for\; i = 0 \in \aleph. \quad (0.5)
$$

Since for the optimum β state, NPV of the benefit rate $E(P_{br, \delta(l), \aleph^c})$ of the repaired system during its expected extended life $E(\tau_{\delta(l),\aleph^c})$ is a concave function of the state $\delta(l) = j \in \wp \subseteq \aleph$ in which it has been placed in expected repair time $E(\mu_{l,\delta(l)})$. Therefore its optimum value would be:

$$E(P^*_{br,\,\delta(l),\,\aleph^c}) \;=\; \max_{\delta(l)=j} \left[\frac{E(P_{cosb,\,\delta(l),\,\aleph^c})}{E(\mu_{l,\delta(l)}) + E(\tau_{\delta(l),\,\aleph^c})} \right] \tag{0.6}$$

NPV of the long-run optimum benefit rate $E(P^*_{br})$ for limiting transition probability $\Pi(l)$ [2] where $l \in \aleph^c$ and $\delta(l) = j \in \subseteq \aleph$:

$$E(P^*_{br}) \;=\; \sum_{l\,\in\,\aleph^c} \Pi(l)\, E(P^*_{br,\,\delta(l),\,\aleph}) \tag{0.7}$$

Discussion and Conclusions

The choice of a production system from many alternatives with various degrees of accuracy, reliability, production capacities and differing purchase and running costs is not an easy one to make. Higher accuracy often results in higher production costs and lower accuracy often results in higher quality costs. Higher production capacity may result in unutilized production capacity and high purchase cost, whereas lower production capacity may result in less profit and loss of customer good will.

Here optimum production and inspection rates are determined keeping in view demand, allowable percentage defectives, job tolerance, process capability and production capacity in order to optimise the long run benefit rate and control the average outgoing quality level. Parts of a multi-level production system are categorised into five basic groups and separate optimum multi-level policies are designed based on transition states and technology developments. The optimum production system can be selected by comparing long run optimum rates for various alternatives.

Capital recovery for a production system is taken to be equal to the drop in value considering infation or deflation, which will enable mature as well as premature replacement by the same or its modern alternative without additional investment. This model is the back bone of authors' series of papers [1,2,3,4], and is universally applicable in machine investment and maintenance policy decisions.

References

1. Williams, G. B. and Hirani, R. S., "Multi-level Maintenance Model for Parts of a Production System with a Self Announcing Present State," Publication in Preparation.
2. Williams, G. B. and Hirani, R. S., "Multi-level On-condition Preventive Maintenance Inspection Model Based on Constant Base Interval Risk – When Inspection Detects Pending Failure, " Submitted for Presentation in $7th$ IFAC Symposium on Information Control Problems in Manufacturing Technology – INCOM 92 Canada.
3. Williams, G. B. and Hirani, R. S., "Multi-level On-condition Preventive Maintenance Inspection Model Based on Constant Base Interval Risk – When Inspection Detects Present State," Publication in Preparation.
4. Williams, G. B. and Hirani, R. S., "Optimal Replacement Policy for Non-repairable Parts of a Production System That Fail Suddenly – Undetectable Deterioration," Publication in Preparation.

The Manufacture of Optical Fibres

P.Wilmott, J.N.Dewynne and J.R.Ockendon

1 Introduction

The motivation for this work is the industrial manufacture of optical fibres. (See figure 1 for a typical geometry.) Within the industrial application the fluid flow may be considered to be slow and Newtonian. The Stokes equations are thus the relevant governing equations. We shall therefore begin with an examination of the slow flow of an axisymmetric jet under tension. We shall exploit the large aspect ratio in this problem, the ratio of the length of the fibre to a typical radius, in order to solve the problem to leading order by asymptotic means. This problem then leads us to consider more general flows, including inertia, surface tension and gravity. Finally we examine the non-axisymmetric fibre. Problem 1 is thus the axisymmetric Stokes flow, Problem 2 is the axisymmetric jet with inertia, surface tension and gravity and Problem 3 the non-axisymmetric Stokes jet.

2 Problem 1

Denote by x and r the axial and radial distances, $u(x, r, t)$ and $v(x, r, t)$ the axial and radial velocities, $p(x, r, t)$ and $A(x, t)$ the pressure and cross-sectional area. Time is t. The small parameter ϵ is the ratio of a typical radius of the fibre to a typical length. All of the above variables are assumed to have been non-dimensionalized and scaled in the obvious ways.

It is a very straightforward matter to expand the dependent variables in an asymptotic series in powers of ϵ, and by solving the Stokes equations (also suitably expanded) to arrive at

$$u \sim u_0(x, t) + \epsilon^2 \left(\frac{r^2}{4\mu} (-3(\mu u_{0_x})_x + \mu u_{0_{xx}}) + a(x, t) \right) + O(\epsilon^4)$$

$$v \sim -\frac{r u_{0_x}}{2} + O(\epsilon^2)$$

$$p \sim -\mu u_{0_x} + O(\epsilon^2),$$

where we have used the zero normal stress boundary condition. $a(x, t)$ is an unknown function. Applying the second stress and kinematic conditions on the free surface leads to

$$A_t + (u_0 A)_x \;=\; 0 \tag{1}$$

and

$$(3\mu A u_{0_x})_x \;=\; 0. \tag{2}$$

The points to note about the above analysis is that to leading order the flow is extensional, that is, the axial velocity is a function of x and t only. Second, we had to go to second order in u in order to close the system. The above analysis is valid for μ a function of x and t. We have retained the factor 3 in (2) for historical reasons, the coefficient 3μ is known as the Trouton viscosity. Details of the analysis may be found in Dewynne, Ockendon and Wilmott (1989).

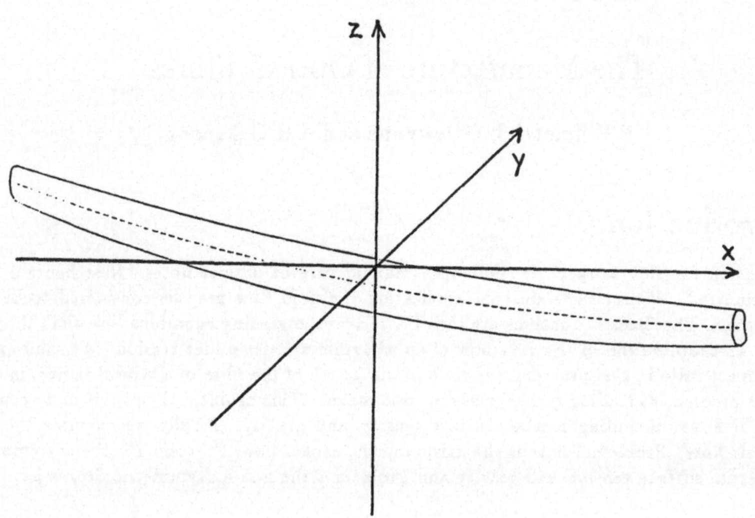

Figure 1: Schematic diagram of the fibre geometry

It is possible to solve (1) and (2) explicitly when μ is spatially piecewise constant by utilizing a semi-hodograph transformation. With A and t as the independent variables we can write (1) and (2) as

$$A(x_{At} - x_{AA}) = x_A$$

and hence

$$x_A = \frac{1}{AF(A+t)},$$

where F is given by the initial shape of the fibre. It is then possible to show, for example, that a fibre can only break if there is a cusp in the initial shape.

3   Problem 2

Staying with the slender axisymmetric fibre we can further introduce the physical phenomena of inertia, surface tension and gravity (acting along the axis of the fibre). We shall only consider the constant viscosity case here. The small ϵ expansion shows that the flow is again extensional to leading order and the equations of motion become

$$A_t + (u_0 A)_x = 0$$

and

$$-3(Au_{0_x})_x + ReA(u_{0_t} + u_0 u_{0_x}) + \frac{Re}{Fr} - \sigma A^{1/2} = 0,$$

where Re is the Reynolds number, Fr the Froude number and σ a surface tension coefficient. Although these equations are not as elegant as (1) and (2) above and cannot be solved explicitly, there do exist interesting similarity solutions. Note that it is also possible to include compressibility within the slender geometry model, this introduces a third equation for the fibre density. Further details may be found in Dewynne and Wilmott (1991).

4   Problem 3

Returning to a Stokes flow in the absence of surface tension and gravity we shall for our third problem consider the stretching and twisting of an asymmetric fibre with constant viscosity. The small ϵ expansion shows that the flow is still extensional to leading order and the pressure and radial velocities ($v(x, y, z, t)$ and $w(x, y, z, t)$ with obvious notation) are given by

$$p \sim -u_{0_x} + O(\epsilon^2)$$

$$v \sim \left(-y\frac{u_{0_x}}{2} + a(x, t)z + b(x, t)\right) + O(\epsilon^2)$$

and

$$w \sim \left(-z\frac{u_{0_x}}{2} - a(x, t)y + c(x, t)\right) + O(\epsilon^2).$$

Here a, b and c are to be determined (along with u_0 as before). Note that the terms on the right-hand sides can be identified as shrinkage, rotation and translation. This is because the leading order problem in the cross-section is that of plane Stokes flow with stress free boundaries but with a uniform mass sink.

Unfortunately, we cannot explicitly find the $O(\epsilon^2)$ term in the expansion of u as we did in the first two problems and we therefore find closure of the problem more complicated. So, rather than continue with the direct asymptotic expansion of the dependent variables, we shall choose to adopt an alternative approach.

We can show directly from the Stokes equations and its boundary conditions that the following global properties must be true.

$$A_t + (u_0 A)_x = 0,$$

$$\left(\int\int_A -p + 2u_{0_x} \, dy dx\right)_x = 0,$$

$$\left(\int\int_A (v_x + u_{1_x}, w_x + u_{1_x}) \, dy dx\right)_x = 0$$

and

$$\left(\int\int_A yu_{1_x} - zu_{1_x} + yw_x - zv_x \, dy dx\right)_x = 0, \tag{3}$$

where u_1 is the $O(\epsilon^2)$ term in the expansion of u and the integral is taken over the cross-sectional area. The above are valid up to the obvious orders of ϵ. The physical interpretations of the above are conservation of mass, axial momentum, transverse momentum and angular momentum.

Some manipulation of these equations is possible without having to solve for u_1. It can be shown that the first three reduce to

$$A_t + (u_0 A)_x = 0,$$

$$(Au_{0_x})_x = 0$$

and

$$b_{xx} = c_{xx} = 0.$$

The first two of these can be solved exactly as shown in Problem 1 and the last simply shows that the centreline is straight. Without loss of generality we can align the x-axis with the centreline so that $b = c = 0$. We have thus shown that the fibre shrinks (a known amount) and rotates about a fixed line (through the centre of mass of each cross-section). It only remains to find the amount of rotation which is determined by eqn. (3).

We shall not include here any details of the analysis which leads to the final equation for the rotation but shall simply give a brief recipe.

The first stage is to recast the problem for u_1 in terms of Lagrangian variables. Hence we write

$$t = \tau, x = X(\xi, \tau), y = Y(\xi, \eta, \zeta, \tau), z = Z(\xi, \eta, \zeta, \tau)$$

where

$$X_\tau = u_0(X, \tau), \ X(\xi, 0) = \xi,$$

$$Y_\tau = v_0(X, Y, Z, \tau), \ Y(\xi, \eta, \zeta, 0) = A(\xi, 0)^{1/2}\eta$$

and similarly for Z_τ, and

$$v_0 = -\frac{yu_{0_x}}{2} + az + b'$$

and similarly for w_0. Note that in these variables the shape of the fibre is independent of time.

It is now possible to write the problem for u_1 in these new variables and, because of the similarity of the shape transformation, show that u_1 can be written as the sum of several terms each of which is the product of two terms; one a function of ξ and τ and the other a function of ξ, η and ζ. With the notation $\theta(\xi, \tau) = \frac{\partial}{\partial \tau}a(x, t)$ we find that eqn. (3) becomes

$$\frac{\partial}{\partial \xi}\left(3\hat{A}\left(u_{0_\xi}\theta_\xi - \theta_{\xi\tau}X_\xi\right)\int\int_{A_0}\eta\hat{u}_{2_\zeta} - \zeta\hat{u}_{2_\eta}d\eta d\zeta\right.$$

$$+3\hat{A}u_{0_\xi}\int\int_{A_0}\eta\hat{u}_{1_\zeta} - \zeta\hat{u}_{1_\eta}d\eta d\zeta$$

$$+ \hat{A}\theta_{\xi\tau}X_\xi\int\int_{A_0}\eta^2 + \zeta^2 d\eta d\zeta\Bigg) = 0. \tag{4}$$

Where $\hat{A} = A^2/X_\xi^2$. Here $\hat{u}_1$ and $\hat{u}_2$ both satisfy Laplace's equation in η and ζ with boundary conditions which depend on ξ, η and ζ only. The integrals in (4) are taken over the initial cross-sectional area of the fibre. Once the double integrals in eqn. (4) have been evaluated once-and-for-all as functions of ξ (at $\tau = 0$) this becomes the final equation from which to determine the amount of twist in the fibre. Details of the analysis and some examples are given in Dewynne, Ockendon and Wilmott (1991).

5   References

Dewynne,J.N., Ockendon,J.R. & Wilmott,P. 1989 On a mathematical model for fibre tapering. SIAM J.Appl.Math.49 983-990.
Dewynne,J.N. & Wilmott,P. 1991 Slender Fluid Jets. Submitted to SIAM J.Appl. Math.
Dewynne,J.N., Ockendon,J.R. & Wilmott,P. 1991 A systematic derivation of the leading order equations for extensional flows in slender geometries. Submitted to J.Fluid Math.

Paul Wilmott
Mathematical Institute,
Oxford, U.K.

UNIT COMMITMENT PROBLEM BY SIMULATED ANNEALING AND AN ANALYSIS OF NEIGHBOURHOOD STRUCTURE

Guobiao Zhou

Glasgow Polytechnic, Glasgow, G4 0BA, UK

ABSTRACT. Unit Commitment Problem (UCP) is a double optimization problem. This paper develops a simulated annealing (SA) algorithm for UCP. The SA solution to UCP involves two key techniques, the design of neighbourhood structure and the optimization of outputs of units of system. One salient feature of the proposed method is that it has no prior bias to any cycling unit. Sample calculations show that SA solution to UCP could give a good compromise between the accuracy of final solutions and the computation time.

1. Introduction

Unit Commitment Problem (UCP) is one of the most difficult problems in electrical power systems analysis. UCP studies when and which unit to be started up or quitted off such that total cost of the power system is minimized under a set constraints during the investigated period ranging from one day to more. UCP is a high dimensional mixed integer, non-linear programming problem with unique constraints. Five kinds of approaches have been proposed to solve UCP [2]: partial enumeration method, dynamic programming, Benders partitioning method, Lagrangian relaxation and heuristic method. UCP with realistic size have been solved so far by heuristics, such as the priority order method and decoupled method [1], [3]. The research of this paper is based on Zhang Chuan-Model [1]. Simulated annealing (SA) is an improved local search method to solve difficult combinatorial optimization problems. Since its introduction in 1983, SA has been applied successfully to various fields. The local search algorithms impose a neighbourhood structure on the solution space such that a comparison of the cost within an artificial local area can be conducted. But local search algorithms will always terminate at one of local minimums, which may be of bad accuracy. SA adopts a new strategy that accepts a neighbour solution with non-reduction in cost with certain probability. Thus SA has an ability to avoid being trapped in a local minimum. Detailed discussions can be found in reference [4]. This paper presents a novel method for UCP: simulated annealing solution. In section 2, the details including the reformulated model of UCP, U-Decision, neighbourhood structure and its features, and P-Decision are discribed. The implementation and results of SA algorithm are involved in section 3. Last section contains a short conclusion.

2. SA Solution to UCP

2.1 UCP MODEL

The Zhang Chuan-Model consists of one object function and several groups of constraints. According to the viewpoint of SA, UCP is a double optimization problem. The first is the optimization of system structure (system configuration) which belongs to combinatorial optimization. Corresponding to each system configuration, the second optimization is to find one set of optimal outputs of units of system, which belongs to a continous optimization. Here the optimization of system configuration is called as U-Decision, while the later P-Decision. The Zhang Chuan-Model of UCP is reorganized into the following:

(I) U-Decision:

$$\sum_{j=1}^{T-1} (u_{i,j+1} - u_{ij})^2 \leq k \tag{2.1}$$

$$\sum_{j=1}^{T} (1 - u_{ij}) \geq \beta_{i1} \qquad \text{or} \qquad \sum_{j=1}^{T} (1 - u_{ij}) = 0 \tag{2.2}$$

$$\sum_{j=1}^{T} u_{ij} \geq \beta_{i2} \qquad \text{or} \qquad \sum_{j=1}^{T} u_{ij} = 0 \tag{2.3}$$

$$\forall\, i: N_2+1,..., N_1, \quad j: 1,2,..., T: \qquad u_{ij} = 1 \tag{2.4}$$

(II) P-Decision:

(II-A) Penalty functions: $\forall j := 1,2,\dots,T$:

$$DH_j = (D_j + R_j) - \sum_{i=1}^{N1} u_{ij} P_j^H \tag{2.5}$$

$$DL_j = \sum_{i=1}^{N1} u_{ij} P_j^L - D_j \tag{2.6}$$

(II-B) Minimize:

$$F(\mathbf{U},\mathbf{P},\boldsymbol{\lambda}) = \sum_{j=1}^{T} \sum_{i=1}^{N1} F_i(P_{ij})\, u_{ij} + \sum_{j=1}^{T} \lambda_j \Big(D_j - \sum_{i=1}^{N1} P_{ij}\Big) + \sum_{i=1}^{N2} S_i(u_{ij})$$

$$+ \sum_{j=1}^{T} (\psi 1_j \times \mathrm{Max}(DH_j) + \psi 2_j \times \mathrm{Max}(DL_j)) \tag{2.7}$$

where

$$\psi 1_j = (D_j + R_j) \times K1 / t_k \tag{2.8}$$

$$\psi 2_j = D_j \times K2 / t_k \tag{2.9}$$

$$\mathrm{Max}(x) = x \text{ if } x.>0, \quad \mathrm{Max}(x) = 0 \text{ if } x \le 0 \tag{2.10}$$

t_k is the temperature parameter of SA algorithm, K1, K2 are constants.

$\forall\ i:\ 1,2,\dots,N_1,\ j:\ 1,2,\dots,T$:

$$u_{ij} P_j^L \le P_{ij} \le u_{ij} P_j^H \tag{2.11}$$

where T: time period length investigated, one hour as one time piece;

k: change times of unit operation state during the time investigated;

N_1, N_2: number of system units and cycling units respectively;

P_{ij}: power output of i-th unit during j-th time piece, positive variables;

u_{ij}: operation state of i-th unit during j-th time piece, 0-1 variables;

P_i^H, P_i^L: upper and lower bound of the output of i-th unit respectively;

β_{i1}, β_{i2}: minimum stop and run time of i-th cycling unit respectively;

D_j, R_j: load and spinning reserve of the system during j-th time piece respectively;

$F_i(P_{ij}) = a_{i1} + a_{i2} P_{ij} + a_{i3} P_{ij}{}^2$: fuel cost function of i-th unit;

$S_i(u_{ij}) = b_{i1}(1 - b_{i2} \exp(-b_{i3} L(U_i)))$: cost function of unit staratup and shutdown of i-th unit, $L(U_i)$ denotes the length of stop time of i-th unit.

Thus, SA algorithm for UCP consists of itreations. The independent combinatorial variables $u_{ij}(\forall i,j)$ which are determined by (2.1)-(2.4) form a possible (instead of feasible) system configuration, or say, a solution of UCP. Corresponding to such a configuration, P-Decision becomes a nonliear programming problem whose variables are $P_{ij}(\forall i,j)$ with a large number of inequqlity constraints. Hence, one of the main problems of applying SA method to UCP is how to search effectively through the configuration space. The iteration process of SA is implemented along with the following line: (see Figure 1)

2.2 U-DECISION

U-Decision in each SA iteration generates a possible, instead of feasible, solution, say, U matrix. The set of possible U matrix includes the sets of feasible and infeasible ones. To eliminate the infeasibility of final solutions, all of infeasible U matrix will be subject to a set of penalty functions (2.5) (2.6), then the penalty parts are added in the cost function proportionally. As t_k is decreased monotonically, the punishment for infeasible U solution will become larger and larger. Thus infeasible solutions are likely to be entirely removed out as the SA algorithm proceeds. When k=2, there only exist six pattern of operetion state each unit:

U_1 {0}-type: $\{0,0,\dots,0\}_{1 \times T}$; U_2 {1}-type: $\{1,1,\dots,1\}_{1 \times T}$;

$$U_3 \ \{0\text{-}1)\text{-type}: \ \{0,0,...,0,1,1,...,1\}_{1\times T}; \qquad U_4 \ \{1\text{-}0\}\text{-type}: \ \{1,1,...,1,0,0,...,0\}_{1\times T}; \qquad (2.12)$$
$$U_5 \ \{0\text{-}1\text{-}0\}: \ \{0,...,0,1,...,1,0,...0\}_{1\times T}; \qquad U_6 \ \{1\text{-}0\text{-}1\}: \ \{1,...,1,0,...,0,1,...,1\}_{1\times T}.$$

where in U_r (r=3,4,5,6), number of 1's, Y, satisfies: $\forall i: 1,...,N_2$: $\quad \beta_{i2} \le Y \le 24 - \beta_{i1}$ $\quad (2.13)$

Thus, one possible solution, U matrix, is comprised by these patterns row by row combinatorially.

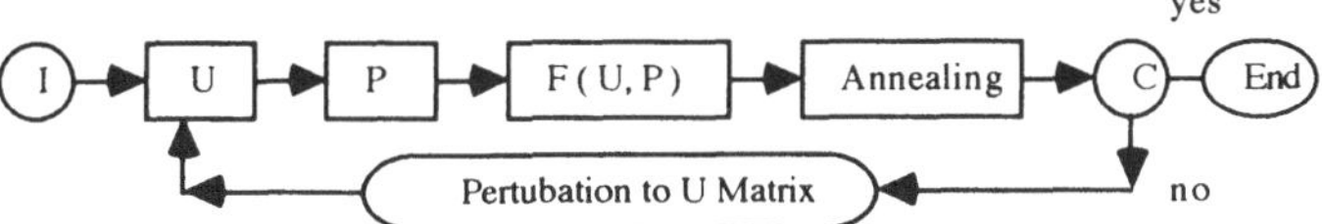

where I: initiation of the SA algorithm;   C: criterion

Figure 1.   Iteration of SA algorithm of UCP

2.3 NEIGHBOURHOOD STRUCTURE

To solve a combinatorial optimization problem is to search over a large, discrete solution space. The cutting-planes and the branch-and bound methods employ the stategies that delete more unnecessry solutions and shrink the scope of the solutions to be searched. The local search methods use neighbourhood structure to search the space. According to the existing theory of SA algorithm, if the neighbourhood structure R satisfies the reversibility, which means that there is a probability distribution α on the solution space S, necessarily the equilibrium distribution for R such that $\alpha(i) \ R(i,j) = \alpha(j) \ R(j,i)$ for all $i,j \in S$, then the homogeneous SA algorithm will converge asymptotily to an equilibrium distribution, which is independent of the size of the neighbourhood structure. Furthermore, the condition on the neighbourhood struvture in inhomogeneous SA algorithm can be reduced as weak reversibility[5]. However, asymptotic convergences for two kinds of SA algorithms are only of theoretical interests and tell us nothing on the effectiveness and the efficiency of SA algorithm. There is a big gap between the existing theory and practical running of SA algorithms. Up to now, the design of neighbourhood structure for a special problem still relies on intuition. Recently, G.B.Sorkin proposaed his research that connects the efficiency of SA algorithm with the landscope of the solutions.[7]. His theory says that when the cost landscope of the solutions of the problem satisfies so-called fractal condition, SA algorithm will run very efficiently. That means that SA algorithm is able to find a good solution within finite time with a very high probability for a kind of instances of problems.

The cost of UCP mainly equals the sum of fuel cost of each unit in every time piece, it is possible to design a neighbourhood structure such that the differences among the costs of most solutions within a neighbourhood structure keep small. It can be estimated that small change of operation state of one or two units only increase or decrease a small amount of cost. Thus the fractal condition is likely satisfied. Therefore, the neighbourhood structure of UCP can be designed by means of the small perturbation, i.e. small change of operation state of the current U matrix. The size of neighbourhood structure is controled by the pertubing parameters, MaxΔY and Nc, where MaxΔY is Maximum number of change of 1's in one row to be perturbed, Nc: number of units to be perturbed. The perturbation is defined as follows (setting T=24):

$$|\Delta Y| \le Inc, \qquad Y_{new} = Y_{old} - \Delta Y, \qquad Flag = (\ \beta_{i2} \le Y \le 24 - \beta_{i1}) \qquad (2.14)$$

------------------ * ------------------ * ------------------

Perturbation to U_1:	Perturbation to U_2:	Perturbation to U_r:
if Flag: U_1->U_r(r=3,4,5,6)	if Flag: U_2->U_r(r=3,4,5,6)	if Flag: U_r->itself with Y_{new};
in probability 0.25;	in probability 0.25;	if Y_{new}>24-β_{i1}: U_r->U_2;
if Y_{new}>24-β_{i1}: U_1->U_2;	if Y_{new}<β_{i2}: U_2->U_1;	if Y_{new}<β_{i2}: U_r->U_1.
if $0 \le Y_{new}$<β_{i2}: U_1->U_1;	if Y_{new}>24 -β_{i1}: U_2->U_2.	
if Y_{new}<0: Y_{new}=-Y_{new},		
perturbating again.		

------------------ * ------------------ * ------------------

The calculations show that when the perturbating parameter MaxΔY and Nc are chosen smaller, the fractal condition can be satisfied. The prososed neighbourhood structure is of several particular features: i) | Si | $\neq$ constant, where | Si | denotes the size of neighbourhood structure of solution i; ii) if $j_1, j_2 \in R(i)$, and the rows of j_1, j_2 which are perturbed do not belong to U_3 and U_4 types at same time, then generation probabilities $G_{i,j_1} \neq G_{i,j_2}$; iii) (S,R) of UCP is irreducible [5].

2.4 P-DECISION

Usually, Kuhn-Tucker conditions can be applied to solve P-Decision in order to avoid too many slack variables in Largrangian relaxation. Let P_{ij}^+, λ_j^+ be minima of the cost function without the constraints. According to K-T conditions, if P_{ij}^+ violate the constraints, then some minima of P_{ij}^* of original problem are comfirmed at certain constraint bounds. But there are lots of combinatorial ways to setting bounds, K-T conditions do not tell any details on the ways and the priority order of setting bounds. In the strict sense, it is a multi-stage decision problem, and may take longer CPU time. A fast approach is developed which is suitable for the iterations of SA algorithm. Let DD be the remnant load of the remnant system, and let MH, ML be maxmal and minmal possible outputs of remnant system. Then this approach is based on two conditions: i) Setting upper or lower bound is feasible if and only if ML$\leq$DD $\leq$ MH; ii) If both of setting upper bound and lower bound are feasible at the same time, then the priority order is given to one whose total amount of exceeding bound is small. Note that the second condition is heuristic, hence the final iteration P_{ij}^* are not sure to be optimal. But the numerical computation of lots instances have shown its high accuracy and fast.

3. Implementation and Results of SA Algorithms

A SA algorithm is coded in Pascal. The cooling schedule employed is the geometry schedule which is faster than the logarithm schedule. In homogeneous version, the iteration times in each temperature is equal to the size of the neighbourhood structure emploted. The SA algorithm with smaller size of neighbourhood structure such as setting Nc=1 and smaller MaxΔY obtains most of better final solutions. The initial U matrix should have a high initial cost value.

30 instances of UCP are run on Prime 9955. About 95% final SA solutions obtained are better than those that are obtained by priority sequence method and decoupled method. But the CPU time of SA algorithm is longer (about 1.5-2.5 times).

4. Conclusions

i) SA is a powerful tool to solve Unit Commitment problem with general applicability and the ability to give satisfactory final solutions.

ii) A major shortcoming of SA solution to UCP is larger CPU time. One way to speed up SA is to use parallel implementation of SA. The choice of better neighbourhood structure for UCP and cooling schedules requires further investigation.

References

[1] Zhang Chuan et al (1988), 'The Decoupled Approaches to Unit Commitment of Power System', Proceedings of Chinese Society of Electric Engineering, vol 8, No6, pp26-31.

[2] A. Merlin et al (1983), 'A New Method for Unit Commitment at Electricite De France', IEEE Transactions, vol PAS 102, No5, pp1218-1225.

[3] Zhang Chuan (1984), 'The Decoupled Ordering Method for Optimization of Unit Commitment of Power System', M.sc dissertation, EIEPE, China.

[4] E.Arrts et al (1989(, Simulated Annealing and Boltzmann Machines, John Wiley & Sons Ltd.

[5] B. Hajek (1988), 'Cooling Schedules for Optimal Annealing', Mathematics of Operations Research, vol 13, No2, pp311-329.

[6] Guobiao Zhou (1991), 'Simulated Annealing Solution to UCP', report, Glasgow Polytechnic.

[7] G.B.Sorkin (1990), 'Simulated Annealing on Fractals: Theoretical analysis and Relevance for Combinatorial optimization', Advanced Research in VLSI Proceedings of the 6th MIT conference, Cambridge, USA, 2 April, pp331-351.

Industrial Design Papers

A SIMPLE MODEL FOR THE SHAPE OF FIBRE TAPERS

T. A. BIRKS, Y. W. LI, R. P. KENNY and K. P. OAKLEY
Lightwave Technology Research Centre
University of Limerick
Plassey Technological Park
Limerick
Ireland.

ABSTRACT.   A model for the shape of optical fibre tapers is presented which avoids any need for the techniques of fluid mechanics. A theoretical procedure for producing any given shape of taper is described, together with a practical system capable of realising the model's predictions. A complete procedure for the formation of fibre tapers with any reasonable shape is thus presented.

1.   Introduction

The fibre taper, formed by stretching a piece of silica optical fibre in a heat source such as a small flame, is the basis for many important components[1]. The performance of a taper component is critically dependent on the shape of the taper[2], drawn schematically in fig.1. This shape $r(z)$ depends on the temperature distribution of the heat source, and particularly on its length L. It is therefore important to understand how a heat source determines the shape of a tapered fibre, if high performance fibre components are to be manufactured.

We consider a simple idealised model in which the heat source raises a length L of fibre to a uniform temperature, while leaving the rest of the fibre cold. Such a model with fixed L is well known[3], but we treat the general case where L may vary. No knowledge of fluid mechanics is needed to develop this model.

2.   The Model

At an instant t, a cylindrical section AB of length L and radius r_w in the taper waist is uniformly heated, while outside this uniform hot zone the glass is cold and solid (fig.2). At time $t+\delta t$ the hot glass has stretched to form a narrower cylinder AB of length $L+\delta x$ and radius $r_w+\delta r_w$, where δx is the increase in extension. The hot zone length is changed to $L+\delta L$ in the same time, where δL may be negative. The extremities AA' and B'B leave the hot zone and solidify, forming new elements of the taper transitions. Thus the local radius $r(z)$ at a point z in the taper transition is equal to the waist radius r_w as z was pulled out of the hot zone.

L is changed by the appropriate control of a heat source and may vary arbitrarily, subject to two constraints $L \geq 0$ (obvious) and $dL/dx \leq 1$ (ensuring that the hot zone does not overtake the retreating taper transitions). The length l_w of the taper waist is equal to the instantaneous hot zone length.

The analysis following from the model is based on two fundamental equations. The first arises from the conservation of glass volume in the stretched taper waist AB, and governs the variation of the waist radius r_w with extension x

$$\frac{dr_w}{dx} = -\frac{r_w}{2L} \tag{1}$$

The second equation determines the extension x at which the point z in the taper transition leaves the hot zone. Comparing the distance PQ as z leaves the hot zone with the initial distance PQ (fig.3), we obtain

$$2z + L = x + L_o \tag{2}$$

where L can be a function of x, and L_o is its initial value, at $x=0$. The solution $x(z)$ of this equation depends on how L varies with x. The time t does not appear explicitly in either equation. Dependent quantities such as r_w and L can thus be expressed as functions of x, regardless of the rate of elongation or of any variations thereof.

Two particular situations are now considered; the "forward problem" in which the elongation conditions $L(x)$ are specified and the resultant taper shape $r(z)$ is to be found, and the "reverse problem" in which a particular taper shape is specified and the conditions to produce it are to be found.

3. The Forward Problem

$L(x)$ is given and must satisfy the constraints identified above, and the taper shape $r(z)$ is to be found. The variation of r_w with x is obtained by integrating eq.1 with the initial condition $r_w(0)=r_o$, giving the general expression

$$r_w(x) = r_o\, exp\left[-\frac{1}{2}\int_0^x \frac{dx'}{L(x')}\right] \tag{3}$$

Eq.2 gives $z(x)$ immediately. Inversion of this expression gives $x(z)$, which can be substituted into $r_w(x)$ from eq.3 to give the taper shape $r(z) = r_w(x(z))$.

In the simplest case, a constant hot zone $L(x)=L_o$ produces an exponential taper shape $r(z) = r_o\, e^{-z/L_o}$ as found before[3]. As another example, a linearly varying hot zone $L(x) = L_o + \alpha x$ gives

$$r(z) = r_o\left[1 + \frac{2\alpha z}{(1-\alpha)L_o}\right]^{-1/2\alpha} \tag{4}$$

which approaches the exponential shape in the limit $\alpha \to 0$. If $\alpha = -1/2$, with the hot zone shrinking at half the rate of taper elongation, a straight-line taper shape results.

4. The Reverse Problem

In this case, the desired taper shape $r(z)$ is given, and $L(x)$ is required. Various manipulations of eqs.1 and 2 give the somewhat artificial function

$$L(z) \;=\; \frac{r_w^2}{r^2(z)} \, l_w \;+\; \frac{2}{r^2(z)} \int_z^{z_o} r^2(z') \; dz' \tag{5}$$

With this expression for L, eq.2 gives $x(z)$ immediately. Inversion gives $z(x)$, which can be substituted into eq.2 again to give the required hot zone variation $L(x) = x + L_o - 2z(x)$. The constraints on L turn out to be equivalent to $l_w \geq 0$ and $dr/dz < 0$, which are automatically satisfied by any reasonable taper shape.

5. Real Heat Source

To imitate the assumptions of the model, a small point-like flame is traversed across and back along the fibre between limit switches, at constant speed. Every hot element of fibre is heated and stretched identically, as required by the model. The travel distance therefore corresponds to L, and can be varied as tapering proceeds simply by moving the limit switches, thus determining the function $L(x)$. Such a fabrication system has indeed been found to produce tapers with shapes corresponding to the predictions of the theory[4].

6. Conclusion

An idealised model for the shape of tapered fibres is developed which avoids using fluid mechanics. The shape resulting from given fabrication conditions can be predicted, as can the conditions required to produce a given required taper shape. A travelling flame heat source satisfies the assumptions of the model, and so a complete procedure for the control of fibre taper shape has therefore been presented.

The authors would like to thank Sumicem Optoelectronics (Ireland) Limited for their support.

References

1. Kawasaki, B.S., Hill, K.O. and Lamont, R.G. (1981) "Biconical-taper single-mode fiber coupler", Optics Letters, **6**, 327-328.

2. Stewart, W.J. and Love, J.D. (1985) "Design limitation on tapers and couplers in single mode fibres", Proc. ECOC '85, Venice, 559-562.

3. Eisenmann, M. and Weidel, E. (1988) "Single-mode fused biconical couplers for wavelength division multiplexing with channel spacing between 100 and 300 nm", IEEE J. of Lightwave Technology, **6**, 113-119.

4. Kenny, R.P., Birks, T.A. and Oakley, K.P. (1991): "Control of optical fibre taper shape", Electronics Letters, **27**, 1654-1656.

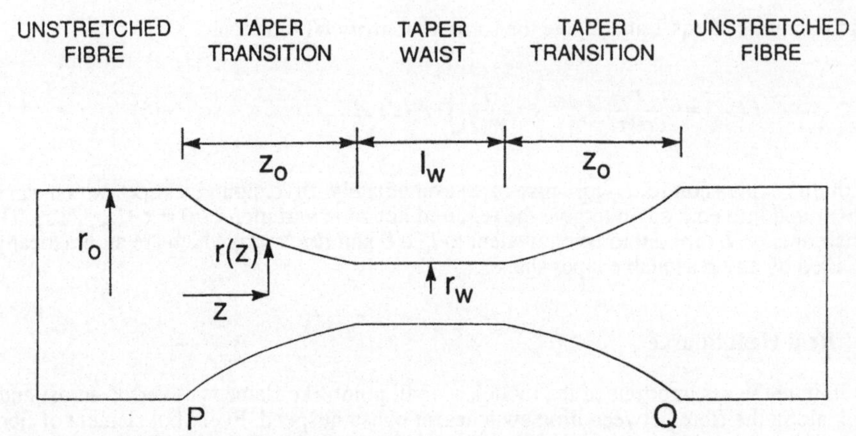

1. The structure of a fibre taper, indicating the terminology used in the text.

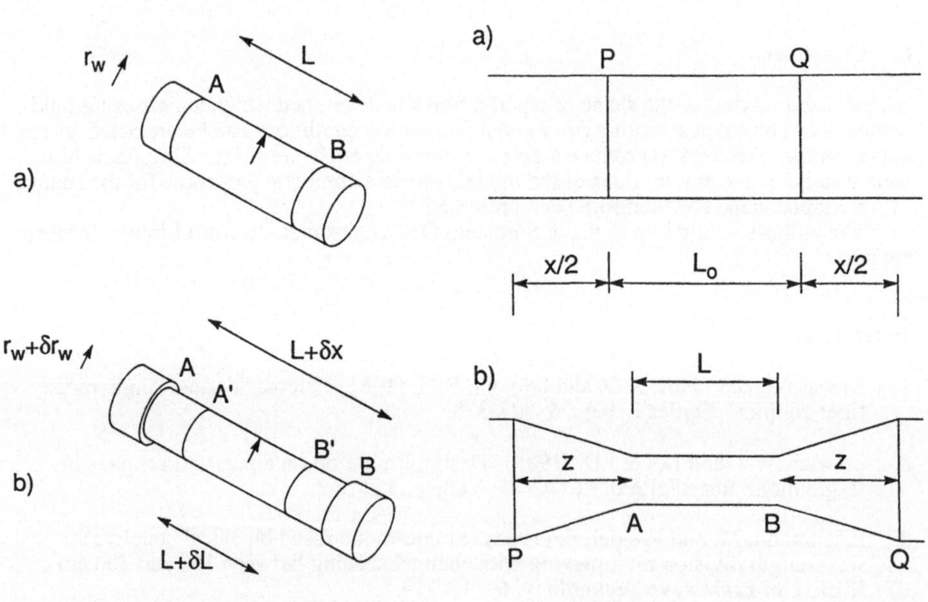

2. a) A cylindrical taper waist at time t, a part AB of which is uniformly heated.
    b) At time $t+\delta t$, AB has been stretched through a distance δx to form a narrower cylindrical taper waist, a part A'B' of which is still heated.

3. a) A fibre at the start of tapering. A section PQ of length L_o is heated.
    b) The fibre as point z is pulled out of the hot zone. P and Q are separated by an additional distance x.

DISPERSION CONTROL IN MULTIPLE LAYER OPTICAL FIBRES

Y. W. LI and C. D. HUSSEY
Lightwave Technology Research Centre
University of Limerick
Plassey Technological Park
Limerick
Ireland

ABSTRACT. Dispersion in single mode optical fibres is due to the material composition of the fibre and to the distribution of that material (the refractive index profile). The two contributions are compensatory, permitting a negligible total dispersion over an appreciable wavelength band. This paper describes how to design a multiple layer fibre with the minimum number of layers (three) that will have zero total dispersion at the telecommunications wavelengths of 1300 nm and 1550 nm, without compromising the other waveguiding properties of the fibre.

1. Introduction

The history of fibre optics is of the quest for bandwidth, in essence the control of dispersion. In multimode fibres, the step index waveguide with its severe intermodal (multipath) dispersion gave way to the graded index (parabolic profile) fibre, which minimises such dispersion by equalising the optical path lengths for the different modes. Multimode fibres were then transcended with the leap to single mode silica fibres, where intermodal dispersion was eliminated altogether, and where chromatic dispersion (a few orders of magnitude smaller) became the limiting factor in transmission bandwidth. In turn, the step index single mode fibre, with a zero of total dispersion in the 1300 nm low loss window, was supplemented with the triangle index fibre, with the zero of total dispersion shifted to the lower loss window at 1550 nm. Multiple layer single mode structures with alternate high and low indices then emerged having two zeros of dispersion, which could be located at both of the operating wavelengths of 1300 nm and 1550 nm. Such fibres can have minimal dispersion between these two wavelengths[1].

Multiple layer fibres have been designed in an *ad hoc* manner, and while they have good dispersion performance, other operating characteristics such bend loss are compromised in the design. Bend loss is related to the optical confinement of the guided mode and is a minimum in those fibres which are operated at wavelengths close to the cutoff wavelength of the second mode[2]. It is therefore required that the second mode cutoff wavelength be less than, but close to, the lower operating wavelength at 1300 nm. In order to minimise stress in the fibre the maximum refractive index difference between successive layers is kept below 1%[3], in which case three is the minimum number of cladding layers required. Fig 1 is a schematic index profile of such a triply clad fibre and defines the parameters used. This paper introduce a design procedure for triply clad fibres which satisfy the above dispersion and bend loss requirements.

2.   Chromatic Dispersion

The chromatic dispersion T in single mode fibres can be broken down into two main components : material dispersion T_M and waveguide dispersion T_W viz[4]

$$T = \delta\lambda\,(T_M + T_W)$$

where $\delta\lambda$ is the spectral width of the source.

$$T_M = -\frac{\lambda}{c}\left[\frac{d^2n_0}{d\lambda^2} + \sum_i\left(\frac{d^2n_i}{d\lambda^2} - \frac{d^2n_0}{d\lambda^2}\right)\Gamma_i\right]$$

where n_0 is the refractive index of the uniform silica outer cladding layer, n_i is the refractive index of the ith layer, Γ_i is the fraction of total modal power propagating in the ith layer, and c is the speed of light in free space. The spectral behaviour of the doped silica layers is obtained from the Sellmeier coefficients[5]. Also

$$T_W = -\frac{n_0\Delta}{c\lambda}\,V\,\frac{d^2(VB)}{dV^2}$$

where Δ is the relative refractive index difference between the core and the uniform outer cladding, V is the waveguide parameter defined as

$$V = \frac{2\pi a_1}{\lambda}\sqrt{(n_1^2 - n_0^2)}$$

where n_1 is the core refractive index and a_1 is the core radius, and B is the normalised propagation constant defined as

$$B = \frac{\beta^2 - k^2 n_0^2}{k^2 n_1^2 - k^2 n_0^2}$$

where β is the modal propagation constant and k is the free space wavenumber. B is found from the solution of the scalar wave equation for the multiple layer structure.

3.   Dispersion Compensation at Two Wavelengths

The material dispersion T_M, for a given dopant concentration in silica fibres, is very similar from one fibre to the next irrespective of the refractive index distribution. Over the wavelength range of interest to us (1100 nm to 1600 nm), the T_M curve is a decreasing function with a zero close to 1300 nm. A typical curve for T_M in a triply clad fibre is drawn in Fig 2. Waveguide dispersion T_W in triply clad fibres can be zero in the single mode region, at a wavelength governed mainly by the parameters (radius and refractive index) of the core. By choosing a suitable core radius a_1, the zero in T_W can be made to coincide with the zero in T_M, giving the required zero total dispersion at 1300 nm. The behaviour of T_W at longer wavelengths then depends mainly on the structure of the second and third layers. Fig 3, for instance, shows the curves of T_W for a series of triple clad fibres, where the values at 1300 nm are approximately the same but the values at 1550 nm can be significantly different. With a suitable choice of the parameters of these outer layers, a fibre with a value of T_W which exactly compensates T_M at 1550 nm can be found. Similarly, the cutoff wavelength of the second mode depends mainly on the parameters of the outer layers, and we have found that fibres with a suitable set of outer layer parameters can be found which simultaneously have zero dispersion at 1300 nm and 1550 nm and have cutoff wavelengths close to 1300 nm.

4.   Procedure

Our procedure is to choose suitable core and first cladding layer materials, thereby fixing D_1, making sure that the refractive index difference between the layers is less than 1%. The second cladding layer is omitted for the moment, so that the next layer is the third uniform cladding layer of pure silica.

The initial thickness of the first cladding layer S_1 is chosen arbitrarily and T_M and T_W are calculated. The zero of T_W is calculated and the core radius a_1 is chosen so that this zero occurs at 1300 nm. The value of T_W is then evaluated at 1550 nm and compared with T_M at 1550 nm to check if T_W cancels T_M. The thickness of the first cladding layer is then changed and the procedure is iterated until cancellation occurs at 1550 nm.

The D_1 and S_1 parameters of the first cladding layer are now known, and the second cladding layer can be added. The refractive index difference between this layer and the outer (third) cladding layer is chosen (a difference about half that between the core and outer cladding tends to be a good starting point), and the thickness of this layer is chosen arbitrarily. T_M and T_W are calculated as before and the second layer thickness is now iterated for the required zeros of dispersion, but now with the additional requirement that the second mode cutoff wavelength occurs at 1200 nm i.e. close to 1300 nm. If a suitable cutoff wavelength cannot be found, the second cladding layer refractive index is then iterated until a solution is found.

Fig. 4 shows the total dispersion curves at different stages in our procedure for the parameters indicated. Curve (a) is the total dispersion curve after optimising the first cladding layer and curve (b) is the final total dispersion curve with the second cladding layer. In case (a) the second mode cutoff is 887 nm while in (b) it is 1225 nm.

References

1.   Cohen, L.G., Mammel, W.L. and Jang, S.J. (1982) "Low-loss quadruple-clad single-mode lightguides with dispersion below 2 ps/km nm over the 1.28 µm - 1.65 µm wavelength range", Electronics Letters, **18**, 1023-1024.

2.   Weling, F., Tjaden, D.L.A. and Van Steenwijk, J.A. (1988) "The design of dispersion flattened single-mode fibres", ECOC'88, 457-460.

3.   White, K.I. (1982) "Design parameters for dispersion-shifted triangular-profile single-mode fibres", Electronics Letters, **18**, 725-727.

4.   Francois, P.L. (1982) "Tolerance requirements for dispersion free single-mode fibre design : influence of geometrical parameters, dopant diffusion, and axial dip", IEEE Journal of Quantum Electronics, **18**, 1490-1499.

5.   Kobayashi, S. *et al.* (1978) "Characteristics of optical fibres in infrared wavelength region", Review of ECL, **26**, 453-467.

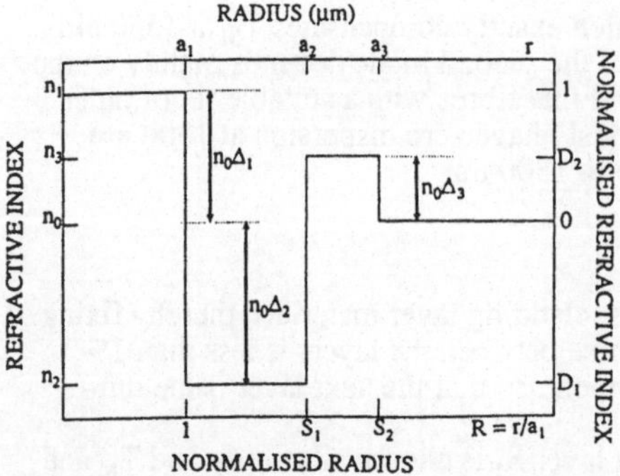

Fig 1. Schematic diagram of a triple clad fibre showing unnormalised and normalised parameters.

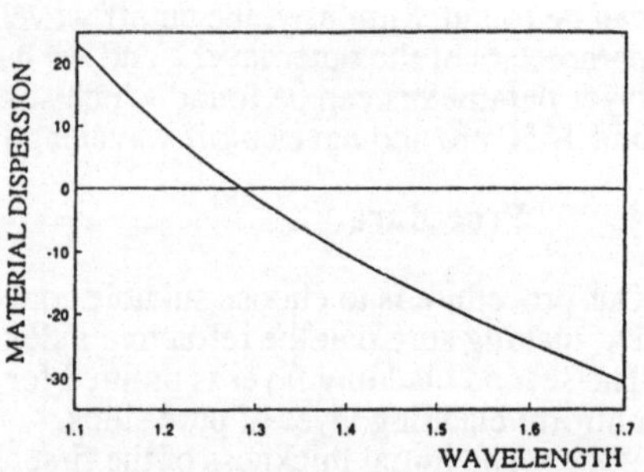

Fig 2. Material dispersion curve for a triply clad fibre with a 3.1% germania-doped silica core, showing the zero at approximately 1300 nm.

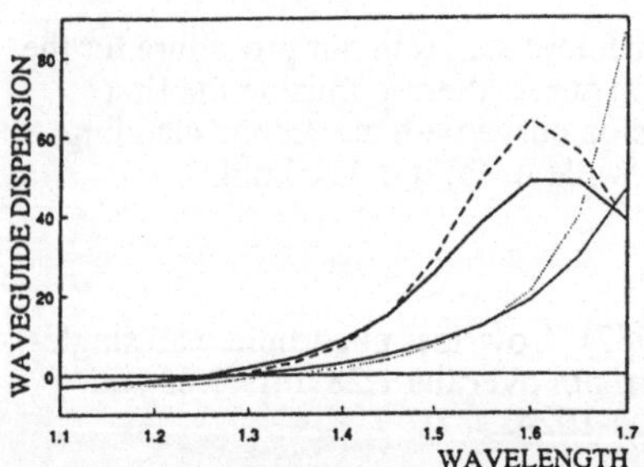

Fig 3. Waveguide dispersion for four different triply clad fibres, showing that large values can be obtained at long wavelengths without significantly altering the values at shorter wavelengths.

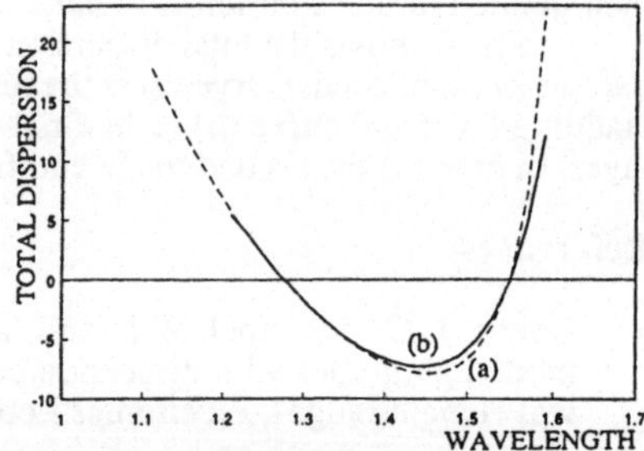

Fig 4. Total dispersion for (a) optimised core and first cladding layer with $a_1 = 3.9$ µm, $\Delta_1 = 0.324\%$, $D_1 = -2$, $S_1 = 1.63$; and (b) optimised core, first cladding and second cladding layers with $a_1 = 4$ µm, $\Delta_1 = 0.324\%$, $D_1 = -2$, $S_1 = 1.63$, $D_2 = 0.11$, $S_2 = 3$.

Mathematical Education
and Industry Papers

THE EXPERIENCE OF MATHEMATICS UNDERGRADUATES DURING INDUSTRIAL PLACEMENT

W J CROMIE and E S GILLESPIE
Department of Mathematics
University of Ulster
Jordanstown
Co Antrim   N Ireland

ABSTRACT. In this paper we investigate the type of Mathematics used by undergraduates during their placement year in commerce and industry. The students in the study are from a BSc (Hons) course in Mathematics, Statistics and Computing. Three cohorts of students with approximately 25 students in each cohort are considered.

1. INTRODUCTION

At the University of Ulster we run a BSc (Hons and Ordinary) degree in Mathematics, Statistics and Computing. This is a thick sandwich course where the third year is spent in paid employment in Industry or Commerce. We have approximately 25 students in each year of the course.

The University is responsible for finding placements for the students. Employers either come to the University to interview or invite a short list of students to visit the organisation. To date we have three cohorts placed and these locations are as follows (Table 1).

Location	1989-90	1990-91	1991-92
Northern Ireland	8	13	12
Southern Ireland	1	2	1
South East England	9	8	6
North England	4	–	–
Canada	1	–	–
Total	23	23	19

Table 1   Placement Locations

The type of placement experiences can be roughly broken down into the following areas (Table 2).

Main Subject Area	1989-90	1990-91	1991-92
Computing	12	10	5
Statistics	5	7	10
Operational Research	4	6	3
Mathematics	2	–	1
Total	23	23	19

Table 2   Main Subject Area During Placement

While we have listed the main subject areas there were overlaps and in all cases students used computers to a lesser or greater degree. Placements involving traditional Applied Mathematics are few and far between. The quality of the placement varies considerably from place to place - with some large organisations having a scheduled training and experience programme drawn up, while in others the firm is looking for cheap labour and the student is put in at the deep end with other employees. Some of our most successful placements have been in the second category where job satisfaction has ranked high.

2. BREAKDOWN OF PLACEMENT YEAR

2.1 PRE-PLACEMENT PREPARATION

This starts really from the beginning of our course but it is concentrated on from the beginning of year II. In addition to the main subjects of Mathematics, Statistics and Computing a slot is time tabled for Professional Studies and Modelling.
Its aims are:-

(i)   to initiate and develop group study of simulated real-life problems involving all areas of Mathematical Modelling;

(ii) to develop skills in oral, visual and written communication.,

In addition to the case studies which are worked on in groups of four or five, lectures and seminars are given on communication skills, interview techniques and report writing. This is put into practice in the case study presentations.

We have found that this is a very useful area of discussion during interviews for placement. In addition we show a video of students on placement, and have talks by the students who have just returned. There is continuous feedback between the class and the placement tutor regarding the current situation at any time.

2.2 PLACEMENT SUPERVISION

During their placement year students are allocated an Industrial Tutor and an Academic Tutor. The role of the Industrial Tutor is to supervise the day to day work of the student and to agree the programme of work experience with the Academic Tutor. The Academic Tutor makes at least two visits to the student in his/her place of work and discusses progress with the student and the Industrial Tutor. Telephone contact is also maintained.

2.3 ASSESSMENT OF PLACEMENT

This is based on three elements:-
(a) The Placement Report and Log Book- the placement report is approximately 5000 words in length and we look for a report giving an overall view and flavour of the year's experience other than detailed technical information. The log book is just a record of tasks undertaken etc.

(b) The Employers Report -this covers all aspects of the student's performance.
(c) The Academic Tutor's Report based on his visits and discussions with the student and his industrial supervision.

2.4 DE-BRIEFING AND FEEDBACK

The students are asked to give a short talk on their year's experience to the second year students. This will cover the nature of the work, salary paid, tips about accommodation etc.

3. REQUIREMENTS OF INDUSTRY

A survey was carried out with a number of firms who have employed mainly Graduates in Computing Studies and in Mathematics, Statistics and Computing. They were asked to tick those areas which were of particular interest to them. The results are given in Table 4.

Computing	%	Computer Languages	%
Business Data Processing	73	Pascal	14
Operating Systems	55	BASIC	39
Hardware	36	Fortran	11
Software Lotus 1-2-3	43	Cobol	25
Symphony	25	Kenetic Algors	0
Wordstar	30	Neural Networks	0
Word Perfect	25	Expert Systems	16
DBASE III	34		

Statistics	%	Operational Research	%
Statistical Modelling		Linear Programming	12
Techniques	11	Dynamic Programming	0
Quality Control Packages	30	Simulation Techniques	12
Survey Sampling	8	Stock Control	55
Forecasting Methods	30	Forecasting	30
Design of Experiments	10	Critical Path Analysis	25
Statistical Software	12	Transportation	

Mathematics	%	General Subjects	%
Mathematical Modelling	5	Communication Skills	77
Discrete Mathematics	0	Report Writing	70
Numerical Methods	0	Business Organisation	59
Mathematical Software	2	European Language	30

Table 3: Industrialists Response to Survey

4.   PREPARATION FOR FINAL YEAR

In our limited experience to date we have found that the year's
placement has had a great maturing influence on the students. The
discipline of working a 40 hour week with deadlines to meet has been a
useful experience. Ideas for honours projects have arisen from the work
during placement, jointly supervised by the industrial and an academic
supervision.

5.   FEEDBACK FROM STUDENTS

The students who have returned from placement have been given a
questionnaire. This was designed to get feedback from students
regarding various aspects of their placement experience. This includes
the pre-placement preparation, the University placement procedure and
various aspects of the actual placement. A summary of the response of
the first cohort of students is given in Table 4.

Question	Response % Satisfactory - Very Good
1. Academic Preparation	67
2. Placement Procedures	100
3. Match between Placement and Preference	87
4. Working Conditions	100
5. Advice and Assistance from Employer	93
6. Accessibility of University during Placement	60
7. Quality of Placement Experience	100
8. Importance of Log book/Diary	67
9. Relevance of Experience to Final Year	40
10. Improvement of Employment Prospects	100

Table 4: Student response to Questionnaire

6.   SUMMARY

Our experience has shown that sandwich placement has many more
advantages than disadvantages. It gives relevant experience to the
students, extending their knowledge and expertise as well as providing
inter-personal skills, maturity and motivation. It is also a very
useful way of obtaining and maintaining links with industry and
commerce.

Ref: A Survey of Industrial Training in Degree Courses in Computing and
Mathematics in Polytechnics. Department of Education and Science, HMI
Report, 1989.

HOW CAN EDUCATION RESPOND TO INDUSTRY'S NEEDS?

Adrian Oldknow   MA, MTech, CEng, FIMA, FBCS
Head of Mathematics, West Sussex Institute of Higher Education
Bognor Regis, Sussex, PO21 1HR, UK

There has been a rapid increase in the range and extent of the applications of mathematics in modelling and solving industrial and commercial problems. Central to this has been the rapid increase in the power and availability of computing systems. Many large companies employ mathematics graduates as mathematical modellers and form inter-disciplinary teams involving, for example, mathematicians, informaticians, scientists and technologists in tackling particular problems. Many smaller companies share common technical problems and buy 'off-the-shelf' mathematical software packages from a range of software houses which themselves employ mathematical modellers and inter-disciplinary teams.

In the UK there is an organisation, called the Confederation of British Industry (CBI), which represents the interests of industrial and commercial employers, and which has considerable influence with government. For a number of years it has been monitoring industry's increasing needs for skilled personnel, particularly in the field of Information Technology (IT) - the term used in the UK for Informatics. In 1989 it published a report 'Towards a skills revolution', pointing to the need for a rapid and massive increase in the numbers of well-trained, skilled employees to meet the predicted demands of industry for a skilled workforce at the end of the 1990s. The three major political parties in the UK have promised to develop policies to make the education and training of 16-19 year olds a major priority in the present decade.

The issue has been complicated by conflicting administrative structures. There is a government department, called the Department of Trade and Industry, which has responsibility for industrial development. There is another, called the Department of Employment, which includes responsibility for vocational training. There is a third, called the Department of Education and Science, which has responsibility for education in schools (primary 5-11 and secondary 11-16), in tertiary education (16-19) and in higher education (universities, polytechnics and colleges of higher education), as well as for national science policy.

The branch of the Department of Employment concerned with training has itself undergone considerable changes in recent years. At its height, in the mid-1980s, it worked through a massive central agency called the Manpower Services Commission (MSC). In the late 1980s this was scaled down considerably and restructured as the Training Agency (TA). In 1991 this has effectively been wound down to a skeleton central structure, with the control of training being devolved to the regions through a network of autonomous bodies called Training Enterprise Councils (TEC) which are intended to be substantially funded by contributions from local industries.

This is also a time of considerable structural change at all levels in the UK educational system. A national curriculum is being introduced for pupils from age 5 to 16. Within this curriculum mathematics is a compulsory element and one of its five components is specifically concerned with Using and Applying Mathematics. The control and funding of Higher Education has been reorganised and undergraduate intakes have risen substantially in each of the last three years. The government has published new proposals for the control and funding of tertiary institutions, within which the majority of the 16-19 year old students are educated. A politically sensitive issue currently is what curriculum and qualifications should be on offer at this level. After the age of 16 education is not compulsory and the UK has had one of the lowest participation rates for post-16 education in the developed world. The system of academic education is currently based upon the Advanced level of the General Certificate of Education, known as A-level, and the usual pattern has been for students to specialise on three subjects for their A-level studies (16-19). Much pressure has been put on the government, from both employers and education, to broaden this curriculum but is being resisted by ministers. In general the increase in the participation rate for these academic courses at age 16 over the last three years has more than compensated for the demographic decline in the numbers of 16 year-olds, but in a few subjects, particularly mathematics and physics, the actual entries have fallen in successive years.

There are a variety of syllabuses in mathematics at A-level but they have all to conform to a 'common core' which specifies at least 40% of their content. The actual content, and its teaching and examining style, has not in general changed much over the past 30 years and is largely concerned with technical manipulation without context. The widespread usage of computers has yet to make an impact at this level.

Over the past few years (until the recession of 1991) there has been a rapid growth in the demand for numerate, well-educated young people, particularly in the field of finance, where material rewards have been very high. This has had the effect of diminishing considerably the numbers of graduates going on to take research and post-graduate courses in mathematics, and, in particular, those entering mathematics teaching.

At various times there have been schemes to involve industry more closely with education. In the climate of recent structural changes in education these might have been expected to have become strengthened. Ironically two recent external factors have countered that development. The recent political changes in Eastern Europe have provided opportunities for reductions in domestic military expenditure. Many technologically advanced companies which have been major defence contractors have not had contracts renewed and have had to shed skilled staff. The economic pressures of high interest rates and the current recession have led to considerable lay-offs both within the industrial and the financial sector. The prospects for graduate employment for the 'class of 1991' are currently very poor in the short term, and a shadow hangs over many of those recruited in 1990.

In summary, then, there is a predicted demand from industry for better educated employees with a greater range of skills, within which mathematics is recognised as having major importance. To meet that demand requires (a) a policy based on a better identification of industry's specific needs, and (b) a national strategy to carry it out. The opportunity for educational change at both tertiary (16-19) and higher (19-21) levels is there, and has probably never been more propitious. However the lack of political will (and consequent lack of availability of resources), the bureaucratic labyrinth of government and the current economic downturn of industry are non-trivial obstacles to this process. It is for others to address the question "How can education respond to industry's needs?" at the national level. The remainder of this paper summarises our attempts to address the question locally with respect to mathematics.

Through a project, funded by the UK government's Training Agency at WSIHE on Flexible Learning Approaches in Sixth Form Mathematics, we have been studying the changing needs of industry and commerce for skilled and numerate recruits who have taken mathematical specialisms to A-level, graduate and post-graduate levels. The identification of particular, general and personal skills appropriate to the mathematical needs of a variety of industrial and commercial sectors is not an easy task - particularly at a time of increasing sophistication and use of IT products. The willingness of employers to cooperate in such a project has also been subject to market forces - initially the dominant factor was a medium-term concern about a shortage of skilled recruits at a time of reducing population cohorts, but this has been overtaken somewhat by reductions in activity in some sectors, first as part of the 'peace dividend' and then through general economic recession.

Despite these difficulties we have been able to establish a reasonable contact with about 90 relevant employers in our region of the UK (central South). Through this we have been able to identify a number of important problems in establishing effective communication. A major problem has been to identify whereabouts in a company there are employees whose work is largely mathematical, and to establish contact with them. Many companies refer educational requests to staff in their personnel departments who may well be unaware of the specific skills used by the more technical employees. Again, an employee using mathematical skills may not see it as a legitimate use of her/his time (and hence the employer's money) in engaging with education in this way.

Another problem is to agree on what 'mathematical' means in this context. A frequent response, perhaps conditioned by their own educational experience, is for employees to say "I don't actually use any mathematics now - when I need to I just use a software package", which reflects the view that 'mathematics' consists of performing technical manipulations. Another understandable first reaction is: "I was educated in the old way and I have been able to respond to the needs of this job - it was good enough for me so why should anyone consider changing it?" Our experience is that each of these common responses makes an excellent opening for a constructive and informative dialogue provided both parties are fairly open-minded. Unfortunately some employees just feel that they cannot spare the time to engage in the dialogue, and others have such entrenched opinions that all changes in education are for the worst that constructive dialogue is not possible.

The other major problem has been encountered at the management level in considering future policies for recruitment. There is still an entrenched view that education should be training young people from the earliest possible opportunity for the very specific tasks that they will undertake when they start their careers. This is of course in stark conflict with the acknowledged problems of adaptation to new equipment and methods, and for continuing re-skilling. Our experience has been that our industrial contacts have been quite surprised to learn that anyone in education is interested in finding out their needs. This has frequently caught them unprepared and produced rather conservative first reactions. Given their interest and good-will we have been able to take this forward and some pointers have emerged.

In as much as there is a common need it seems to be for 'flexible thinkers' who can keep themselves up to date, use computer technology with confidence and who can communicate effectively. These general skills can certainly be given explicit attention in both tertiary and undergraduate courses and, in some cases, are receiving attention now. When it comes to the need for content the three most frequent responses were (a) a familiarity with ways of handling and representing data, (b) a familiarity with the use and notation of matrices and vectors and (c) experience in the process of mathematical modelling. Again these specific areas can also be given due weight in course design, and there are courses in which this already happens.

At a local level WSIHE has been involved in reflecting these identified needs in developing course both at A-level and in Higher Education. For the past three years students in local schools and colleges have been following an A-level in mathematics with in-course assessment, developed through the Department of Education and Science funded project: Raising AChievement in Mathematics, which was directed from WSIHE by Afzal Ahmed. The results of our industrial contacts have been fed into design and continuing development of that course. A similar process has also influenced the approach taken in the undergraduate (BEd) and postgraduate (PGCE) programmes of initial teacher training at WSIHE.

At a national level WSIHE is collaborating with the Nuffield Advanced Mathematics Project, directed by Hugh Neill, in the design of a new A-level course. Although the path for A-level developments in the short term in general remains cloudy, the designers have had to make some major decisions in principle, and have put data-handling, modelling and matrices at the core of the course, which encourages students to develop flexible study skills, to gain confidence in the use of computing systems and to become effective communicators of mathematics.

We hope that our local experience may be encouraging to others and may contribute to national developments should circumstances allow.

SOFTWARE-ORIENTED MATHEMATICAL CONTINUING EDUCATION

M. Schulz-Reese
*Zentrum für Techno-
und Wirtschaftsmathematik
Universität Kaiserslautern
Postfach 30 49
D-6750 Kaiserslautern*

Without doubt mathematics is nowadays an indispensable and important tool in almost all fields of science, industry and economy. But a lot of people working with mathematical tools are neither sufficiently mathematically educated nor are they familiar with modern developments of mathematics. It follows that there must exist significant demand for more training in mathematics, in particular for continuing education courses.

The European Consortium for Mathematics in Industry (ECMI) has chosen mathematical continuing education as one of its main educational activities. In the academic year 1990/91 more than 20 courses were organized by institutions belonging to the ECMI educational system. Table 1 shows the organizing institutions, the titles of the courses and the number of participants from industry. With respect to the number of participating industrial scientists the effect of these courses is certainly not too bad but not at all satisfactory. What are the problems? Why could not more industrial scientists be attracted in spite of the advertising efforts made by the ECMI institutions? It seems that not enough attention is paid to the question what mathematics in industry really is.

After more than ten years of intensive cooperation with industry we finally came to the conclusion that mathematics in industry mainly means computer simulation. Above all it means descriptions on how to use software-packages for special technical problems. Sometimes also how to improve these packages, rarely how to develop new software. In many cases the mathematical background of the simulation is even forgotten. These statements may be illustrated by an example:

We have asked staff from the chemical corporation BASF and the car producer Mercedes-Benz to present the mathematics they use in practice in a series of lectures to our students. Almost all lectures dealt with the use of special software-packages for certain applications, e.g. how to use NASTRAN, ASKA or PERMAS for acoustic calculations in car industry.

These observations have obvious implications for our concept of industrial mathematics. An important consequence concerning continuing education offered by university institutions is that such courses must become more software-oriented. Specifically, software packages used in industry must be dealt with.

Let me try to outline a way how such software-oriented courses could be developed and organized:

1. Find out which scientific software-packages are used by which firms. Attention should be paid not only to big companies but also to small and medium-size firms.

TABLE 1.   ECMI Continuing Education Courses in 1990/91

Organized by	title	number of partici-pants from industry
Eindhoven	Finite Element Methods	2
	Heat and Mass Transfer	5
	Design of Experiments	4
Glasgow	Spline Approximation	1
Helsinki	Cryptology	10
	Free Boundary Problems and Industrial Applications	4
	Differential Equations in Industry	0
Kaiserslautern	Quantitative Methods for Production Planning	6
	Practice of Computer Aided Geometric Design	37
	Finite Element Methods in Industrial Applications	10
Linz	Numerical Methods for Eddy Currents	0
	Mathematical Modelling	0
Oxford	Modelling, Analysis, Computation of Fluid Flow	9
	Image Analysis and Pattern Recognition	7
	Dynamical Systems, Fractals and Chaos	17
	Computational Fluid Dynamics	13
SASIAM/Bari	Experimental Design	9
	Computer Simulation	1
	Information Theory and Image Processing	9
	Reliability and Quality Control	8
	Inverse Problems	1
	Free Boundary Problems and Industrial Applications	3
	Optimal Strategies in Finding Allocation	3
	Theoretical and Computational Aerodynamics	4

2. Try to get familiar with the most important packages and/or try to find good experts who know them.

3. Make the mathematical background of the packages transparent and determine those mathematical methods the user has to know to guarantee an efficient and appropriate application.

4. Build up an efficient organization. In general there will not be enough experts at one university (not even in one country) to lecture those courses. It would be wasteful to develop parallel courses on the same topics at different universities. ECMI is likely to be a good forum for information transfer to avoid such duplication.

In Kaiserslautern we have started to develop courses as mentioned above. First we sent out more than 2,000 questionnaires to mainly small and medium size enterprises in fields where we thought scientific software is used: To chemical industry, mechanical engineering industry, woodworking and paper industry, consulting bureaus, etc. We asked in which applications which software-packages are used, e.g. in fluid mechanics, acoustics, heat transfer, structural analysis, CAD/CAM. We also asked whether the users are content with the packages or not. We received almost 300 replies (about 13 %). The results listed in table 2 do not allow representative quantitative statements but give some qualitative indications.

TABLE 2.   Some results of the evaluation of the questionnaires

Application field	Number of firms using software packages
CAD/CAM	118
Construction engineering/statistics	54
Production planning	48
NC	44
FEM	30
Project management	28
Quality control	22
Statistics/data analysis	17

It can of course not be sufficient to make such an inquiry only in a small region as we did it. To identify fields of common interest of different firms the whole network of ECMI has to be used.

The Centre of Applied Mathematics, an institution run by the Technical University Darmstadt and the University of Kaiserslauern, has organized a first software-oriented course in May 1991. The programme of the course with the title "Practice of Computer Aided Geometric Design" is listed in table 3.

TABLE 3.   Programme of the course "Practice of Computer Aided Geometric Design", Lambrecht, 13.-17.05.1991

Monday:	Lectures
	▪ Freeform requirements for CAD systems
	▪ Integration of physical phenomena into CAD
	▪ Report on interfaces
	▪ General matrix representation for NURBS curves and surfaces for interfaces
	▪ Approximation conversion and marging of spline surface patches
Tuesday:	Presentation EUCLID
	by Matra Datavision, Paris/Munich
	Presentation ICEM
	by CDC/ICEM, Hannover
Wednesday:	Presentation CATIA
	by Dassault, France, and users
	Presentation UNIGRAPHICS
	by McDonnel Douglas Germany
Thursday:	Presentation SYRKO
	by Mercedes Benz, Stuttgart
	Presentation STRIM
	by cisigraph, France/Germany
Friday:	Presentation MENTAL RAY
	by mental images, Berlin
	Panel discussion

The audience of this course consisted of producers of commercially successful CAD packages, users of CAD software, potential users, and scientists, mainly university mathematicians but also computer scientists.

I am sure that such courses are advantageous for all participating groups: Maybe the advantage for the producer is not obvious at first sight. After all he has to face the discussion with rival firms and with critical users. He cannot give an advertising show but has to reveal even the weak points of his software. On the other side there is a lot of feedback which should be useful for further developments.

The user, or potential user, who is looking for appropriate software which fits best for his problem finds detailed information. Those already using some packages know weak and strong points and will make suggestions for improvement.

The scientist, in this case the mathematician, learns the problems of the users and the trends of software and in this way he gets impulses for promising research problems. Concerning continuing education he can identifiy mathematical methods essential for an efficient application of scientific software. It is obvious that not all users have to know the whole mathematical background of a software package. But in special fields, e.g. in the research and development departments of high-tech industry, the user wants to know details in order to have the possibility to manipulate the programme. Such users are certainly a target group for mathematical continuing education.

Altogether this is a very sensitive and complicated field of educational activities, not only in mathematics and not only for universities. In 1991 in the May issue of the German Manager Magazin several articles on continuing education were published. One of these was entitled: "Much money for nothing". It was claimed that often only the organizers profit by continuing education courses and that the participants often come away empty-handed. Millions of marks are wasted because the market is not transparent and the intramural continuing education is badly managed.

To earn at least a small part of this amount of money by organizing successful and efficient courses should be perhaps an incentive for mathematical institutes to engage in this educational field.